Springer Series on Touch and Haptic Systems

More information about this series at http://www.springer.com/series/8786

Matteo Bianchi · Alessandro Moscatelli
Editors

Human and Robot Hands

Sensorimotor Synergies to Bridge the Gap Between Neuroscience and Robotics

Editors
Matteo Bianchi
Department of Advanced Robotics
Istituto Italiano di Tecnologia
Genoa
Italy

and

Research Centre "E. Piaggio"
Università di Pisa
Pisa
Italy

Alessandro Moscatelli
Department of Cognitive Neuroscience and Cognitive Interaction Technology Centre of Excellence (CITEC)
University of Bielefeld
Bielefeld
Germany

and

Department of Systems Medicine and Centre of Space Bio-Medicine
Università di Roma "Tor Vergata"
Rome
Italy

ISSN 2192-2977 ISSN 2192-2985 (electronic)
Springer Series on Touch and Haptic Systems
ISBN 978-3-319-26705-0 ISBN 978-3-319-26706-7 (eBook)
DOI 10.1007/978-3-319-26706-7

Library of Congress Control Number: 2015957224

Printed on acid-free paper

This Springer imprint is published by SpringerNature
The registered company is Springer International Publishing AG Switzerland

Series Editors' Foreword

This is the 11th volume of 'Springer Series on Touch and Haptic Systems', which is published as a collaboration between **Springer** and the **EuroHaptics Society**.

Human and Robot Hands provides a comprehensive introduction to human hand sensorimotor synergies and the manner in which this approach is applied to robotic hands and their manipulation of objects. This book is organized into two parts and 15 chapters. The first part is devoted to neuroscience topics, which are mainly focused on understanding the neuroanatomical, physiological and behavioural mechanisms that are related to the sensorimotor control of the human hand. The second part shows developments and guidelines for the replication of sensory and motor features of the human hand using robotic and haptic devices.

More than 30 well-known researchers from the field of neuroscience, psychology, robotics and computer science have contributed to this book edited by Matteo Bianchi and Alessandro Moscatelli. Most of the results included in this issue have been developed within 'THE Hand Embodied' project (http://www.thehandembodied.eu/project; duration: 1 March 2010–28 February 2014) coordinated by Prof. Antonio Bicchi (Università di Pisa), which was funded by the European Commission under CP grant no. 248587, within the FP7-ICT-2009-4-2-1 programme 'Cognitive Systems and Robotics'.

September 2015

Manuel Ferre
Marc O. Ernst
Alan Wing

Contents

Part II Robotics, Models and Sensing Tools

Contributors

Arash Ajoudani Department of Advanced Robotics, Istituto Italiano di Tecnologia, Genoa, Italy

Charalampos P. Bechlioulis National Technical University of Athens, Zografou, Greece

Wouter M. Bergmann Tiest Department of Human Movement Sciences, MOVE Research Institute, Vrije Universiteit Amsterdam, Amsterdam, BT, The Netherlands

Matteo Bianchi Department of Advanced Robotics, Istituto Italiano di Tecnologia, Genoa, Italy; Research Center "E. Piaggio", Università di Pisa, Pisa, Italy

Antonio Bicchi Department of Advanced Robotics, Istituto Italiano di Tecnologia, Genoa, Italy; Research Center "E. Piaggio", Università di Pisa, Pisa, Italy

Manuel Bonilla Research Center "E. Piaggio", Università di Pisa, Pisa, Italy

George I. Boutselis Georgia Institute of Technology, Atlanta, USA

Claudio Castellini Institute of Robotics and Mechatronics, DLR - German Aerospace Center, Oberpfaffenhofen, Germany

Manuel G. Catalano Department of Advanced Robotics, Istituto Italiano di Tecnologia, Genoa, Italy; Research Center "E. Piaggio", Università di Pisa, Pisa, Italy

Marc O. Ernst Department of Cognitive Neuroscience and Cognitive Interaction Technology Centre of Excellence (CITEC), University of Bielefeld, Bielefeld, Germany

Edoardo Farnioli Research Center "E. Piaggio", Università di Pisa, Pisa, Italy; Department of Advanced Robotics, Istituto Italiano di Tecnologia, Genoa, Italy

Qiushi Fu School of Biological and Health Systems Engineering, Arizona State University, Tempe, AZ, USA

Marco Gabiccini Research Center "E. Piaggio", Università di Pisa, Pisa, Italy; Department of Advanced Robotics, Istituto Italiano di Tecnologia, Genoa, Italy; Department of Civil and Industrial Engineering, Università di Pisa, Pisa, Italy

Manolo Garabini Research Center "E. Piaggio", Università di Pisa, Pisa, Italy

Guido Gioioso Department of Advanced Robotics, Istituto Italiano di Tecnologia, Genoa, Italy; Dipartimento di Ingegneria dell'Informazione e Scienze Matematiche, University of Siena, Siena, Italy

Sasha B. Godfrey Department of Advanced Robotics, Istituto Italiano di Tecnologia, Genoa, Italy

Giorgio Grioli Department of Advanced Robotics, Istituto Italiano di Tecnologia, Genoa, Italy; Research Center "E. Piaggio", Università di Pisa, Pisa, Italy

Giacomo Handjaras Laboratory of Clinical Biochemistry and Molecular Biology, Department of Surgery, Medical and Molecular Pathology, and Critical Care, Università di Pisa, Pisa, Italy

Henrik Jörntell Neural Basis of Sensorimotor Control, Department of Experimental Medical Science, Lund University, Lund, Sweden

Astrid M.L. Kappers Department of Human Movement Sciences, MOVE Research Institute, Vrije Universiteit Amsterdam, Amsterdam, BT, The Netherlands

Kostas J. Kyriakopoulos National Technical University of Athens, Zografou, Greece

Andrea Leo Laboratory of Clinical Biochemistry and Molecular Biology, Department of Surgery, Medical and Molecular Pathology, and Critical Care, Università di Pisa, Pisa, Italy

Minas V. Liarokapis Mason Laboratory, Yale University, New Haven, CT, USA

Monica Malvezzi Dipartimento di Ingegneria dell'Informazione e Scienze Matematiche, University of Siena, Siena, Italy

Hamal Marino Research Center "E. Piaggio", Università di Pisa, Pisa, Italy

Alessandro Moscatelli Department of Cognitive Neuroscience and Cognitive Interaction Technology Centre of Excellence (CITEC), University of Bielefeld, Bielefeld, Germany; Department of Systems Medicine and Centre of Space Bio-Medicine, Università di Roma "Tor Vergata", Rome, Italy

Abdeldjallil Naceri Department of Cognitive Neuroscience and Cognitive Interaction Technology Centre of Excellence (CITEC), University of Bielefeld, Bielefeld, Germany

Martin Nilsson Swedish Institute of Computer Science (SICS), Kista, Sweden

Cristina Piazza Research Center "E. Piaggio", Università di Pisa, Pisa, Italy

Pietro Pietrini Laboratory of Clinical Biochemistry and Molecular Biology, Department of Surgery, Medical and Molecular Pathology, and Critical Care, Università di Pisa, Pisa, Italy

Domenico Prattichizzo Department of Advanced Robotics, Istituto Italiano di Tecnologia, Genoa, Italy; Dipartimento di Ingegneria dell'Informazione e Scienze Matematiche, University of Siena, Siena, Italy

Emiliano Ricciardi Laboratory of Clinical Biochemistry and Molecular Biology, Department of Surgery, Medical and Molecular Pathology, and Critical Care, Università di Pisa, Pisa, Italy

Paolo Salaris Laboratoire d'Analyse et d'Architecture des Systèmes (LAAS) – CNRS, Toulouse Cedex 4, France

Gionata Salvietti Department of Advanced Robotics, Istituto Italiano di Tecnologia, Genoa, Italy

Marco Santello School of Biological and Health Systems Engineering, Arizona State University, Tempe, AZ, USA

Alessandro Serio Department of Advanced Robotics, Istituto Italiano di Tecnologia, Genoa, Italy; Centro di Ricerca E. Piaggio, Università di Pisa, Pisa, Italy

Nikos Tsagarakis Department of Advanced Robotics, Istituto Italiano di Tecnologia, Genoa, Italy

Vonne van Polanen Department of Kinesiology, KU Leuven, Movement Control and Neuroplasticity Research Group, Leuven, Belgium

Chapter 1
Introduction

Matteo Bianchi and Alessandro Moscatelli

Abstract The human hand is our preeminent and most versatile tool to explore and modify the external environment. It represents both the cognitive organ of the sense of touch and the most important end effector in object manipulation and grasping. Our brain can cope efficiently with the high degree of complexity of the hand, which arises from the huge amount of actuators and sensors. This allows us to perform a large number of daily life tasks, from the simple ones, such as determining the ripeness of a fruit or drive a car, to the more complex ones, as for example performing surgical procedures, playing an instrument or painting. Not surprisingly, an intensive research effort has been devoted to (i) understand the neuroanatomical and physiological mechanisms underpinning the sensorimotor control of human hands and (ii) to attempt to reproduce such mechanisms in artificial robotic systems.

This book reports relevant issues in robotics and neuroscience of the hand, which were investigated within the international cooperation project "THE Hand Embodied". The leading idea of THE project was the concept of synergies, intended as "a functional property of a multi-element system performing an action, whereby many elements of the system are or can be constrained to act as a unit through a few coordination patterns to execute a task" [15, 16]. There is extensive evidence in neuroscience for

M. Bianchi (✉)
Department of Advanced Robotics (ADVR), Istituto Italiano di Tecnologia, via Morego 30, 1613 Genoa, Italy
e-mail: matteo.bianchi@iit.it

M. Bianchi
Research Center "E. Piaggio", Università di Pisa, Largo Lucio Lazzarino 1, 56126 Pisa, Italy

A. Moscatelli
Department of Cognitive Neuroscience, University of Bielefeld, Universitätsstrasse 25, 33615 Bielefeld, Germany
e-mail: alessandro.moscatelli@uni-bielefeld.de

A. Moscatelli
Department of Systems Medicine and Centre of Space Bio-Medicine, Università di Roma "Tor Vergata", Via Montpellier 1, 00133 Rome, Italy

M. Bianchi and A. Moscatelli (eds.), *Human and Robot Hands*, Springer Series on Touch and Haptic Systems, DOI 10.1007/978-3-319-26706-7_1

the organization of the sensorimotor system in functional or structural "synergies", at different levels, such as neural (e.g. [19]) and muscular (e.g. [20]) (see [15] for an exhaustive review). Accordingly, recent studies have demonstrated strong covariance patterns in the control of the hand, in both the kinematics and force domains (see e.g. [17, 22]). At the same time, the idea has become popular in robotics of exploiting these reduction mechanisms to better control and design robotic hands and haptic systems using a reduced number of control inputs, with the goal of pushing their effectiveness close to the natural performance. Central to this point of view is the concept that merely mimicking the architecture of the hand is an unfeasible and daunting task: our belief is that the modelling of the synergistic organization and its translation into a mathematical language can represent an effective step forward to advance the state of art of artificial systems [2, 4].

Research in neuroscience can provide the theoretical and experimental foundations to describe the hierarchical organization of the human hand. Then such foundations can be suitably translated into a language understandable by artificial systems and used to drive the design of more effective robotic devices (see e.g. [5, 8]). At the same time, robotic and technological systems can represent useful tools to perform neuroscientific investigations on human hand and to offer new insights to better understand human grasping and manipulation behavior. With this book, we propose to bridge the gap between neuroscience and robotics with the twofold goal of increase the comprehension of functional and neuroanatomical organization of the human hand and to derive the guidelines for a more effective development of robotic and haptic devices. This book is organized into two parts to mirror this dual approach.

Part I of the book deals with the functional and behavioral aspects of the sensorimotor control of the hand. In all chapters, the theoretical framework of synergies provides a coherent solution to reduce the degrees of freedom in motor control and to organize the rich sensory information provided by cutaneous touch.

Chapter 2 analyzes dexterous manipulation in two-finger grasp. The authors propose a theoretical framework based on high-level representation of the task that can be learned in an effector-independent fashion. These results are discussed in relation to the concept of motor equivalence and sensorimotor integration of grasp kinematics and kinetics as well as low-level coordination of digit force and position. Chapter 3 analyzes force synergies in unconstrained hand grasping, examining how humans stabilize an external object in response to external perturbations [14]. Chapters 4 and 5 discuss the possible neural bases of synergies in subcortical and cortical structures.

Specifically, Chap. 4 reviews recent findings from neurophysiology showing the synergistic organization of the subcortical circuitry [15, 18]. Synergies appear to be a natural solution due to the diverging organization of the nervous system, where each neuron is connected to multiple motor units. Chapter 5 proposes a functional Magnetic Resonance Imaging (fMRI) study [9], where novel encoding techniques are employed to determine whether regional brain activity during grasping movements can be predicted by the kinematic combination of hand synergies. Chapter 6 models the control of the hand-arm system through building blocks consisting of neuronal populations. These blocks can be regarded as neuronal operation amplifiers (opamps) that implement an efficient adaptive feedback control that could be profitably applied in robotics for the identification of unknown sensors on-the-fly. Chapter 7 evaluates the hypothesis that cutaneous touch from the interaction with external object provides information on the hand displacement. This cutaneous contact information is fused with "classical" proprioceptive cues from musculoskeletal system and skin stretch to produce a robust estimation of the displacement. The correlation between finger movements and skin deformation suggests the existence of sensorimotor synergies [4, 13].

Part II of the book focuses on the definition of exploitable guidelines for the replication of the sensory and motor aspects of the human hand through robotic and haptic/sensing devices. These guidelines are devised from the neuroscientific results reported in Part I. Furthermore, Part II describes tools and procedures that can be used to perform more effective behavioral investigations, even providing novel inspirations to better understand natural systems. Chapter 8 presents the Pisa/IIT SoftHand, a novel robot hand prototype designed with only one motor with the purpose of being robust and easy to control as an industrial gripper, while exhibiting high grasping versatility. The design is inspired by the synergistic control of human hand along the first most common actuation pattern [7]. Chapter 9 proposes a learn-by-demonstration approach to learn anthropomorphic robot motions for reach to grasp movements towards different positions and objects in 3D space. Exploiting principal component analysis to extract a lower dimensional manifold for human-like robot data, Navigation Functions (NF) based models [10] are defined, which operate in a synergistic manner. A methodology for robust grasping with tactile sensing is also used to relax uncertainties and increase robustness of the final grasp. Chapters 10 and 11 deal with electromyography (EMG) based synergistic control of robotic and prosthetic hands. Chapter 10 presents an overview of the teleimpedance control concept [1] and provides two application examples. In this teleimpedance control, the user's postural and stiffness synergy references are tracked in real-time by using surface EMG signals acquired from one pair of antagonistic muscles on the forearm. In the first example, an electromyography based model is developed to estimate the operator's arm endpoint stiffness in real-time, while in the second example the teleimpedance concept is translated to control the Pisa/IIT SoftHand described in Chap. 8. Chapter 11 analyzes synergistic muscle patterns for the control of a dexterous hand prosthesis [6, 20] and for the restoration of a missing hand function by introducing the concept of "incremental learning" as the main feature of modern machine learning for the amputees. Chapters 12 and 13 propose mathematical tools to model and analyze the grasp and control of under-actuated synergistic robotic hands. Chapter 12 identifies a mapping strategy [12] to transfer human hand synergies onto robotic hands with dissimilar kinematics and presents a novel software tool "SynGrasp", which includes functions for the definition of hand kinematic structure, the coupling between joints induced by a synergistic control, compliance at the contact, joint and actuator levels and graphical functions. Chapter 13 describes new, general approaches for the analysis of grasps with synergistic underactuated robotic hands [11]. Two different approaches to the analysis are presented: the first one is based on a systematic combination of the quasi-static equations, while the second one focuses on the strategies to determine the feasibility of the pre-defined tasks, operating a systematic decomposition of the solution space of the system. Chapter 14 presents a model of the human hand calculated from data obtained from a small number of sensors, which can be used for movement analysis in object exploration [21] and contact point analysis.

Finally, Chap. 15 describes how the concept of hand synergies, which has been used to control and design robotic hands with a reduced number of control inputs and actuators, can be also exploited to optimize the performance of hand pose reconstruction systems in terms of estimation accuracy and optimal design [3].

Overall, this book provides a comprehensive overview of the complex topic of sensorimotor synergies. The outstanding results of the studies described in the book demonstrate the advantages of our integrated approach in neuroscience and robotics.

References

1. Ajoudani A, Tsagarakis NG, Bicchi A (2012) Tele-impedance: teleoperation with impedance regulation using a body-machine interface. Int J Robot Res p 0278364912464668
2. Bianchi M (2012) On the role of haptic synergies in modelling the sense of touch and in designing artificial haptic systems. PhD thesis, Pisa, Italy
3. Bianchi M, Salaris P, Bicchi A (2013) Synergy-based hand pose sensing: optimal glove design. Int J Robot Res 32(4):407–424
4. Bicchi A, Gabiccini M, Santello M (2011) Modelling natural and artificial hands with sinergie. Phil Trans R Soc B 366:3153–3161
5. Brown C, Asada H (2007) Inter-finger coordination and postural synergies in robot hands via mechanical implementation of principal component analysis. In: IEEE-RAS international conference on intelligent robots and systems, pp 2877–2882
6. Castellini C, Fiorilla AE, Sandini G (2009) Multi-subject/daily-life activity emg-based control of mechanical hands. J Neuroeng Rehabil 6(1):41
7. Catalano MG, Grioli G, Farnioli E, Serio A, Piazza C, Bicchi A (2014) Adaptive synergies for the design and control of the pisa/iit softhand. Int J Robot Res 33:768–782
8. Ciocarlie MT, Goldfeder C, Allen PK (2007) Dimensionality reduction for hand-independent dexterous robotic grasping. In: IEEE/RSJ international conference on intelligent robots and systems, pp 3270–3275
9. Ehrsson HH, Kuhtz-Buschbeck JP, Forssberg H (2002) Brain regions controlling nonsynergistic versus synergistic movement of the digits: a functional magnetic resonance imaging study. J Neurosci 22(12):5074–5080
10. Filippidis IF, Kyriakopoulos KJ, Artemiadis PK (2012) Navigation functions learning from experiments: application to anthropomorphic grasping. In: 2012 IEEE international conference on robotics and automation (ICRA), pp 570–575. IEEE
11. Gabiccini M, Farnioli E, Bicchi A (2013) Grasp analysis tools for synergistic underactuated robotic hands. Int J Robot Res p 0278364913504473
12. Gioioso G, Salvietti G, Malvezzi M, Prattichizzo D (2013) An object-based approach to map human hand synergies onto robotic hands with dissimilar kinematics. Robotics p 97
13. Moscatelli A, Bianchi M, Serio A, Al Atassi O, Fani S, Terekhov A, Hayward V, Ernst M, Bicchi A (2014) A change in the fingertip contact area induces an illusory displacement of the finger. In: Haptics: neuroscience, devices, modeling, and applications, pp 72–79. Springer

14. Naceri A, Moscatelli A, Santello M, Ernst M (2014) Coordination of multi-digit positions and forces during unconstrained grasping in response to object perturbations. In: Haptics symposium (HAPTICS), 2014 IEEE, pp 35–40
15. Santello M, Baud-Bovy G, Jorntell H (2013) Neural bases of hand synergies. Front Comput Neurosci 7
16. Schieber MH, Santello M (2004) Hand function: peripheral and central constraints on performance. J Appl Physiol 96(6):2293–2300
17. Soechting JF, Flanders M (1997) Hand synergies during reach-tograsp. J Comput Neurosci 4:29–46
18. Spanne A, Jorntell H (2013) Processing of multi-dimensional sensori- motor information in the spinal and cerebellar neuronal circuitry: a new hypothesis. PLoS Comput Biol 9(3):e1002979
19. Taylor AM, Enoka RM (2004) Quantification of the factors that influence discharge correlation in model motor neurons. J Neurophysiol 91(2):796–814
20. Valero-Cuevas FJ (2000) Predictive modulation of muscle coordination pattern magnitude scales fingertip force magnitude over the voluntary range. J Neurophysiol 83(3):1469–1479
21. van Polanen V, Tiest WMB, Kappers AM (2014) Target contact and exploration strategies in haptic search. Sci Rep 4
22. Zatsiorsky VM, Li Z-M, Latash ML (2000) Enslaving effects in multi-finger force production. Exp Brain Res 131(2):187–195

Part I
Neuroscience

Chapter 2
Dexterous Manipulation: From High-Level Representation to Low-Level Coordination of Digit Forces and Positions

Qiushi Fu and Marco Santello

Abstract The ability to perform fine object and tool manipulation, a hallmark of human dexterity, is not well understood. We have been studying how humans learn anticipatory control of manipulation tasks to characterize the mechanisms underlying the transformation from multiple sources of sensory feedback to the coordination of multiple degrees of freedom of the hand. In our approach, we have removed constraints on digit placement to study how subjects explore and choose relations between digit forces and positions. It was found that the digit positions were characterized by high trial-to-trial variability, thus challenging the extent to which the Central Nervous System (CNS) could have relied on sensorimotor memories built through previous manipulations for anticipatory control of digit forces. Importantly, subjects could adjust digit forces prior to the onset of manipulation to compensate for digit placement variability, thus leading to consistent outcome at the task level. Furthermore, we found that manipulation learned with a set of digits can be transferred to grips involving a different number of digits, despite the significant change in digit placement distribution. These results have led us to propose a theoretical framework based on high-level representation of manipulation tasks can be learned in an effector-independent fashion and transferred to some, but not all that contexts. We discuss these findings in relation to the concept of motor equivalence and sensorimotor integration of grasp kinematics and kinetics.

2.1 Introduction

Goal-directed dexterous manipulation is accomplished by controlling the distribution of digit forces among multiple digits that grasp the object to generate or resist object motions. It should be emphasized that most hand-object interactions humans encounter in activities of daily living do not constrain where each digit is placed on

---

Q. Fu · M. Santello (✉)
School of Biological and Health Systems Engineering, Arizona State University, Tempe AZ, USA
e-mail: Marco.Santello@asu.edu

Q. Fu
e-mail: qiushifu@asu.edu

M. Bianchi and A. Moscatelli (eds.), *Human and Robot Hands*,
Springer Series on Touch and Haptic Systems, DOI 10.1007/978-3-319-26706-7_2

the object, or how many digits are used. Therefore, for a given manipulation a given object can be grasped in many different ways. Mechanically, both digit positions and forces contribute to meet the manipulation task requirement. Therefore, studying grasping kinematics or digit forces in isolation limits the extent to which we can advance our understanding of how the CNS resolves the classic problem of redundant degrees of freedom in dexterous manipulations [1]. Specifically, focusing on either grasping kinematics or kinetics overlooks the underlying sensorimotor transformations from multiple sources of sensory feedback to multiple motor commands necessary to modulate multiple digit forces to positions.

The vast majority of grasping studies over the past 30 years has examined either digit positions or forces. For instance, studies of grasp kinematics have shown that subjects tend to maximize the end-state comfort by choosing sub-optimal hand locations for the initial grasp when the object has to be transported to a different location [2]. Hand shaping can be also changed according to the task goals [3] and learned object dynamics [4, 5], even though the hand degrees of freedom are controlled through synergistic motion patterns [16, 17]. However, these studied did not measure digit forces. In contrast, studies of grasp kinetics have examined the control and coordination of digit forces by using objects that constrained digit contacts, hence removing the natural digit placement variability that occurs when subjects can choose where to grasp an object. For constrained grasping studies, subjects are usually required to grasp force/torque sensors mounted at fixed locations on the object. These studies have revealed that, similar to what had been reported for reach-to-grasp kinematics, digit forces are also modulated as a function of object properties and task goals, such as object weight [6], surface friction [7], and shape [8]. Furthermore, it has been shown that digit forces can be controlled in the form of synergies, effectively reducing the number of independent degrees of freedom [9, 10]. Although research of grasp kinematics and kinetics in isolation has revealed important insights, a major question remained: *how are digit forces and positions synergistically controlled as a unit?*

To illustrate the above scenario of grasping at unconstrained contacts, let us consider the task of two-digit grasping, lifting, and balancing an object whose mass distribution is asymmetrical. This particular object would tilt towards the heavier side if one does not actively produce a torque to counterbalance the external torque, i.e., a compensatory torque. In order to prevent the tilt, one has to first position the digits on the object, then gradually exert forces through the digit-object contacts to generate a torque at object lift onset in the direction opposite to the torque caused by object mass distribution (Fig. 2.1a). Whereas the task space in this example is only three dimensional (i.e., translations and rotation in the x-y plane; Fig. 2.1a), the actions of the hand lie within a space that consists of the grip forces (normal to the surface, f_x), load forces (tangential to the surface, f_y), and positions of the contacts from at least two digits $C_i = (x_i, y_i)$. Mathematically, a two-digit planar grasp can be represented as

$$F_0 = Gf_c \quad f_c \in FC \tag{2.1}$$

where F_o is the task space consisting of net forces exerted in the horizontal and vertical directions as well as the torque in the x-y plane, G is a 3×4 'grasp matrix'

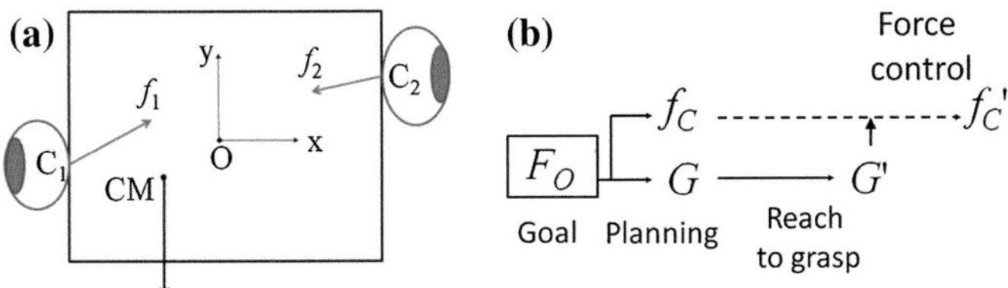

Fig. 2.1 Redundancy in two-digit dexterous manipulation. **a** Example of two-digit planar manipulation. **b** Temporal evolution of a manipulation task. *G* and *G'* denote planned and actual digit contact distributions, f_C denotes planned digit forces, and f_C' denotes the digit forces required to attain the manipulation task goal

determined by digit positions (x_1, y_1) and (x_2, y_2), f_c is a vector consisting of digit forces [f_{x1}, f_{y1}, f_{x2}, f_{y2}], and *FC* is the frictional constraints.

This redundancy poses three major challenges to the CNS. First, without physical constraints on the objects' contact locations, the CNS has to determine where to position the digits among many possible locations. Note that some contact locations may enable a grasp matrix *G* that is more suitable (e.g., less force is required) for the upcoming manipulation than other contact locations. Second, the actual contact sites may be quite variable from trial to trial due to noise in motor planning and/or execution [11] (Fig. 2.1b). Variability in digit placement implicitly requires the CNS to select appropriate digit forces to ensure task completion. This means that simply reproducing the digit force distribution used in the manipulation performed in the previous trial (i.e., stored as sensorimotor memory [12]) may not generate the same outcome at the task level if the current and previous digit contact distribution are significantly different. In this case, the digit forces required for the manipulation will not match planned digit forces (f_C and f_C', respectively; Fig. 2.1b). Lastly, once the manipulation is successfully learned, are the neural representations of learned hand-object actions independent from the effectors used during learning? If so, one would predict that humans should be able to perform the same manipulation task by using a set of effectors, e.g., digits or hand, that differs from that used to learn the manipulation.

In this chapter we review our work addressing the above three question and highlight open questions for future research. Specifically, we investigated the coordination of digit positions and forces by using a novel apparatus that allows subjects to choose where they grasp while still providing measurement of digit forces. The results reviewed here are the first direct evidence about how the CNS exploits the sensorimotor redundancy in both grasp kinematics and kinetics.

2.2 Materials and Methods

Subjects. Twenty-four right-handed subjects (12 females and 12 males, ages 20–26) and ten right-handed volunteers (4 males and 6 females, ages 19–24) participated in the first and second study, respectively. They had normal or corrected-to-normal

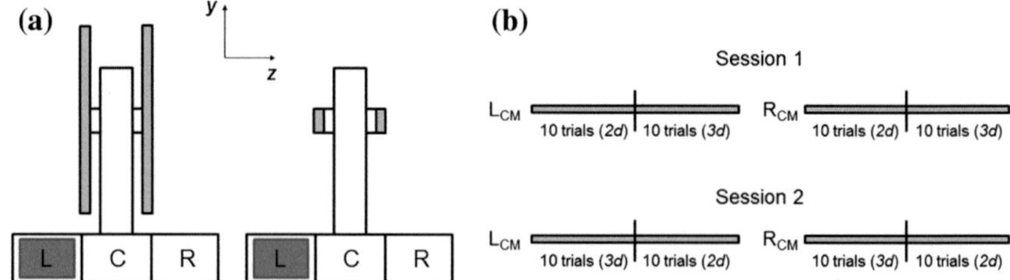

Fig. 2.2 Experimental setup. **a** The unconstrained (*left*) and constrained (*right*) devices used in Experiments 1 and 2. **b** The experimental sequence used in Experiment 2. The labels "2*d*" and "3*d*" denote two- and three-digit grip, respectively, whereas L_{CM} and R_{CM} denote left and right center of mass, respectively

vision, no previous history of orthopedic, neurological trauma, or pathology of the upper limbs and were naive to the purpose of the study. Subjects gave their informed consent according to the declaration of Helsinki and the protocols were approved by the Office of Research Integrity and Assurance at Arizona State University.

Apparatus. We used a custom-made grip device to measure digit forces and their points of application during manipulation tasks (Fig. 2.2a; see [13] for details). Forces and torques exerted by the thumb and fingers were recorded by two six-component force/torque (F/T) transducers (Nano-17, ATI Industrial Automation) mounted collinearly on each side of the object handle. Force and torque data were sampled at 1 kHz by 12 bit analog-to-digital converter boards (PCI-6225, National Instrument). To allow unconstrained placement of the digits, the grip surfaces consisted of two parallel long PVC plates (length and width: 140 mm and 22 mm, respectively) each mounted vertically on an F/T transducer and were covered with 100-grit sandpaper. These two plates were replaced by two small circular plates (diameter: 22 mm) during constrained grasping experiment (Fig. 2.2a). The distance between the two grip surfaces (grip width) was always 6.07 cm. A Plexiglass box attached underneath the grip apparatus was used to change the mass distribution to the left, right or center of the grip device by inserting a mass (400 g) into one of three compartments. The total mass of the grip device and load was 790 g. A torque in the frontal plane of −255 or +255 N/mm is introduced when the load was placed in the left or right compartment (L and R), respectively. For Experiment 1, we also placed the mass in the center such that no compensatory torque was required to lift the object straight. The visual identification of the actual object center of mass (CM) was blocked from view by a lid.

For Experiment 1, object kinematics was recorded with a magnetic tracker at 120 Hz. For Experiment 2, hand and object kinematics were recorded using an active marker 3D motion capture system at a sampling rate of 480 Hz. Subjects were outfitted with active markers on the fingernails of thumb, index, and middle fingers.

Experimental Procedures. Subjects sat comfortably with the hand resting on a table with the elbows flexed. The apparatus was placed at a distance of 30 cm from the hand start position, and the midpoint of the apparatus was aligned with subjects' right shoulders. For each trial, after a verbal signal from the experimenter, subjects reached from this start location, grasped the grip surfaces with the tip of a required set of digits of the right hand, lifted the grip device at a natural speed to a height of ~10 cm, held it for ~1 s, and replaced it. Subjects were instructed to extend the non-involved fingers throughout the task. At the beginning of each block of trials, we instructed subjects to minimize object roll during the lift. The between-trial interval within a block was ~10 s.

Experiment 1. Subjects were assigned to one of two groups ($n = 12$ for each group): the unconstrained group used the apparatus with long graspable surfaces, whereas the constrained group used the apparatus with small circular graspable surfaces (left and right object, respectively; Fig. 2.2a). Both subject groups were given the same task instructions. After three practice trials (with center CM location), subjects performed three blocks of ten consecutive trials per CM location for a total of 30 experimental trials. Subjects were informed that CM location would remain the same for the entire block of trials, but they could not anticipate CM location at the beginning of each block of trials as changes of object CM across blocks of trials were performed out of view. The consecutive presentation of a given object CM location was used to allow subjects to learn implicitly the magnitude and direction of the external torque caused by the added mass. The order of CM blocks of trials was counterbalanced across subjects. On average, the time between blocks of trials was 1 min, respectively.

Experiment 2. Each subject performed the task the unconstrained grasp surfaces with two grip types: (1) two-digit grasping (thumb and index finger; 2*d*) and (2) three-digit grasping (thumb, index, and middle fingers; 3*d*). Subjects performed 10 2*d* trials followed by 10 3*d* trials ($2d \rightarrow 3d$) with left CM (L_{CM}). After a short break (~ 20 s), subjects performed the $2d \rightarrow 3d$ experimental condition with right CM (R_{CM}). Each subject was tested again two weeks later but on a trial sequence opposite to that experienced on his/her first experimental session, i.e., $3d \rightarrow 2d$ on the L_{CM} condition followed by $3d \rightarrow 2d$ on the R_{CM} condition (Fig. 2.2b). The break between the two experimental sessions was designed to minimize potential positive or negative learning transfer effects from one sequence to the next. Prior to the experiment, subjects lifted the object once with each grip type (2*d* and 3*d*) with the load placed in the center compartment to familiarize with the task, texture, and weight of the grip device. Subjects were informed that the load could be placed either in the left or right compartment of the Plexiglass box and would remain the same for two blocks of trials. At the beginning of each block, subjects were told the number of digits to be used for the upcoming block of trials.

Data Processing. Force and position data were temporally aligned offline and analyses were performed using MATLAB. We analyzed the following variables (see [13, 14] for details): (1) Object lift onset: the time at which the vertical position of the grip device crossed and remained above a threshold for 200 ms; (2) Object roll: the angle between the gravitational vector and the vertical axis of the grip device, and peak roll is the peak of object roll shortly (~150 ms) after object lift onset; (3) Digit forces: force perpendicular (grip force, GF) and parallel (load force, LF) to the grip surface; (4) Digit center of pressure (CoP): the vertical coordinate of the point of resultant digit force application, calculated for each digit using the force and torque output of each sensor (positive and negative values CoP denoted higher and lower CoPs relative to the center of transducer, respectively). Note that GF, LF, and CoP recorded on the finger side of the grip device are the resultant net forces and net center of pressure of both index and middle finger when subjects performed the task using the 3*d* grip. To quantify the modulation of individual digit position, we used the fingertip marker position defined as the vertical position of the marker on the nail of the thumb, index, and middle fingers.

We then used digit forces and CoP to compute the following performance variables: (a) the average of the digit grip forces (F_{GF}), (b) the difference between load forces exerted on the thumb and finger side of the grip device (ΔF_{LF}), and (c) the vertical distance between the thumb and index finger CoP on the thumb and finger side of the grip device (d_y). How these three variables are coordinated dictates whether the compensatory torque (T_{com}) necessary for minimizing object roll is attained before the object is lifted to balance the external torque caused by the added mass (see [15]). We use the following equation to approximate the relationship among these variables

$$T_{com} = F_{GF} \times d_y + \Delta F_{LF} \times \frac{w}{2} \tag{2.2}$$

where *w* is the distance between the graspable surfaces. To further understand how digit placement changed when there is a change in grip type (Experiment 2), we also computed the vertical distance between thumb and index finger markers (d_{tip}). Note that all of these performance variables were computed at object lift onset. Our analysis focused on object lift onset because this event allows for an unbiased estimation of *anticipatory* modulation of digit forces as a function of digit positions, i.e., prior to experiencing the external torque that occurs as soon as the object is lifted (for more details on this rationale, see [13]).

2.3 Experiment 1: Digit Force and Position Coordination in Unconstrained Grasping

In the first experiment, we studied how digit positions and forces are modulated and coordinated during learning the manipulation as well as throughout stable performance of our manipulation task. We also compared the results obtained using unconstrained grasps with the same task performed with constrained contacts as this model has been used by most grasp studies. Overall, subjects learned to generate compensatory torque equally well and at similar rates in both grasp conditions, even though the underlying control mechanisms differed significantly between the two subject groups.

Learning Constrained Versus Unconstrained Grasping. At the task level, compensatory torque generation (i.e., object roll minimization) was learned within the first three trials by both constrained and unconstrained subject groups (Fig. 2.2). Therefore, only data from the unconstrained group is described here. On trial 1, subjects exerted little or no compensatory torque (mean ± SE: −26.6 ± 16.9 N·mm, −14.1 ± 33.8 N·mm, and 22.5 ± 17.5 N·mm for left, center and right CM, respectively). Unlike the compensatory torques developed after trial 1, the direction and magnitude of these torques were not correctly scaled to the external torque. On trials 2 and 3, however, the compensatory torque gradually approached the external torque and settled at a mean value of −188.3 (±13.9) N·mm and 189.7 (±17.0) N·mm (right and left CM, respectively) from trial 4 through 10 (Fig. 2.3). The compensatory torque at lift onset for the center CM condition changed little after trial 1, reaching a mean value of 11.3 (±13.1) N·mm. Despite the fact that subjects' performances were different for center vs. left and right CM locations (CM × Trial interaction $p < 0.001$), post hoc comparisons between neighboring trials (1 vs. 2, 2 vs. 3, and so forth) revealed that subjects learned to generate anticipatory compensatory torque to minimize object roll early in the trial sequence, the only significant

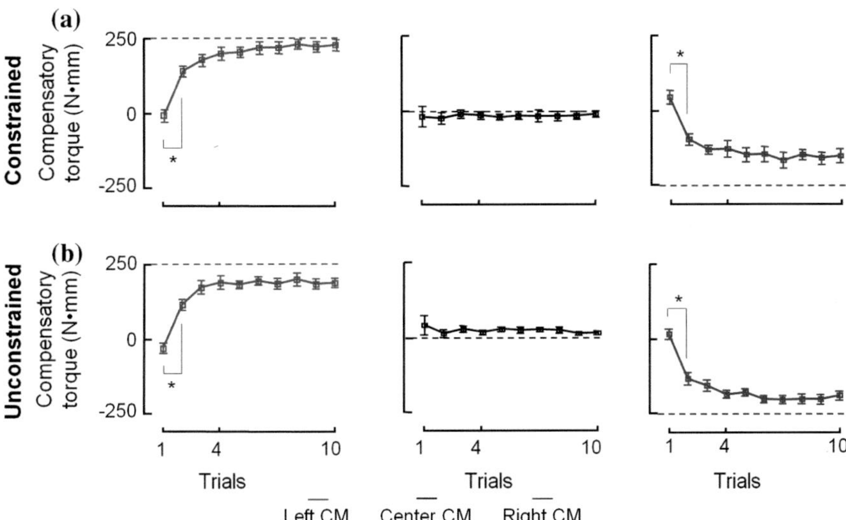

Fig. 2.3 Anticipatory control of compensatory torque as a function of trial. Data in **a** and **b** denote compensatory torque for the constrained and unconstrained grasp conditions, respectively. *Dashed horizontal lines* denote the external torque caused by the added mass. For graphical purposes, the external and compensatory torques are plotted with the same sign. All data are means averaged across subjects (±S.E.). *Asterisks* indicate significant differences ($p < 0.05$) between trials. Adapted from [13]

difference in compensatory torque occurring between trial 1 and trial 2 ($p < 0.05$ for both right and left CM in both groups). The relation between peak object roll and trials paralleled the relation between compensatory torques and trials shown in Fig. 2.3.

For digit positions, the centers of pressure (CoP) of thumb and index finger at object lift onset were modulated as a function of trial and object CM location. Figure 2.4a shows d_y averaged across all subjects as a function of trial for each CM location and subject group. On the first trial, subjects tended to position the digits collinear to each other regardless of CM location. After trial 1, when lifting the object during the center CM condition, thumb and index finger CoP tended to remain collinear across all subsequent trials in both groups. In contrast, left and right CM locations elicited opposite patterns of digit CoP modulation. Specifically, the thumb CoP tended to be positioned progressively higher and lower relative to the index CoP for the left and right CM locations, respectively, and for both subject groups (CM × Trial interaction, $p < 0.001$; Group × Trial interaction, $p > 0.05$). Similar to the above results on compensatory torque, post hoc comparisons between neighboring trials revealed that the only significant change in dy occurred between trial 1 and 2 but only for right and left CM in both groups ($p < 0.05$).

For digit forces, grip forces (F_{GF}) tended to increase as a function of trial for left and right CM conditions and decrease for center CM in both subject groups (Fig. 2.4b; CM × Trial interaction, $p < 0.001$, Group × Trial interaction, $p > 0.05$). However, post hoc analyses showed that these trends were significant only for the left CM condition in both groups ($p < 0.001$ and $p < 0.005$ for unconstrained and constrained groups, respectively). Subjects also used different patterns of load force

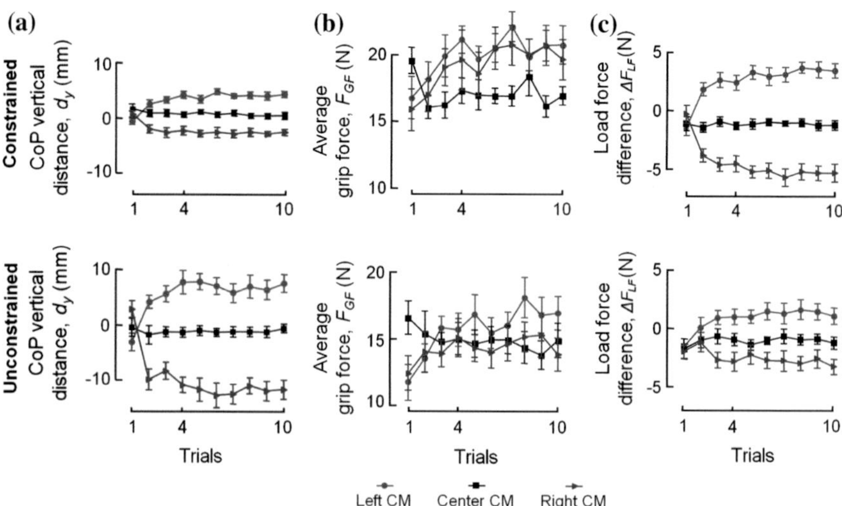

Fig. 2.4 Learning of digit placement and forces. All data averaged across subjects (±S.E.). *Top* and *bottom rows* show data from constrained and unconstrained groups, respectively. Adapted from [13]

distribution across CM locations (Fig. 2.4c). Specifically, thumb and index finger load forces remained symmetrical across all trials in both subject groups in the center CM condition. In contrast, the difference between thumb and index finger load forces (ΔF_{LF}) tended to be modulated as a function of trial early in the trial sequence to then remain relatively constant for left and right CM conditions for both subject groups. On the first trial, subjects tended to use nearly symmetrical load forces for both CM and subject groups. After trial 1, load forces applied by the thumb and index finger were applied asymmetrically to counteract the CM asymmetries. Specifically, the thumb load force tended to be progressively larger and smaller relative to the index load force for the left and right CM conditions, respectively, in both subject groups (CM × Trial interaction, $p < 0.001$; Group × Trial interaction, $p > 0.05$). However, post hoc comparisons between neighboring trials revealed that only the constrained group modulated ΔF_{LF} significantly from trial 1–2 for both CM conditions (both $p < 0.05$).

Since stable level of task performance was attained within the first 3 trials (Fig. 2.3), we used trial 3 as the cut-off after which (trial 4 through 10) we defined subjects' performance as stable, i.e., where further practice with the manipulation task did not lead to statistically significant improvements in compensatory torque at object lift onset and object roll minimization. Therefore, we examined the magnitude of digit forces and CoP during the last seven trials of each block. The two subject groups exhibited significant differences in their overall strategy. Specifically, constrained grasp trials relying mostly on modulation of grasp kinetics (force application), and unconstrained grasp trials relied primarily on modulation of grasp kinematics (digit placement on the object). Digit position modulation in the constrained group was significantly smaller than that exhibited by the unconstrained group in left and right CM conditions (Fig. 2.4a; 3-way analysis of variance, ANOVA, on factors Group, CM, and Trial; CM × Group interaction, $p < 0.001$; post hoc tests on Group effects within right CM: $p < 0.05$; non-significant Group effects within left CM). Furthermore, the constrained group used larger grip force than the unconstrained group across trial 4–10 (Fig. 2.4b; main effect of CM, $p < 0.01$; main effect of Group, $p < 0.001$). Post hoc tests also revealed that subjects used significantly larger grip force only for left and right CM conditions ($p < 0.05$). Lastly, the constrained group showed larger asymmetry of digit load forces than the constrained group in left and right CM conditions (Fig. 2.4c; CM × Group interaction $p < 0.001$; post-hoc tests on Group effects within left and right CM: both $p < 0.05$). These results suggested that, despite inter-subject variability, the subjects had common strategies according to the object physical properties. We speculate that such strategies were implemented to optimize the motor output by reducing the total digit force, thus minimizing the energy cost and motor noise [13]. Note that unlike the result obtained from random perturbation trials presented in the Chap. 3 by Naceri and colleagues, all subjects adapted similar preferred strategies in an anticipatory fashion in our experiment. This is because the perturbation induced by the added mass following object lift is predictable.

Covariation of Digit CoPs and Digit Forces. The above analysis revealed that digit forces and positions at object lift onset were controlled differently depending

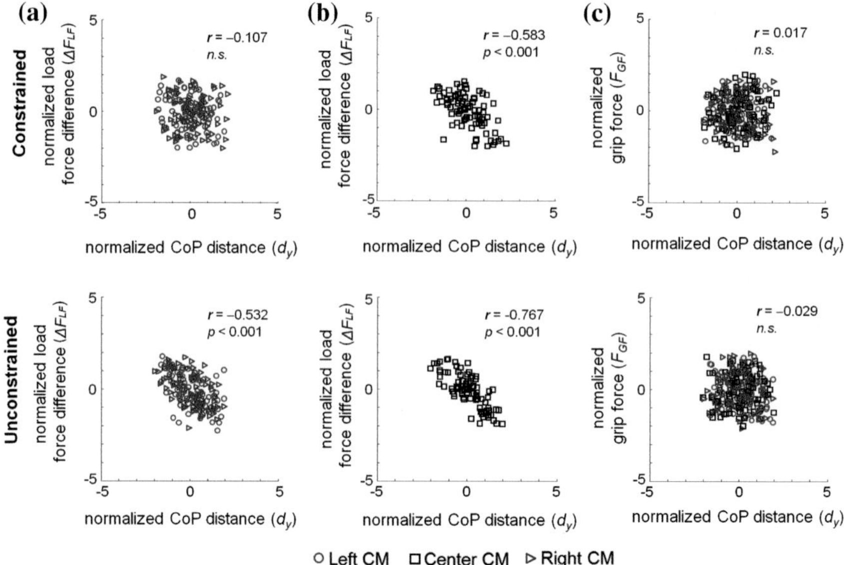

Fig. 2.5 Relations between digit centers of pressure, grip force, and load force. Data are from constrained and unconstrained grasp trials 4 through 10 from each subject and CM condition (*top* and *bottom row*, respectively) and are shown in normalized form. Data in **a** and **c** are from *left* and *right* CM conditions, whereas data in **b** are from the center CM condition. The Pearson's r-value and corresponding p-value are shown in each panel (n.s. = not significant, $p > 0.05$). Adapted from [13]

on whether or not the grip device constrained digit placement. Surprisingly, however, subjects from the unconstrained and constrained groups learned to generate compensatory torques with similar consistency (Fig. 2.3). This was confirmed by a lack of a significant Group effect on the standard deviation of T_{com} of the mean compensatory torque averaged from trial 4–10 ($p > 0.05$). This result is remarkable particularly when considering that the variability of digit placement at object lift onset of the unconstrained group was significantly larger than the constrained group. With regard to standard deviation of individual digit CoP, we found only a significant main effect of Group ($p < 0.001$). The standard deviation of d_y was significantly different across subject groups and CM (both $p < 0.001$).

In contrast, there was no significant difference between the two groups with regard to the standard deviation of either digit load forces or grip forces ($p > 0.05$). As T_{com} is the net result of d_y, F_{GF}, and ΔF_{LF} (Eq. 2.2), the large variability in digit placement in the unconstrained group was effectively compensated by digit force modulation such that trial-to-trial variability of T_{com} was similar to the constrained group. We therefore examined how subjects modulated, on a trial-to-trial basis, digit forces as a function of position. Linear regression analyses on data normalized to zero mean and unit standard deviation (see [13] for details) was performed. We observed significant negative correlations between d_y and ΔF_{LF} in both the unconstrained group (r =

−0.615, $p < 0.001$) and constrained group ($r = -0.263$, $p < 0.001$). Furthermore, the correlation coefficient of the unconstrained group was significantly larger than that of the constrained group ($p < 0.001$). We also found that center CM was different from left and right CM conditions. Specifically, for the center CM condition, both subject groups showed negative correlations between d_y and ΔF_{LF} (Fig. 2.5b). This correlation was significantly larger in the unconstrained than in the constrained group ($p < 0.05$). Interestingly, for left and right CM conditions, the constrained group did not exhibit a significant correlation. In contrast, negative correlations were still found for the unconstrained group (Fig. 2.5a). Lastly, the strength of the correlation between d_y and ΔF_{LF} was significantly larger in the unconstrained than in the constrained group ($p < 0.05$). We found no significant correlation between d_y and F_{GF} or between ΔF_{LF} and F_{GF} in either subject group.

2.4 Experiment 2: Transfer of Learned Manipulation Between Different Grip Types

Experiment 1 revealed that natural trial-to-trial variability occurs when subjects can choose where to grasp an object. Remarkably, such variability persists even after the manipulation has been learned (trials 4–10). An important result was that subjects actively compensate for digit placement variability by modulating digit forces such that the required compensatory torque is attained at object lift onset. As this trial-to-trial variability in digit placement was relatively small, in a second experiment we introduced a large change in digit position after subjects had learned the manipulation task. This was achieved by asking subjects to use a different grasp configuration by adding or removing one digit relative to the grasp configuration used to learn the task. The ability to transfer learned manipulation across grasp types was examined at both task and digit level. Overall, we found complete transfer of task performance despite significantly different digit position and force patterns.

Learning and Learning Transfer of Compensatory Torque. Subjects were able to learn the manipulation task with both two-digit and three-digit grasp, in a similar fashion to that described for Experiment 1. Specifically, T_{com} averaged across all subjects changed significantly as a function of consecutive practice in the pre-switch block (main effect of Trial, $p < 0.001$; Fig. 2.6). On average, all subjects learned to anticipate the T_{com} necessary to minimize object roll within the first 3 trials regardless of grip type and CM location, after which T_{com} did not change any further on subsequent trials (all tests on trials 4–10: $p > 0.05$).

Following a change in grip type, i.e., at the beginning of the second block (trial 11, Fig. 2.6), we performed ANOVA for each CM location to examine the immediate effect of changing grip type on T_{com}. Subjects were able to generate T_{com} whose magnitude was statistically indistinguishable from that generated on the pre-switch trial (trial 10) for all but one experimental condition ($3d \rightarrow 2d$, L_{CM}, $p = 0.035$).

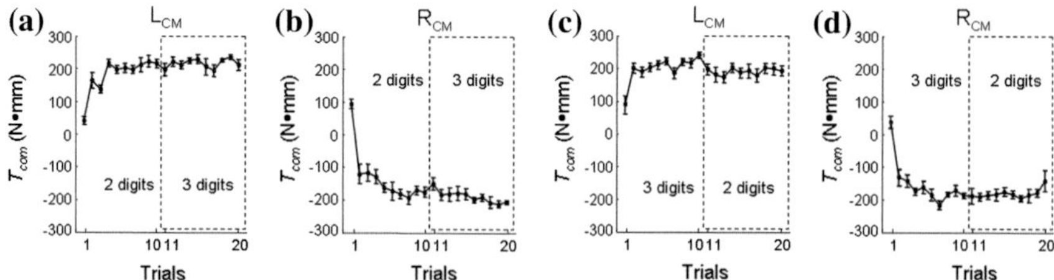

Fig. 2.6 Learning curves of compensatory torque: pre- and post-grip type switch. Trials within the *dashed box* (11–20) indicate the grip type subjects switched to after learning the manipulation with a different grip type. T_{com} denotes compensatory torque, whereas L_{CM} and R_{CM} denote left and right center of mass, respectively. Data in **a** and **b** are from the $2d \rightarrow 3d$ condition, whereas data in **c** and **d** are from the $3d \rightarrow 2d$ condition. "L_{CM}" and "R_{CM}" denote left and right center of mass, respectively. Data are averages of all subjects (±S.E.). Adapted from [14]

However, no significant differences were found when comparing peak object roll on trial 10 versus 11 on any of the four experimental conditions. This indicates that the statistically significant difference for the $3d \rightarrow 2d$ L_{CM} condition did not have significant behavioral consequences on the manipulation, thus suggesting that anticipatory control of T_{com} in the pre- and post-switch trials was equally appropriate to attain the task goal. Furthermore, we examined average differences between 7 trials pre- versus post-switch in grip type using ANOVA with repeated measures for each CM location with within-subject factors of Trial (7 levels; 7 pre- and 7 post-switch) and Grip type (2 levels, pre- and post-switch). We found no significant main effect of Trial, Grip type, or interaction ($p > 0.05$ for each experimental condition). This suggests that no further learning of T_{com} occurred before and after the switch in grip type, the average T_{com} being statistically similar for the two grip types. Similarly, no significant main effect of Trial, Grip type, or interaction were found on peak object roll as well ($p > 0.05$ for each experimental condition).

Change of Digit Positions and Forces Following a Change of Grip Type. The positive learning transfer of T_{com} to a different grip type implies that subjects were able to coordinate, in an anticipatory fashion, the components that generate the T_{com}: d_y, ΔF_{LF}, F_{GF}. Since there are infinite number of combination of these components (see Eq. 2.2), we analyzed each T_{com} component separately.

Digit Center of Pressure. Subjects used significantly different vertical separations between digit center of pressure (d_y) after switching grip type on all but one experimental condition ($3d \rightarrow 2d$, R_{CM}; Fig. 2.7). For the left CM location, subjects significantly increased d_y when adding middle finger to the grip and decreased d_y when removing one finger from the grip, whereas an opposite pattern was found for the right CM location. In addition to the change found in the net center of pressure on the finger side when adding or removing a finger, we found that the distance between thumb and index finger marker (d_{tip}) was significantly modulated such that the index finger was positioned higher when adding the middle finger and lower when remov-

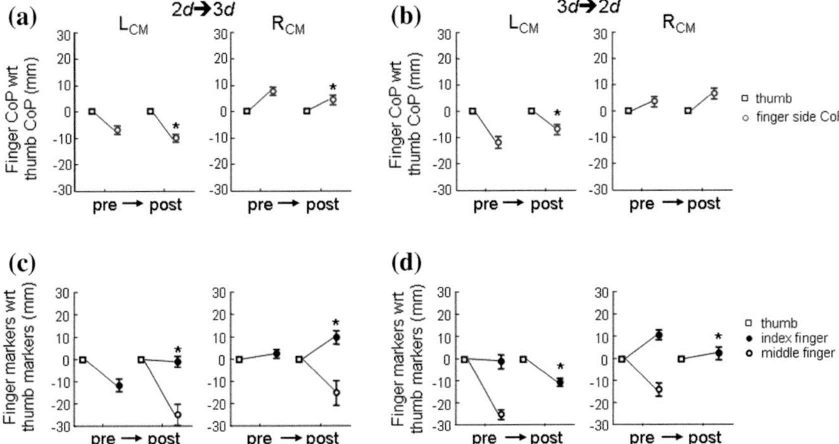

Fig. 2.7 Immediate learning transfer of digit center of pressure and position from pre- to post-grip type switch trials. *Asterisks* denote a statistically significant difference ($p < 0.05$) between pre- and post-switch trials. **a** and **b** are digit center of pressure measured by F/T sensors. **c** and **d** are digit position measured by motion tracking. Data are averages of all subjects ($\pm$ S.E.). Adapted from [14]

ing the middle finger (Fig. 2.7). Therefore, subjects *immediately* used significantly different digit placement distribution when changing grip type.

After the immediate adaptation (i.e., trial 11) following a change in grip type, there were no further significant modulation of d_y (no significant Trial effect or Trial $\times$ Grip interaction, $p > 0.05$). Specifically, the new digit placement was maintained for all but one experimental condition (significant Grip effect for three conditions and non-significant Grip effect for one condition, $3d \rightarrow 2d$, R_{CM}; Fig. 2.8a). With regard to the relative fingertip positions, d_{tip} was significantly modulated in a similar fashion to that observed in the immediate adaptation in all conditions (significant Grip type effect only).

Digit Load Forces. Subjects used significantly different load force sharing (ΔF_{LF}) after switching grip type on only one experimental condition ($2d \rightarrow 3d, L_{CM}$). However, unlike the immediate adaptation in the ΔF_{LF}, three out of four experimental conditions showed long term modulation of ΔF_{LF} throughout the first post-switch trials in response to the modulation of d_y (significant Grip effect; $p < 0.05$). ΔF_{LF} remained unchanged only in $3d \rightarrow 2d$, R_{CM} condition in which d_y was not modulated significantly (Fig. 2.8b). However, there were no significant Trial effect or Trial $\times$ Grip type interaction on ΔF_{LF}. In general, subjects tend to use larger load force difference in $2d$ than $3d$ grip.

Digit Grip Force. No significant main effect of Grip type ($p > 0.05$ for each experimental condition) were found on the immediate post-switch trial, indicating that subjects exerted similar net grip forces regardless of the number of digits used for

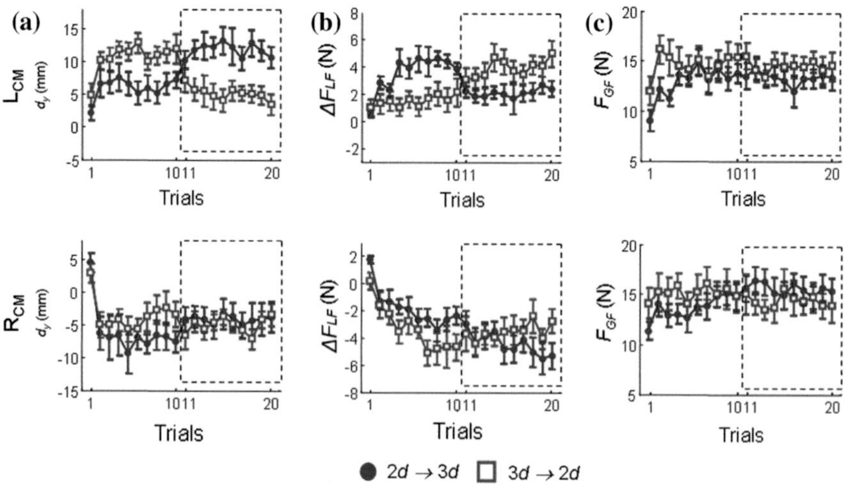

Fig. 2.8 Learning curves of digit center of pressure and forces: pre- and post-grip type switch. Each plot shows data from two-digit grip trials followed by three-digit grip trials (*squares*) and trials performed in the reverse order (*circles*). **a–c** are dy, ΔF_{LF}, F_{GF}, respectively, across pre- and post- transfer blocks. Data are averages of all subjects (±S.E.). Adapted from [14]

the grasp. However, whereas grip force is provided by index finger only in 2*d*grip on the finger side, the middle finger may contribute substantially in 3*d* grip for left CM condition. In the trials following the switch in grip type, subjects also exerted similar grip forces to those exerted before the switch (Fig. 2.8c), with lack of significant main effects of Grip type, Trial, or interaction ($p > 0.05$ for each experimental condition).

2.5 Discussion

This chapter reviewed recent evidence describing how the CNS addresses the redundancy of available kinematic and kinetic solutions to perform dexterous manipulation. It should be noted that the control of grasp kinematics and kinetics are part of a continuum, i.e., digit force distribution must take into account digit contact distribution. Nevertheless, until recently most of the literature on grasping has focused on either kinematic or kinetic synergies. Specifically, biomechanical analyses of multi-digit grasping have characterized prehension synergies that emerge from 'chain effects' through which obligatory and non-obligatory relations among digit forces and torques emerge (for review, see [10]). Similarly, studies of hand shaping during reach-to-grasp have characterized kinematic synergies, i.e., systematic covariation patterns of digit joint angles that can be described by a very small number of principal components across a wide variety of object shapes and sizes ([16, 17]; for review see [18]). Both of these approaches have revealed significant insights into motor control strategies that result in dimensionality reduction across the available degrees of freedom (i.e., joints, forces). However, and given the above-mentioned continuum of grasp kinematic and kinetics, a major gap remains: *How does the CNS control synergies that combine the control of digit positions and forces?* Here, the term 'synergy' is used as a broader term that describes the covariation of variables in a high-dimensional space that consists of both kinematic and kinetic components. Kinematic or kinetic synergies are throught to originate from biomechanical and neural constraints. However, the synergy between kinematic and kinetic variables discussed here denotes the ability of the CNS to integrate sensory information in the continuum of reaching-grasping-manipulation. This allows the system to make corrections to accomplish high-level goals that are independent of the configuration and forces of the hand.

2.5.1 Redundancy of Kinematic and Kinetic Solutions Through Digit Force-to-Position Modulation

As described in the Results section, learning of dexterous manipulation appears to occur despite execution and/or planning noise causing trial-to-trial variability in digit placement [13]. These observations were confirmed using a virtual reality environment where digit placement variability was induced by changing object width across trials [19]. Even when differences in digit placement are not accidental, but intentional, i.e., when we grasp an object using a set of digits that was not used to learn a given manipulation, the sensorimotor system can still re-organize digit force distribution to a new digit contact distribution [14]. These two scenarios, characterized by small and large variability in digit placement, respectively, capture a critically important sensorimotor ability, digit force-to-position modulation, which we propose as a hallmark of dexterous manipulation in humans. It should be noted that modulation of digit forces to positions requires a synergy-based mechanisms, whereby multiple sources of sensory feedback are integrated and used to generate a set of forces that can satisfy the task requirements. Remarkably, and despite the large number of degrees of freedom involved in object manipulation, subjects are able to adjust multiple digit forces as a function of digit contact distribution *within a few 100 ms*, i.e., between contact and onset of manipulation.

Equation 2.2 describes the relation between two contacts and their forces for generating a given torque in two dimensions. As mentioned in the Methods, the desired compensatory torque can theoretically be attained through a number of infinite digit force-position relations, i.e., an example of the Bernstein's problem of redundant degrees of freedom [1]. However, we found that subjects respond to trial-to-trial variability in digit placement by (a) exerting similar grip forces, thus reducing a free variable to a constant, and (b) linearly modulating the difference between thumb and index finger load forces as a function of the vertical distance between the digits (Fig. 2.5) [13]. This relatively simple relation describes a family of digit position-force relations that ensure the attainment of the same compensatory torque. Further studies have provided additional supporting evidence on the covariations of digit positions and forces as a way of synergistic control in other dexterous manipulation tasks (see Chap. 3 for details).

Although the mechanisms responsible for digit force-to-position modulation remain to be determined, behavioral evidence suggests that humans are particularly skilled at sensing digit placement during digit force production for large vertical distances between the fingertips [20]. Yet, how this sensory feedback is used to select the appropriate digit force distribution is not known. However, a recent study suggests that feedback-mediated force corrections occur to a greater extent when grasping an object at unconstrained than constrained contacts [21]. This result is compatible with the notion that the control of digit forces for constrained grasps benefits from the fact that the same digit forces can be used across trials, and therefore one would expect a greater reliance on sensorimotor memory of digit forces used in previous manipulations. In contrast, trial-to-trial variability in digit placement found for unconstrained grasps demands that forces are distributed in a way that reflects the actual digit contact distribution. Hence, corrective forces responses occurring between contact and onset of manipulation in unconstrained grasping may reflect such compensatory mechanisms triggered by sensing the mismatch between planned and actual digit placement (Fig. 2.1b).

2.5.2 High-Level Representation of Learned Manipulation

The above-described digit force-to-position modulation has led to the proposition that the sensorimotor system builds a high-level representation of learned manipulation, which for the above-reviewed studies corresponds to the desired compensatory torque [13, 14]. This concept, which is a fundamental component of the theoretical framework underlying motor equivalence [22–25], is based on the notion that the neural representation of learned motor behavior can be dissociated from the effectors (muscles, joints, limbs, etc.) with which the behavior was learned. Therefore, a learned motor behavior could be performed at a similar level of performance even when the effectors engaged in learning the behavior differ from those used to perform it. For the case of dexterous manipulation at unconstrained contacts, the ability to transfer a compensatory torque learned with two digits to three digits, or vice versa, could be interpreted as an example of motor equivalence. It has been pointed out, however, that effector-independence for motor execution is not routinely observed. Specifically, a number of notable exceptions have been reported in the literature, whereby changing the context of a learned manipulation prevents subjects to perform the same manipulation. For example, subjects are unable to transfer a learned compensatory torque after the object is rotated 180° ([15, 26]). Failure to generate the same torque in the opposite direction to that learned through previous manipulations indicates that manipulation learning transfer is sensitive to the congruency between the frame of reference of the learned action and the object. Such congruency is maintained when exerting a learned compensatory torque in the same direction despite a change in grip type [14], but not when rotating the object, in which case subjects have to perform a mental rotation of the action [15, 26]. Additional examples of failure to transfer learned manipulation to different contexts [27–29] indicate that there are circumstances that prevent high-level representations of learned manipulation to be transferred. This experimental evidence, combined with evidence indicating successful transfer of learned manipulation [14], indicate that the sensorimotor system's ability to build synergies for coordinating multiple digit positions with forces and retrieve them is sensitive to several factors, including previous manipulation history and frames of references associated with the 'old' versus 'new' manipulation context.

2.5.3 Open Questions and Future Research

Our understanding of the factors underlying humans' ability to control multiple position or force variables in a synergistic fashion to perform dexterous manipulation has improved significantly over the past two decades, leading to the characterization of kinematic and kinetic synergies. The work reviewed in this chapter is the first attempt at answering the question of how the CNS combines kinematic and kinetic grasp synergies. Nevertheless, further work is needed to characterize the underlying sensorimotor mechanisms. Specifically, behavioral data suggest that unconstrained grasping is mediated by more corrective force responses than constrained grasping [21]. However, it remains to be determined how cortical areas within the so-called 'grasp circuit', which has been characterized in human and non-human primate studies of grasping and manipulation [30], interact when sensory feedback of digit placement must be integrated with digit force control. Similarly, to date there is no comprehensive theoretical framework describing the key features of manipulation tasks that allow or prevent learning transfer to different task contexts. Furthermore, we do not know why a given manipulation task can be transferred to some, but not all contexts. Lastly, another critical open question is: how does the CNS transition from kinematic synergies before contact to kinetic synergies after contact? It remains unclear whether kinetic synergies are simply a result of kinematic synergies interacting with the environment or originate from different neural mechanisms. Future work should combine complementary experimental approaches to study unconstrained grasping and manipulation, including brain imaging, non-invasive brain stimulation, virtual reality environments, and sensorized objects and/or gloves, to gain insight into sensorimotor mechanisms responsible for dimensionality reduction in dexterous manipulation.

Acknowledgments This work was made possible by a National Science Foundation grant BCS-1153034 "Collaborative Research: Sensory Integration and Sensorimotor Transformations for Dexterous Manipulation".

References

1. Bernstein NA (1967) The co-ordination and regulation of movement. Pergamon Press, Oxford
2. Cohen RG, Rosenbaum DA (2004) Where grasps are made reveals how grasps are planned: generation and recall of motor plans. Exp Brain Res 157(4):486–495
3. Ansuini C, Santello M, Massaccesi S, Castiello U (2006) Effects of end-goal on hand shaping. J Neurophysiol 95(4):2456–2465
4. Lukos JR, Ansuini C, Santello M (2007) Choice of contact points during multidigit grasping: effect of predictability of object center of mass location. J Neurosci 27(14):3894–3903
5. Lukos JR, Ansuini C, Santello M (2008) Anticipatory control of grasping: independence of sensorimotor memories for kinematics and kinetics. J Neurosci 28(48):12765–12774
6. Johansson RS, Westling G (1984) Roles of glabrous skin receptors and sensorimotor memory in automatic control of precision grip when lifting rougher or more slippery objects. Exp Brain Res 56(3):550–564
7. Johansson RS, Westling G (1988) Coordinated isometric muscle commands adequately and erroneously programmed for the weight during lifting task with precision grip. Exp Brain Res 71:59–71
8. Jenmalm P, Johansson RS (1997) Visual and somatosensory information about object shape control manipulative fingertip forces. J Neurosci 17(11):4486–4499
9. Santello M, Soechting JF (2000) Force synergies for multifingered grasping. Exp Brain Res. 133(4):457–467
10. Zatsiorsky VM, Latash ML (2004) Prehension synergies. Exerc Sport Sci Rev 32(2):75–80
11. Faisal A, Selen LPJ, Wolpert DM (2008) Noise in the nervous system. Nat Rev Neurosci 9(4):292–303
12. Johansson RS, Flanagan JR (2009) Coding and use of tactile signals from the fingertips in object manipulation tasks. Nat Rev Neurosci 10(5):345–359
13. Fu Q, Zhang W, Santello M (2010) Anticipatory planning and control of grasp positions and forces for dexterous two-digit manipulation. J Neurosci 30(27):9117–9126
14. Fu Q, Hasan Z, Santello M (2011) Transfer of learned manipulation following changes in degrees of freedom. J Neurosci 31(38):13527–13534
15. Zhang W, Gordon AM, Fu Q, Santello M (2010) Manipulation after object rotation reveals independent sensorimotor memory representations of digit positions and forces. J Neurophysiol 2953–2964
16. Santello M, Flanders M, Soechting JF (1998) Postural hand synergies for tool use. J Neurosci 18(23):10105–10115
17. Santello M, Flanders M, Soechting JF (2002) Patterns of hand motion during grasping and the influence of sensory guidance. J Neurosci 22(4):1426–1435
18. Santello M, Baud-Bovy G, Jörntell H (2013) Neural bases of hand synergies. Front Comput Neurosci 7:23
19. Fu Q, Santello M (2014) Coordination between digit forces and positions: interactions between anticipatory and feedback control. J Neurophysiol 111(7):1519–1528
20. Shibata D, Choi JY, Laitano JC, Santello M (2013) Haptic-motor transformations for the control of finger position. PLoS One 8(6):e66140
21. Mojtahedi K, Fu Q, Santello M (2015) Extraction of time and frequency features from grip force rates during dexterous manipulation. IEEE Trans Biomed Eng 62(5):1363–1375
22. Lashley KS (1930) Basic neural mechanisms in behavior. Psychol Rev 37:1–24
23. Cole KJ, Abbs JH (1986) Coordination of three-joint digit movements for rapid finger-thumb grasp. J Neurophysiol 55(6):1407–1423
24. Rijntjes M, Dettmers C, Büchel C, Kiebel S, Frackowiak RSJ, Weiller C (1999) A blueprint for movement: functional and anatomical representations in the human motor system. J Neurosci 19(18):8043–8048
25. Wing AM (2000) Motor control: Mechanisms of motor equivalence in handwriting. Curr Biol 10(6):245–248

26. Bursztyn LLCD, Flanagan JR (2008) Sensorimotor memory of weight asymmetry in object manipulation. Exp Brain Res 184(1):127–133
27. Fu Q, Santello M (2012) Context-dependent learning interferes with visuomotor transformations for manipulation planning. J Neurosci 32(43):15086–15092
28. Fu Q, Santello M (2015) Retention and interference of learned dexterous manipulation: interaction between multiple sensorimotor processes. J Neurophysiol 113(1):144–155
29. Ingram JN, Howard IS, Flanagan JR, Wolpert DM (2011) A single-rate context-dependent learning process underlies rapid adaptation to familiar object dynamics. PLOS Comput Biol 7(9):e1002196
30. Davare M, Kraskov A, Rothwell JC, Lemon RN (2011) Interactions between areas of the cortical grasping network. Curr Opin Neurobiol 21(4):565–570

Chapter 3
Digit Position and Force Synergies During Unconstrained Grasping

Abdeldjallil Naceri, Marco Santello, Alessandro Moscatelli and Marc O. Ernst

Abstract Grasping is a complex motor task which requires a fine control of the multiple degrees of freedom of the hand, in both the position and the force domain. In this chapter, we investigated the coordinated control of digit position and force in the human hand while grasping and holding a moving object. We observed a substantial variability between participants in the hand posture. Instead, digit placement was rather stereotyped for repeated grasps of the same participant. The normal forces applied by the digits co-varied with their placement across trials. Specifically, we observed an exponential relationship between finger placement and normal force applied for the thumb and lateral fingers. For the middle and ring fingers, the force responses co-varied in an approximately linear fashion with digit position. Principal component analysis revealed that more than 97 % of the finger force variance was accounted by the first two components (corresponding to the first and the second force synergy). This is consisted with the framework of motor synergy, since two components successfully explained most of the variability in the data.

A. Naceri (✉) · A. Moscatelli · M.O. Ernst
Department of Cognitive Neuroscience and Cognitive Interaction Technology Centre of Excellence (CITEC), University of Bielefeld,
33615 Bielefeld, Germany
e-mail: abdeldjallil.naceri@uni-bielefeld.de

M.O. Ernst
e-mail: marc.ernst@uni-bielefeld.de

A. Moscatelli
Department of Systems Medicine and Centre of Space Bio-Medicine,
Università di Roma "Tor Vergata",
Via Montpellier 1, 00133 Rome, Italy
e-mail: alessandro.moscatelli@uni-bielefeld.de

M. Santello
School of Biological and Health Systems Engineering, Arizona State University,
Tempe, AZ 85287, USA
e-mail: marco.santello@asu.edu

M. Bianchi and A. Moscatelli (eds.), *Human and Robot Hands*,
Springer Series on Touch and Haptic Systems, DOI 10.1007/978-3-319-26706-7_3

3.1 Introduction

Grasping is a complex motor task which requires a fine control of the multiple degrees of freedom of the human hand, in both the position and the force domain. In order to control digit positions and forces, humans use both anticipatory strategies and sensory feedback from touch, proprioception and vision (see Chaps. 2, 4 and 5). Results of previous studies showed the important role of the anticipatory behavior in grasping [8, 9, 11]. As showed in Chap. 2, anticipatory strategies take into account object geometry, its estimated mass and material, and the planned action and manipulation [8, 11, 12]. Prior knowledge of the center of mass of the grasped object determines the position of the fingers relative to the object and the applied force [16].

Due to the many degrees of freedom of the human hand, a given object can be grasped with several—virtually infinite—postures. Therefore, the central nervous system has to efficiently manage this redundancy in degrees of freedom, which determines the choice of the digit contact point and the hand posture, as well as the forces and torques to be applied [4, 7]. Two main frameworks have been proposed in order to explain how the central nervous system solves this redundancy problem: Motor synergies [18, 19] and task-optimal control [21, 22]. Motor synergies, which are the main focus of the book, are defined as a correlation between a large set of variables in the kinematic (postural synergies) or in the force domain (force synergies), oriented towards a specific motor goal [7, 18, 25]. The co-activation of a specific group of muscles towards a given action is a well-established example of motor synergy [6]. The synergy approach has received broad interest in both the neuroscience (see [19] and Chaps. 2, 4 and 5) and robotic communities (see [5] and Chaps. 8, 10, 11–14). The frameworks of synergies and optimal control are not mutually exclusive; accordingly Latash [10] proposed a combination of these two motor strategies to solve the redundancy problem.

Several studies investigated force synergies in multi-digit grasping [3, 17, 24]. The experimental paradigms used in these studies constrained the contact points at fixed locations, to ensure that the participants grasped the manipulandum on the force sensors. It has been shown that the first two synergies (identified using a Principal Component Analysis) accounted for more than 90 % of the force variance during constrained grasping [15, 25]. Lukos et al. [11] investigated learning and digit placement in unconstrained grasping, without examining the distribution of forces produced by individual digits. Fu et al. [8] investigated anticipatory control in both position and force domain (two-digit grasping; see also Chap. 2). Specifically, the authors investigated the digit forces and torque modulation in tasks that did not constrain contact points. They showed that the coordinated control of digit location and force is crucial for successful grasping and manipulation [8]. During unconstrained grasping, the motor system choose the location of the fingers in order to modulate the control of grip and load force [8].

In this chapter we reviewed two previous studies [13, 14] requiring unconstrained grasping of a sensorized object. In order to measure both position and force of by each of the five digits, we integrated a newly developed force sensor array [20] in an object of cuboid shape connected with a force feedback devices (PHANToM™ device).

Participant were required to grasp and hold the object, while we applied external disturbance forces using the two PHANToM™ devices. Owing to the tactile sensitive object, we estimated the location on the object's surface and the normal forces of all five fingers. We evaluated the stability of the grasp and the force distribution across the fingers.

3.2 Methods

3.2.1 Participants

Five right-handed participants, 29 ± 4 years of age (4 males), took part in the experiment. All participants had no known neurological or sensorimotor deficits and they gave informed written consent in accordance with the Declaration of Helsinki.

3.2.2 Hardware

For this study we developed a new sensorized object (TACtile Object, referred as TACO in the manuscript) that is able to record the position and the normal force exerted by each finger on the object's surface. This setup allowed the participants to freely choose the position of the fingers during grasping. The TACO has a rectangular cuboid shape (length: l = 170 mm, height: h = 85 mm and width w = 55 mm) and is built using high-speed tactile sensors [20] on each of the two sensorized sides of the object. Each tactile sensor has a modular architecture with 16 × 16 modules or tactels. We mounted four sensors on the TACO, providing 64 × 16 tactels. A single tactel has a spatial resolution of 5 mm and a sampling rate of 1.9 kHz. We recorded both the position and of the normal force exerted by each of the five fingers. The TACO was calibrated using a force gauge with a force ranging from 0 to 25 N and we varied also gauge tip across sections from 10 to 50 mm^2 with a step of 20 mm^2.

During the experiments, a virtual rectangular cuboid was displayed between the participants and the TACO so as to remove any visual feedback of their actual finger position. The visual scene was displayed on a 21 CRT-computer monitor (SONY® CPD-G520) with a resolution of 1280 × 1024 pixels (refresh rate 100 Hz). Participants viewed the mirror image of the visual scene via liquid-crystal shutter glasses (CrystalEyes™) providing binocular disparity (Fig. 3.1a). The TACO was attached to two PHANToM™ (SensAble® Technologies) forces feedback devices in order to track its position and to apply force/torques perturbations during grasping and lifting (Fig. 3.1b). The sampling rate of the PHANToM™ was 1 kHz. The total weight of TACO attached to the PHANToM™ arms was m = 0.470 kg. Constrained by the arrangement of the PHANToM™ force feedback devices, TACO has five degrees of freedom of unconstrained motion ($x, y, z, 0$: no pitch rotation, $\alpha : yaw, \beta : roll$).

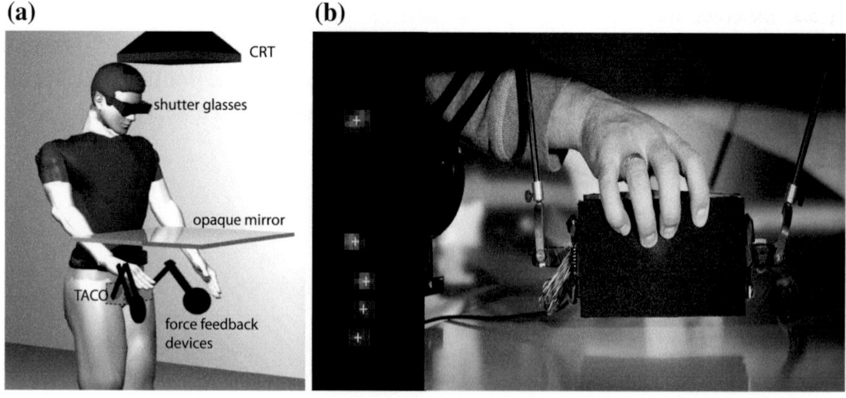

Fig. 3.1 Experimental materials. **a** Participants binocularly view the mirror image of the visual scene. **b** The TACO attached to the PHANToM™ force feedback devices. On the *left*, the TACO output image with *yellow cross* represents digit center of pressures ($CoPs$)

3.2.3 Procedure

Participants sat on a chair of adjustable height. Before the start of the grasping movement, participants forearm rested on a plank with the palm of the hand facing downward. Participants received an auditory "GO" signal prompting them to grasp the TACO (five-fingers grasp) and lift it from 100 to 150 mm. The color of the virtual rectangular cuboid changed when the TACO reached the target height. Thereby, the two PHANTOM arms generated the disturbance forces and the participants were instructed to hold the TACO as stable as possible for 20 s. After having stabilized the TACO for approximately 20 s, participants received another auditory signal prompting to replace the TACO on the table. Perturbation forces and torques where applied using the PHANToM™ force feedback devices. We studied 5 conditions of force/torque perturbations: force of Fy = 2.4 N (Fig. 3.2a) was applied in vertical direction, or torques of 25 N·cm were applied around the y- or z-axis (Fig. 3.2b, c respectively) causing yaw and roll rotations around TACO's center of mass (T_y and T_z). The perturbations were turned on with a duration ranged between 1 to 3.5 s and off with durations ranged between 0.6 to 1 s. Both perturbations were randomly presented. Both torques perturbation (T_z and T_y) were applied in clockwise (CW) and counter-clock-wise (CCW) directions. Thus, there were in total 5 conditions: (F_y, T_y^{CCW}, T_y^{CW}, T_z^{CCW}, T_z^{CW}). The order of conditions was randomly presented to the participants. Participants performed twenty trials conducted for each condition. Each trial lasted approximately 25 s from grasp onset to the end. Before starting the experiments, subjects performed four trials with F_y perturbation in order to familiarize them with the task. Participants could rest as much as they needed between two consecutive trials. The total duration of the experiment was approximately 2 h per subject with a break in half-way through the experiment.

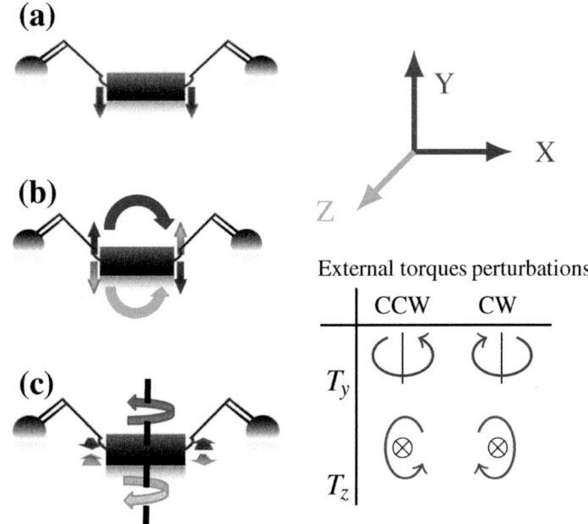

Fig. 3.2 Experimental protocol. **a** Perturbation force F_y. **b** Perturbation torque T_z CW and CCW. **c** Perturbation torque T_y CW and CCW

3.3 Data Processing and Analysis

The normal forces F of the fingers and the x- and y-position of the center of pressures (CoP_x and CoP_y, respectively) were directly read from the force modules of the TACO. The CoP_x and CoP_y were defined as the location of the maximally (one output: global maximum of the activated region) activated tactels for each fingers' region in the output matrix that was converted to force in Newton using the lookup table from the calibration. The calibration table was obtained with a resolution ± 0.2 N. Digit locations, normal forces, and TACO position were recorded and ran through a second order Butterworth low pass filter with 1 Hz cutoff frequency (Fig. 3.3). Digits locations CoP_x and CoP_y were both extracted during the holding phase. The positions and rotations of TACO were tracked using PHANToM™ devices.

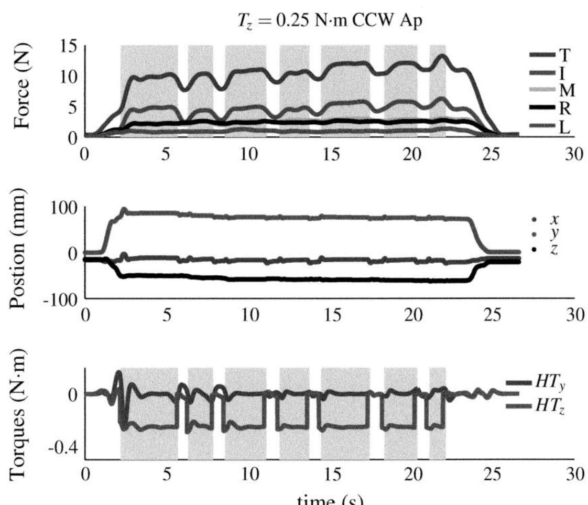

Fig. 3.3 Digit normal forces, TACO position and hand torques for representative subject. *Gray areas* represents intervals when perturbations "on". Legend 1: *T* thumb, *I* index, *M* middle, *R* ring, *L* little. Legend 2: *x*, *y*, *z* are the TACO's coordinates. Legend 3: HT_y, HT_z are the hand net torque

Digit peak normal forces were extracted for perturbations "on" within trial and then averaged. Linear mixed model (LMM; [1]) with repeated measure structure was used to analyze the data.

3.4 Results

3.4.1 Center of Pressure for Individual Participants

First, we investigated the contact locations of each digit on the TACO during the holding phase, to quantify the variability across trials (grasps) and participants. Figure 3.4 shows CoP data for individual participants. We found large variability between participants in digit location, indicating that participants differed in the initial hand posture when grasping the TACO. This is also illustrated in Fig. 3.5. Instead, digit placement was rather stereotyped for repeated grasps within the same participant.

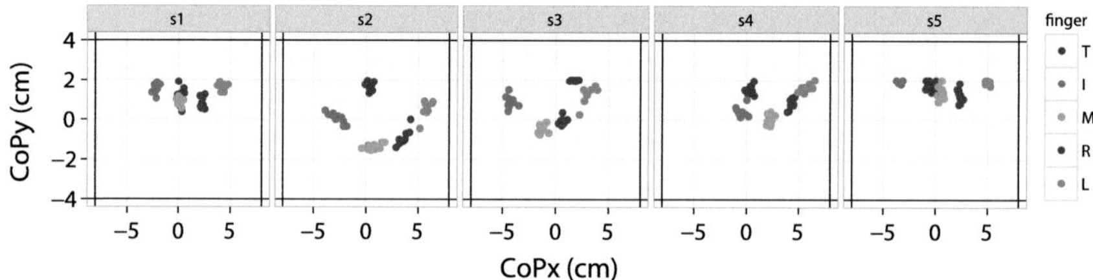

Fig. 3.4 Digit CoP results for individual participants averaged across trials for each condition. The thumb CoPs were plotted at the same plane with other fingers (*T* thumb, *I* index, *M* middle, *R* ring, *L* little)

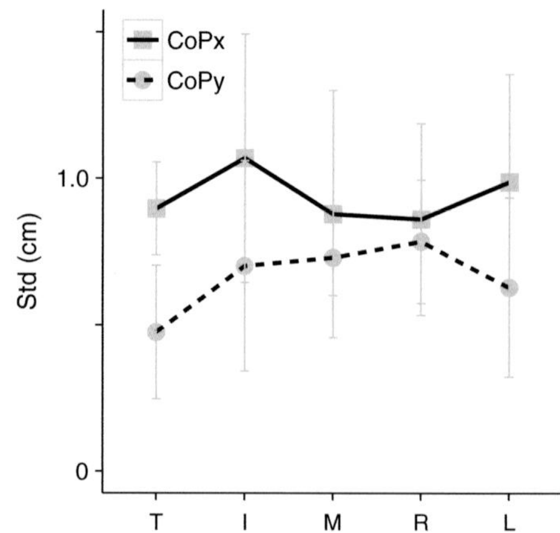

Fig. 3.5 Average variability in digit CoP across participants for both *x* and *y* coordinates. *Error bars* represents the standard deviations of the mean std (*T* thumb, *I* index, *M* middle, *R* ring, *L* little)

3.4.2 Digit Normal Forces Versus CoPs

Next we sought to evaluate the modulation of digit forces as a function of digit horizontal locations CoP_x. LMM was used to fit the data of digit forces as a function of CoP_x. Specifically, four models were tested for each finger: M_1, M_2, M_3 and M_4 where we varied the way of including our main effect CoP_x. In the model M_1, CoP_x was included as a fixed effect. In the model M_2 we included CoP_x as a fixed effect and also as random effect in order to allow the adjustment to the individual participants CoP_x. The models M_3 and M_4 were similar to the models M_1 and M_2, respectively with including CoP_x as a non-linear term ($exp(CoP_x)$). Furthermore, in the above-described four models, participants were included as random effect, whereas perturbation conditions and trials were included as fixed effects.

For the thumb and index fingers, M_4 was selected as best fit based on the obtained AIC values (the smallest AIC indicating the best model) of the tested models (thumb: $AIC_1 = 3066$, $AIC_2 = 3144$, $AIC_3 = 3103$, $AIC_4 = 3019$ and index: $AIC_1 = 2081$, $AIC_2 = 2075$, $AIC_3 = 2074$, $AIC_4 = 2066$). The chosen model M_4 revealed that the normal forces of these two digits decreased exponentially with increasing CoP_x values (Fig. 3.6). For the Middle and ring fingers, the obtained AIC values were approximately similar, thus, we chose the simplest model, i.e. the model with less

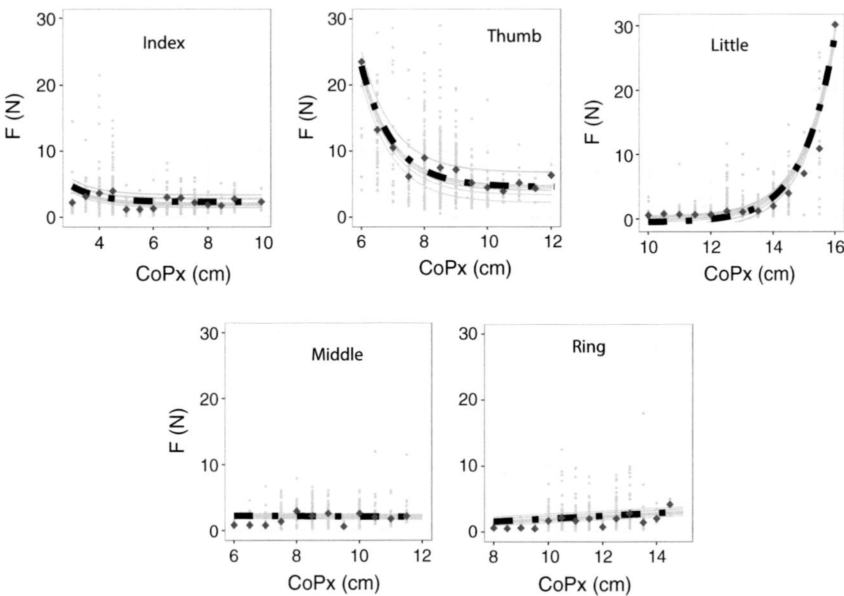

Fig. 3.6 Force and CoP_x relationship for each finger. *Gray points* represents original data for all participants in all conditions. *Gray lines* represents LMM fitting for each perturbation condition. *Red points* represents the average of force data across perturbation conditions. *Black dashed line* represent the average LMM fitting

parameters, M_1 (middle: $AIC_1 = 1528$, $AIC_2 = 1532$, $AIC_3 = 1528$, $AIC_4 = 1532$ and ring: $AIC_1 = 1919$, $AIC_2 = 1923$, $AIC_3 = 1927$, $AIC_4 = 1923$). The chosen model M_1 revealed a decrease in the middle finger force per centimeter of CoP_x (0.03 ± 0.06 N; $Estimate \pm SE$). This result indicate that the middle finger normal force was nearly constant within CoP_x variation (Fig. 3.6). For the ring finger, M_1 revealed that the force increased linearly by (0.21 ± 0.08 N) per centimeter of CoP_x. For the little finger M_4 was selected as best fit based on the obtained AIC values of the tested models ($AIC_1 = 2097$, $AIC_2 = 2017$, $AIC_3 = 1873$, $AIC_4 = 1866$). The chosen model M_4 revealed that the normal force of the little finger increased exponentially with an increase of CoP_x (Fig. 3.6). The selected models for each digit revealed small decrease of forces per trial (thumb: $4.44 \times 10^{-4} \pm 7.26 \times 10^{-3}$, index: 0.003 ± 0.003, middle: 0.003 ± 0.002, ring: 0.004 ± 0.003, little: $7.44 \times 10^{-4} \pm 4.49 \times 10^{-3}$ N). This result indicates that possible fatigue effects that might have occurred across trials were negligible.

3.4.3 Digit Forces Synergies

Finally, we investigated digit force synergies using principal component analysis (PCA). PCA was conducted on the mean peak normal forces, averaged across five perturbations and participants for the four fingers opposing the thumb. The matrix for PCA analysis was constructed by defining digit forces as variables (four columns) and perturbation conditions as entries (five rows). The results of the PCA revealed that the first two PCs accounted for 97 % of the variance of the normal force. Specifically, PC1 accounted for 71.6 % of the variance while PC2 accounted for 25.8 % of the variance. Figure 3.6 shows a biplot of the PCA for the index, middle, ring, and little fingers (I, M, R, L). Each finger force is characterized by 2 loadings, $w1$ and $w2$ ($F_{finger} = w_1\mathrm{PC1} + w_2\mathrm{PC2}$) represented by a vector (Fig. 3.7). PC2 loadings were higher for the index and little fingers compared to the middle and ring (PC2 loading of middle finger was approximately zero). PC1 loadings were all positive for all fingers, while the PC2 loading was negative only for the index finger. Figure 3.7 shows that the middle and ring fingers were more involved in supporting the load task within the perturbation condition F_y. The index finger was more involved in the rotational task within the conditions T_y^{CW}, T_z^{CCW}, T_z^{CW}. The little finger was more involved in the rotational tasks within the conditions T_y^{CCW}, T_y^{CW}, T_z^{CCW}. Finally, the cosine of the angle between arrows plotted in Fig. 3.7 indicates the correlation between the fingers' forces (see also Fig. 3.8 for detailed correlation coefficients of normal force exerted by pairs of digits). All between-digits correlation coefficients were larger than 0.6 with the exception of the index finger that showed weak correlations with normal force exerted by all other fingers. This weak correlation was required to generate a net torque against the external perturbations when the index finger was mainly involved (i.e., T_y^{CW}, T_z^{CCW}, T_z^{CW}). In contrast, little finger normal force was highly correlated with normal force exerted by the middle fingers, indicating that

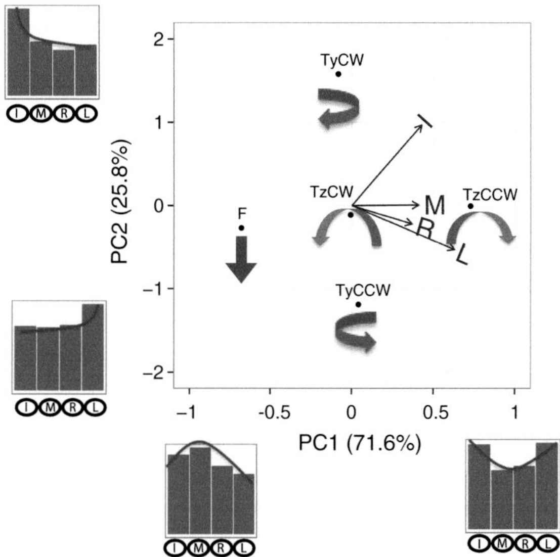

Fig. 3.7 PCA results. Data points with *red*, *blue* and *green arrows* indicate PCs loadings for all conditions. *Brown* vectors represents the PCs loadings for each finger. The cosine angle between vectors indicate the correlation between the fingers. *Blue bars* represent digit normal forces

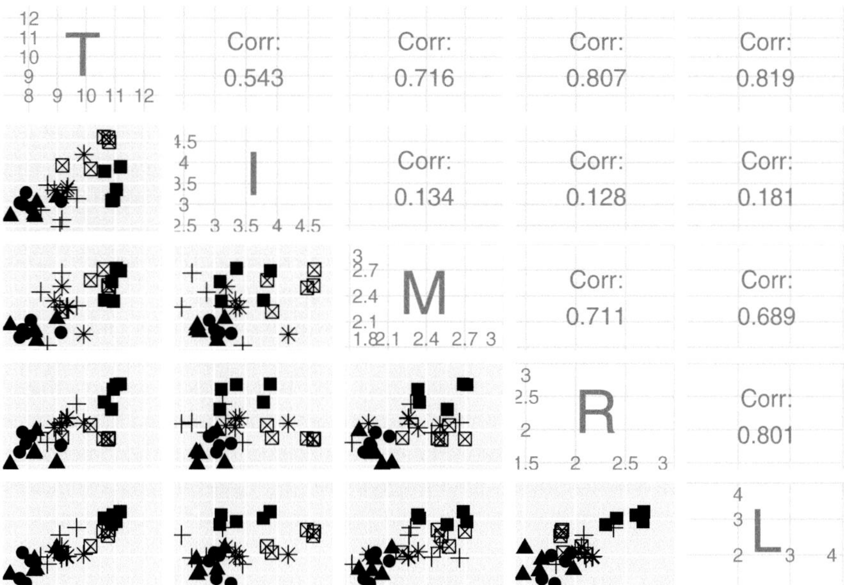

Fig. 3.8 Correlation coefficients between digit normal forces

these fingers were more involved with the little when this latter was mainly involved to compensate external torques (T_y^{CCW}, T_y^{CW}, T_z^{CCW}; Fig. 3.7).

3.5 Discussion

In this chapter we addressed the question of how multiple digit forces are coordinated during grasps at self-chosen contacts on an object in response to external force and torque perturbations. We found a large variability in the contact distributions across participants. Furthermore, digit force responses systematically varied as a function of digit horizontal locations CoP_x. Specifically, thumb, index, and little finger forces exhibited an exponential modulation as function of digit displacement (CoP_x), whereas the normal force exerted by the inner fingers (middle and ring finger) exhibited linear responses, indicating that most of the across-trial variability in digit contacts was compensated by normal force modulation by the outer fingers. Moreover, PCA computed on multi-digit forces revealed that two force synergies accounted for more than 97 % of the normal force variance.

We observed a large variability between-participants in both horizontal and vertical digit location (CoP_x and CoP_y). This confirmed the results of a previous study of unconstrained two-digit grasping [8]. Specifically, in [8] participants modulated the vertical spacing between the thumb and index fingers at grasping in order to generate a torque to compensate an external torque whose direction was changed across blocks of consecutive trials. In contrast to previous studies that investigated multi-digit prehension could not assess modulation of digit contacts because they studied constrained grasping. Moreover, it has been reported that for the case of tripod grasping, the thumb exhibited the largest variability in vertical CoPs (CoP_y) and the index and middle fingers were characterized by the largest CoP variability in both coordinates $CoP_{x,y}$ [2]. Our results showed that the variability between participants was similar in the x- and y-axis, although slightly larger for CoP_x (Fig. 3.5). This high variability can be a hint of idiosyncratic grasping strategies.

Our results revealed that digit forces co-varied with the horizontal and vertical digit locations $CoP_{x,y}$. Specifically, linear and exponential force modulation was observed across digits as a function of CoP_x (Fig. 3.6). Fu et al. [8] reported a linear relation between the load force and vertical distance between thumb and index finger CoP_y (equivalent to CoP_x in our study). Thus, we conclude that the phenomenon of digit force-to-position modulation is not limited to two-digit grasping, but extends to cases where a larger number of contacts is involved as the present five-digit grasping scenario.

PCA revealed that two force synergies accounted for most of the variability in digit normal force. This result is consistent with findings from previous studies on multi-digit force synergies [15, 23, 25] using a manipulandum with fixed digit locations. Despite the large variability in digit $CoP_{x,y}$, the first two force synergies accounted for nearly all the variability in fingers forces. These results supports the above proposition that digit normal forces were modulated to compensate for the variability in digit locations. These findings indicate that the central nervous system solved the redundancy problem by reducing the dimensionality of the force space, i.e., from five to two dimensions, despite the within- and across-subject variability in digit locations. The neural mechanisms underlying the observed digit force-to-position modulation are currently under investigation.

References

1. Baayen R, Davidson D, Bates D (2008) Mixed-effects modeling with crossed random effects for subjects and items. J Memory Lang 59(4):390–412. doi:10.1016/j.jml.2007.12.005. http://linkinghub.elsevier.com/retrieve/pii/S0749596X07001398
2. Baud-Bovy G, Soechting J (2002) Factors influencing variability in load forces in a tripod grasp. Exp Brain Res 143(1):57–66. doi:10.1007/s00221-001-0966-8
3. Baud-Bovy G, Soechting JF (2001) Two virtual fingers in the control of the tripod grasp. J Neurophysiol 86:604–615
4. Bernstein N (1967) The co-ordination and regulation of movements. Pergamon Press, Oxford
5. Bicchi A, Gabiccini M, Santello M (2011) Modelling natural and artificial hands with synergies. Philos Trans Royal Soc Lond Ser B Biol Sci 366(1581):3153–3161. doi:10.1098/rstb.2011.0152
6. D'Avella M, Portone A, Fernandez L, Lacquaniti F (2006) Control of fast-reaching movements by muscle synergy combinations. J Neurosci 26(30):7791–7810
7. Friedman J, Flash T (2007) Task-dependent selection of grasp kinematics and stiffness in human object manipulation. Cortex 43(3):444–460
8. Fu Q, Zhang W, Santello M (2010) Anticipatory planning and control of grasp positions and forces for dexterous two-digit manipulation. J Neurosci 30(27):9117–9126. doi:10.1523/JNEUROSCI.4159-09.2010
9. Johansson RS, Westling G (1984) Roles of glabrous skin receptors and sensorimotor memory in automatic control of precision grip when lifting rougher or more slippery objects. Exp Brain Res 56(3):550–564. doi:10.1007/BF00237997
10. Latash ML (2012) Movements that are Both Variable and Optimal by 34(September):5–13. doi:10.2478/v10078-012-0058-9
11. Lukos J, Ansuini C, Santello M (2007) Choice of contact points during multidigit grasping: effect of predictability of object center of mass location. J Neurosci 27(14):3894–3903. doi:10.1523/JNEUROSCI.4693-06.2007
12. Lukos JR, Ansuini C, Santello M (2008) Anticipatory control of grasping: independence of sensorimotor memories for kinematics and kinetics. J Neurosci: Off J Soc Neurosci 28(48):12765–12774. doi:10.1523/JNEUROSCI.4335-08.2008
13. Naceri A, Moscatelli A, Santello M, Ernst M (2014a) Multi-digit position and force coordination in three- and four-digit grasping. In: Haptics: neuroscience, devices, modeling, and applications. doi:10.1007/978-3-662-44193-0_14. http://faculty.engineering.asu.edu/santello/wp-content/uploads/2014/05/eurohaptics2014_submission_110-.pdf
14. Naceri A, Moscatelli A, Santello M, Ernst MO (2014b) Coordination of multi-digit positions and forces during unconstrained grasping in response to object perturbations. In: 2014 IEEE haptics symposium (HAPTICS), IEEE, pp 35–40. doi:10.1109/HAPTICS.2014.6775430. http://ieeexplore.ieee.org/lpdocs/epic03/wrapper.htm?arnumber=6775430
15. Park J, Zatsiorsky VM, Latash ML (2010) Optimality vs. variability: an example of multi-finger redundant tasks. Exp Brain Res 207(1–2):119–132. doi:10.1007/s00221-010-2440-y
16. Salimi I, Hollender I, Frazier W, Gordon AM (2000) Specificity of internal representations underlying grasping. J Neurophysiol 84:2390–2397

17. Santello M, Soechting JF (2000) Force synergies for multifingered grasping. Exp Brain Res 133(4):457–467
18. Santello M, Flanders M, Soechting JF (1998) Postural hand synergies for tool use. J Neurosci 18(23):10105–10115
19. Santello M, Baud-Bovy G, Jörntell H (2013) Neural bases of hand synergies. Front Comput Neurosci 7(April):23. doi:10.3389/fncom.2013.00023
20. Schurmann C, Koiva R, Haschke R, Ritter H (2011) A modular high-speed tactile sensor for human manipulation research. In: World haptics conference (WHC), 2011 IEEE, pp 339–344. doi:10.1109/WHC.2011.5945509
21. Terekhov AV (2010) NIH Public. Access 61: doi:10.1007/s00285-009-0306-3.An
22. Todorov E, Ghahramani Z (2004) Analysis of the synergies underlying complex hand manipulation. Conference proceedings: annual international conference of the IEEE engineering in medicine and biology society IEEE engineering in medicine and biology society conference 6:4637–4640. doi:10.1109/IEMBS.2004.1404285. http://www.ncbi.nlm.nih.gov/pubmed/17271341
23. Wu YH, Zatsiorsky VM, Latash ML (2012) Multi-digit coordination during lifting a horizontally oriented object: synergies control with referent configurations. Exp Brain Res 222(3):277–290. doi:10.1007/s00221-012-3215-4. http://www.ncbi.nlm.nih.gov/pubmed/22910900
24. Zatsiorsky VM, Latash ML (2008) Multi-finger prehension: an overview. J Motor Behav 40:446–476
25. Zatsiorsky VM, Gregory RW, Latash ML (2002) Force and torque production in static multifinger prehension: biomechanics and control. I. Biomechanics. Biol Cybern 87(1):1–19. doi:10.1007/s00422-002-0321-6.Force

Chapter 4
The Motor Control of Hand Movements in the Human Brain: Toward the Definition of a Cortical Representation of Postural Synergies

Andrea Leo, Giacomo Handjaras, Hamal Marino, Matteo Bianchi, Pietro Pietrini and Emiliano Ricciardi

Abstract The control of the many degrees of freedom of the hand through functional modules (hand synergies) has been proposed as a potentially useful model to describe how the hand can maintain postures while being able to rapidly change its configuration to accomplish a wide range of tasks. However, whether and to what extent synergies are actually encoded in motor cortical areas is still debated. A direct encoding of hand synergies is suggested by electrophysiological studies in nonhuman primates, but the evidence in humans resulted, so far, partial and indirect. In this chapter, we review the organization of the brain network that controls hand posture in humans and present preliminary results of a functional Magnetic Resonance Imaging (fMRI) on the encoding of synergies at a cortical level to control hand posture in humans.

4.1 Introduction

The human hand shows an extraordinary ability to perform a wide range of volitional movements that are adaptable—changing in response to modifications in the task demands—and are, in the meantime, characterized by a great precision. As widely discussed throughout this book, this is made possible only by reducing the impact of the redundancy of effectors that characterizes the hand from an anatomical, functional and kinematic point of view. Indeed, according to the redundancy

---

A. Leo · G. Handjaras · P. Pietrini · E. Ricciardi (✉)
Laboratory of Clinical Biochemistry and Molecular Biology, Department of Surgery, Medical and Molecular Pathology, and Critical Care, Università di Pisa, Pisa, Italy
e-mail: emiliano.ricciardi@bioclinica.unipi.it

H. Marino · M. Bianchi
Research Center "E. Piaggio", Università di Pisa,
Pisa, Italy

M. Bianchi
Department of Advanced Robotics, Istituto Italiano di Tecnologia,
Genoa, Italy

M. Bianchi and A. Moscatelli (eds.), *Human and Robot Hands*,
Springer Series on Touch and Haptic Systems, DOI 10.1007/978-3-319-26706-7_4

principle, motor systems have more components (i.e. bones, muscles or joints) than needed for accomplishing particular tasks, leading to an excess of degrees of freedom. The multiple combinations of these degrees of freedoms lead to many possible strategies for movement execution. For example, all motor tasks (e.g. pinching or grasping) can be achieved with more than one configuration of hand joints. As already described in previous chapters, redundant degrees of freedom can be grouped in weighted combinations called *motor synergies*. Synergistic control represents one of the most interesting theoretical approaches that have been attempted to solve this redundancy problem [1]. Hand muscular and postural synergies may therefore be the way such optimized hand control is achieved. However, neural functional evidences of synergistic control of the hand are very scarce, comprising a few electrophysiological studies performed on monkeys and an even more slender number of reports in humans. Chapter 5 analyzes the theoretical and experimental evidences in support of an encoding of hand synergies at the level of subcortical circuitry. In this chapter, we will push forward those observations, reviewing the structure of the brain network that controls hand posture in humans. We will show how the evidences obtained so far definitely support the idea of synergistic control of human hand at a brain cortical level. We will then discuss the importance that multimodal investigations can take in studies about hand postures, presenting the preliminary results of a conjoint kinematic and functional Magnetic Resonance Imaging (fMRI) experiment about the neural correlates of hand postural synergies in humans.

4.2 Action Processing in the Brain

The ability to grasp or to use objects and tools is one of the most important skills in humans and primates. Grasping and interacting with surrounding items have been essential for evolution, allowing even for the development of many meaningful gestures that contributed to shape the relationships between humans and their external world (see e.g. Chap. 2). The investigation of the neural correlates of grasping and of other hand-mediated movements has been therefore one of the most fascinating topics of research in neuroscience during the last two centuries. Furthermore, along the 20th century, with the development of the electrical and functional recording techniques, the science of grasping has obtained its most striking results, paving the way to a better understanding of such important and meaningful motor acts [2].

The studies regarding hand actions—and, in more detail, grasping—have tried to shed light on the many possible ways in which the brain deals with the different aspects of action processing. They have therefore focused on the neural networks potentially involved in action perception, recognition and execution. Chapter 5 discussed the importance of subcortical circuitry in the synergistic organization of motor control in the brain, with a focus on spinal circuitry and add-on capability provided by the spinocerebellar system for complex synergies.

However, the extent to which these domains of action processing are correlated and whether such a synergistic organization can be observed also in specific brain areas are still fascinating open questions: are our internal representations of actions necessary to understand actions intended or performed by other individuals? And, to what extent our ability to recognize the motor acts of others is needed to be able to perform them on our own? Answering to these questions strongly implies to know the organization and the internal structure and differences of the cortical networks engaged by these aspects of action processing.

The commonalities and differences of these networks have been extensively explored since the earliest years of functional neuroimaging. A highly-cited meta-analysis [3] evaluated the correlates of action execution, observation, verbalization and simulation. The four modalities could achieve a high degree of overlap in a network including the supplementary motor area, dorsal premotor cortex, superior parietal lobe and supramarginal gyrus. However, some regions were activated only in a single modality: inferior parietal lobe, for example, was found activated in action execution and imitation, while it was not recruited by verbalization. The high degree of overlap has suggested the presence, in the commonly recruited regions, of a core of shared representations for both performing actions and recognizing the ones being executed by others [4]. However, this hypothesis is highly disputed and further studies [5] and comprehensive reviews [6] have found differences between the representations of executed, imagined and observed actions in this networks, concluding that the consistency of engagement can be apparent and due only to different factors such as attention or high-level "semantic" processing of the actions.

This chapter will focus on hand action execution, presenting the approaches that have been attempted for studying hand motor control in monkeys and humans and the theories about the way motor commands may be encoded in the cerebral cortex and giving an overview about the possible "languages" used by the brain to control the hand, including the evidences of a neural coding of muscular and postural hand synergies.

4.3 A Cortical Network for Hand Posture Control

Evidence from animal neurophysiology suggests indeed that the control of goal-directed hand motor acts relies only on a network of brain regions. Studies that applied single neuron recording to monkeys during the execution of grasping tasks found three specific grasp-related regions in the monkey: area F1 (primary motor cortex), area F5/PML (premotor cortex) and the anterior intraparietal sulcus (AIP). Area F1 is the output channel of this network: axons originating from this cortical region form the corticospinal tract, which reaches the spinal cord and extends into the peripheral nerves directly connected to hand extrinsic and intrinsic musculature [7]. The other regions of the hand motor control network are related to the processing of grasp-relevant properties: area F5 and area AIP are directly connected to each other and process visuomotor features of action planning and control, thus modulating their activity according to the motor act that is specifically performed (e.g. precision vs. power grasp). While the area AIP, which is structurally and functionally closer to visual areas and to those cortical regions that form the dorsal visual pathway [8], may be related to the processing of visual features of the hand-target interaction, providing the information that is required to adapt the motor act to the actual object size [9, 10], the premotor area F5 may be more related to the motor representation and control, selecting the most appropriate strategy for accomplishing a task consequently to the visual properties of objects [11]. Lesion or inactivation studies in monkeys have substantially confirmed this putative organization of the grasping network [12].

4.4 The Network for Hand Control in Humans: fMRI Evidences

During the last years, fMRI has become the most used technique for in vivo exploration of brain activity. The measurements provided by fMRI are indirect estimations of brain activity: the increased discharge rate of neurons following—for example—the presentation of a stimulus, leads to a greater need of blood supply, provided by the small cortical vessels. Because of the different magnetic properties of hemoglobin, this event is reflected by a local alteration of magnetic field. This phenomenon, called neurovascular coupling, is the basis of BOLD (Blood Oxygenation Level Dependent) signal and expresses the way brain activity is measured by fMRI. Every fMRI voxel samples a portion of brain tissue of some mm^3, thus containing a number of neurons that can reach some thousands [13, 14]. For this reason, BOLD can measure brain activity in an indirect and coarse way, providing a less accurate measurement with respect to single neuron recordings.

The signal variations across a functional run (which may include—for example—the execution of a task) are represented as a time course that expresses the BOLD signal over time. These time series are sampled at a low temporal frequency (very often 0.3 Hz) and need to undergo some preprocessing steps, in order to correct motion and acquisition artifacts, and can be later analyzed with an operation called *deconvolution*, that implies the computation of a measure of fit between the standard time pattern of an activated voxel (Hemodynamic Response Function, HRF) and the time course of the actual voxel. This operation results in a set of scores, usually defined as Statistical Parametric Map, that is expression of the fMRI activity.

The study of hand movements with functional MRI is potentially interesting yet very difficult, due to the artifacts resulting from head motion, which often lead to data loss and hamper the findings. Therefore, functional imaging evidences on the hand network relied mainly on simple finger tapping paradigms that were quite inspiring for investigating the organization of primary motor areas [15–17]. Overall, these studies confirmed the distributed and overlapping maps of fingers inside hand primary motor cortex.

Functional studies with complex hand movements related to hand-tools interaction started quite lately, with an important report of an experiment during which subjects had to pantomime the use of imagined objects triggered by an auditory presentation of their names [18]. Despite some technical shortcomings, the study identified a bilateral activation of the intraparietal sulcus (IPS) while no motor regions resulted activated, probably due to contrast with a moving-hand control condition.

In the years following this first report, the number of studies implying tasks of real movement execution has been fairly low, even if the interest in the neural basis of action planning is surely great. Moreover, many of the fMRI studies on grasping were based on the classic distinction introduced for the first time by Napier [19, 20] that divided grasp gestures into essentially two big classes: power and precision grips. In the former class, the object is "held in a clamp between the partly flexed fingers and the palm", while in the latter the object is "pinched between the fingers and the opposing thumb" [19].

The studies conducted with these grasp categories helped to a great deal to understand the role of parietal cortex in motor planning: superior and intraparietal cortex indeed may be devoted to sensorimotor transformations, gathering information about the target object (e.g. its size or distance) and using them to correctly preshape the hand for the upcoming grasping movement. The anterior part of intraparietal sulcus may correspond therefore to the putative human homologue of monkey area AIP [21–23]. Further studies applied classification algorithms on fMRI activation patterns and demonstrated that different tool-directed movements could be decoded from brain activity in IPS [10, 24] and that this region is sensitive to the difference between precision and power grasp acts [9, 25, 26]. These differences in activations are indeed suggestive of a modulation of activity in hand-related regions induced by object size: indeed precision grips require a prolonged preparatory phase and are likely to be associated to specific activation patterns in the regions integrating object information such as the intraparietal sulcus [27–29].

Due to the low number of studies performed so far, a recent, comprehensive meta-analysis on grip type identified only 28 original papers focused on grasp type and grasp execution using gestures that were actually performed inside the MRI scanner. This meta-analysis [30] identified specific activations for power grip in the postcentral gyrus and unique activations for precision grip in precentral gyrus and in SMA. However, most of the activation clusters overlapped in the aforementioned regions and in frontal and parietal areas, sustaining the idea of a specific network for hand control, activated with great consistency across subjects. When the attention is concentrated on the primary motor cortex only, most of the studies have examined single digit representations, as reported above. Nonetheless, in a recent experiment [31, 32] single meaningful movements could be successfully decoded from their evoked BOLD patterns at ultra-high field, only using the voxels in primary motor cortex.

Functional MRI studies have therefore helped to confirm the organization of the hand related network that result from electrophysiology and clinical reports: the primary motor cortex, organized as a distributed and overlapping set of patterns displaying only moderate preferences for a single digit, is the effector of a prefrontal-parietal system which plans movements and hand-tool interactions on the basis of visual and proprioceptive features and of information about target objects which is conveyed by the occipitoparietal visual pathways.

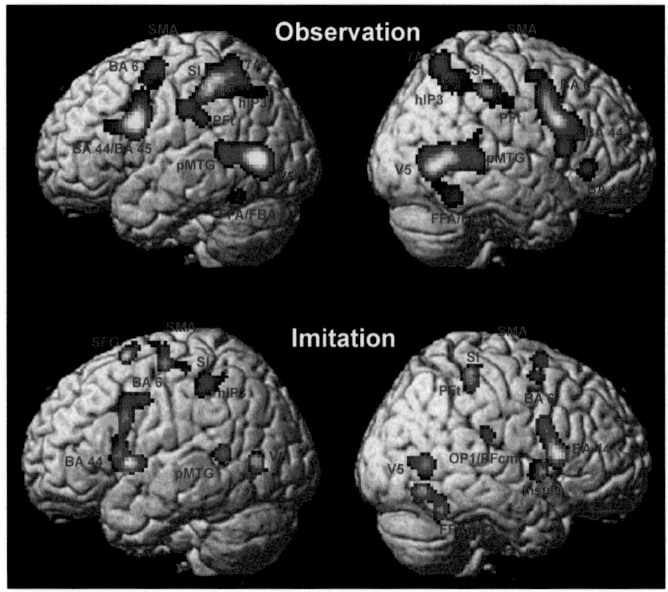

Fig. 4.1 Cortical networks involved in action observation (*top*) and imitation (*bottom*), according to the meta-analysis by [6]

Even if the largest part of these studies focused on transitive hand actions, which contemplate interaction with tools and objects, fMRI studies regarding intransitive actions—that do not require an interaction with objects—found activations in a similar network comprising motor, premotor, inferior parietal and intraparietal regions [33, 34] (Fig. 4.1).

4.5 Somatotopic Control of Hand Muscles

When aiming at characterizing the brain structural and functional organization of hand motor control, one of the main questions inevitably focuses on how motor information is encoded in the primary motor cortex (M1). Actually, some different theories exist about the possible 'format' of how the information on distinct motor acts is represented at a cortical level and then descends from M1 to the hand muscles and joints.

Historically, the first important hypotheses were drawn during the 18th century by John Hughlings Jackson who observed direct relationships between seizures and spreading of muscular jerking, with specific patterns that suggested a strict ordering of body segments in the motor cortex. Later, Clinton Woolsey, Wilder Penfield and other researchers observed that the cortical stimulation of single neurons in M1 elicited distinct motor responses with the contraction of small groups of muscles controlling one or few digits [35, 36]. Those observations introduced the concept of *motor somatotopy*: the motor cortical regions are topographically organized into sub-segments that control a single effector, or a reduced group, of muscles or joints. The subsequent success of this theory arose from many observations in patients or animals that brought to similar conclusions. Nonetheless, while a coarse arrangement of the limbs in the primary motor regions could be demonstrated with few doubts, leading to the concept of separated—for instance—hand, mouth or face areas [37], the intrinsic organization of movement information within such motor regions was harder to assess.

When looking specifically to hand motor control, the almost contemporary reports by Foerster [38] and Penfield and Boldrey [35] contained completely opposing considerations: Foerster claimed to have detected subregions within the M1 hand area that controls specifically single digits, and lesions to these clusters could lead to weakness or deficit in control of single digits [38]. Successively, Penfield himself applied electrical stimulation to a wide number of sites in the monkey primary motor area, without finding unique, segregated clusters of neurons that were specific for single digits [35]. Foerster's observations were quite isolated and, in the following decades, no further claims were made regarding single digit deficits induced by cortical lesions [39], even if this can be due to the scarcity of lesions that respect the neuroanatomical boundaries of single digit-related areas. Penfield's theory was indeed more successful, providing the basis for the new accounts on topographical hand representations that still today are regarded as being the most reliable theoretical framework. The hypothesis of a "distributed and overlapping" hand area, in which neurons pertaining to different digits are intermingled, is highly supported nowadays. Additionally, movements of more than a single digit could be evoked stimulating primary motor areas, thus providing evidence against strict segregation of single finger clusters [39–41]. Two different theories on complex, coordinated hand movements can be derived from each of these accounts: if single digits were represented in segregated spatial clusters, descending motor commands are composed by single effector or single joint elementary movements that should be assembled by a structure in the Central Nervous System (CNS), such as basal ganglia or spinal cord. Overlapping representations covering more than a single finger require a hand control that should occur through complex multidigit commands, projected in a descending way from the primary motor cortex. The spinal cord, in this framework, may simply act as a relay station, refining and enhancing the motor instructions.

It is worth noting that brain functional imaging studies in humans, which typically adopt simple finger tapping tasks to study the representations of single digits in motor cortex, have essentially confirmed a "distributed" model of finger representations. Indeed, no evidence was found about an orderly organized set of digit-specific clusters: hand area is instead more likely to be represented as a patch of partially overlapping foci of activation, which can be more specific for the control of a single digit without, however, being exclusively associated with it [15–17]. In conclusion, somatotopy has been assessed in motor cortical areas at a coarse level (i.e. "hand" or "head" areas), failing to support strongly a finer level coding (i.e. regions specific for single digits or muscles), in favor of a distributed representation of digit-specific neurons, widespread across the "hand" area, and organized in clusters that overlap at least partly.

4.6 "Languages" of Hand Control in Primary Motor Cortex

Despite the evidences supporting somatotopic control of body segments, another hypothesis has raised regarding the 'nature' of motor commands encoded in the cortical regions involved in motor planning. Such information can be represented in a somatotopic or topographical way (i.e. as sets of muscles that need to be controlled) or in a goal-oriented way (i.e. different sets of actions are represented in specific ways according to their goal). While several studies using electrical microstimulation elicited movements of well-defined limb segments, new fascinating theories have emerged in recent years. Among the most significant, the first studies by Graziano and colleagues [42] and later confirmed by other authors [43, 44] are worth mentioning. Applying electrical pulses (intracortical microstimulation, ICMS) to primary motor and premotor areas in monkeys, with the delay associated with motor planning, could evoke complex and often behaviorally relevant motor acts or postures, such as self feeding or grips [40, 45]. Moreover, the evoked movements shared stable end points that were highly specific for the group of neurons that were stimulated. More interestingly, the movements were directed towards a final posture that was independent on the starting position of the limb. Therefore, these results suggested that the stimulation over primary motor and premotor neurons could specifically evoke final limb positions and that the posture to be achieved affected neural activity more than the specific recruitment of hand muscles or joints.

Even if these previous findings hinted at an end-point or movement specific coding within the primary motor cortex, Muir and Lemon's following observation of neurons that were specific for precision grips [7] and the discovery of posture-specific coding nurtured further investigation on the internal nature of the motor commands encoded in the cortex. Recently, a study demonstrated the ability in monkeys to discriminate between a very large number of postures using the neuronal activity in the premotor and motor areas [46]. Schaffelhofer and colleagues showed that the activity patterns were highly specific thus allowing for an object-specific decoding: different accuracies were obtained for the planning phase (higher role of premotor regions) and actual motor execution (higher discrimination in primary motor cortex). This specificity allows for the programming of a robotic device, which was able to achieve the same postures after being trained on the differences between neural patterns. However, despite these results, only a fraction of the total variance of M1 activity has been explained either with end posture or trajectory, suggesting that much has still to be clarified about the neural mechanisms that underlie hand control in nonhuman primates and, consequently, in humans.

4.7 Synergies and Their Brain Correlates

Among the strategies possibly adopted by CNS to deal with motor redundancy, the presence of linear dimensionality reduction strategies or motor synergies is one of the most theoretically successful [47, 48], as discussed in Chaps. 2–7. A synergy is a "collection of relatively independent degrees of freedom that act as a single functional unit" [48, 49]. Synergies have been described both at a muscle level, as co-occurrent patterns of contraction (muscle synergies) (Chaps. 10 and 11) or at a postural level (Chaps. 2, 3, 8 and 12), as groups of degrees of freedom that are controlled together [50]. These synergistic models offer a great computational advantage, since a few synergies (from three to five) can explain a great fraction of the variance associated to posture or muscle activity in a wide range of tasks. However, in order to state that synergies are not only an advantageous and strategic way for controlling artificial and robotic devices, as discussed in Chaps. 8 and 9, but can also represent at least one, or even the main, strategy of motor control in primates, further neurophysiological and functional evidence is needed.

Synergistic motor control hypotheses have been so far corroborated by a great amount of observations, regarding gait and stance [51], arm movements [52] or hand control [53, 54]. In all these reports, a small number of synergies can be extracted from behavioral data to explain the majority of variance. However, these data can only confirm that the motor output (i.e. posture or muscle activity) can be described through synergies, without informing about the possibility that such model has a neural correlate in the encoding activity of motor regions. For this reason, despite these behavioral and kinematic descriptions of synergies, the presence of synergies as a neuronal underpinning for motor commands in the brain has been posited only indirectly.

In fact, the studies that have been performed, both in monkeys and humans, to verify the presence of muscle or postural synergies in the brain relied on the use of integrated, multimodal paradigms: generally, these investigation require both a technique for recording posture or muscle activity (e.g. postural tracking or Electromyography -EMG-) and a technique for studying brain activity (e.g. intracortical potentials, Electroencephalography -EEG- or fMRI. Combining these pieces of information, it is possible to verify whether the synergy-based descriptions of behavioral data are consistent with the brain activity patterns.

Important evidence for synergistic control of the hand was provided by studies on monkeys in which intracortical microstimulation (ICMS) was used along with EMG to evoke and record hand movements, extracting a small set of synergies from the EMG patterns and analyzing the covariation with neuronal activity. In these experiments [55, 56], it was possible to confirm both that evoked movements tend to converge towards particular positions and that a low number of synergies could be extracted from the EMG activity patterns. However, those authors argued that a cortical coding of synergies is unlikely and that motor primitives could be instead represented in the spinal cord, as suggested by earlier evidence obtained in animals [57].

In humans, the number of reports is even more slender: a pioneering study [58] applied a quite similar paradigm to that adopted by Overduin by using Transcranial Magnetic Stimulation (TMS) over M1 to elicit hand movements and optical tracking to record the joint angle patterns of the evoked postures. The results showed that TMS-induced postures have low complexity, with a small set of synergies able to explain most of the postural variance. Other confirmations were provided by stroke patients in which the synergies in the affected and unaffected arm were compared [59, 60]. In those patients, the performance in the affected and unaffected arms differed, but the synergy patterns were highly consistent. As also described in Chap. 5, where the importance of subcortical circuitry in synergistic organization is pointed out, these data suggest that the spinal cord may actually encode muscle synergies and that motor cortex with its descending projections could act as a controller that selects and tunes the synergies encoded downstream. The temporal distance from stroke onset has a role on the synergy patterns, providing useful hints for rehabilitation.

So far, a neuroimaging confirmation of a synergistic control of hand posture is totally lacking: the only results provided so far came from a simple binary comparison between synergistic/dexterous and non-synergistic hand movements and assessed a differential recruitment in the well-known premotor and parietal net-work associated to control of hand posture [61].

Apart from these very partial indications, no studies have so far unveiled whether and to what extent the cerebral cortex can adopt synergy-based models to control the hand: a clear, direct evidence of the presence of dimensionality reduction strategies in motor planning has yet to be achieved.

4.8 Alternative Hypotheses: A Revised Somatotopy?

Despite the growing support to the motor control theories based on synergies, alternative models of hand motor control have been suggested. In fact, additional hypotheses originating from the models based on somatotopic control imply the presence of a strictly topographical organization of motor system based on cortical neurons directly connected to spinal motoneurons with monosynaptic projections. Thus, a cortical motor neuron would directly control a single joint or muscle [35–37]. This organization has been assessed at a coarse scale (i.e. "hand" or "leg" areas) and it is still debated whether a finer level representation for finger or joints is also present. Actually, the existence of horizontal connections and distributed representation of fingers, along with the end point tuning of hand motor neurons, hint at a coding of hand movements based on multi-joints and multi-muscles modules. However, recently, the firing patterns of primary motor neurons were compared both with components extracted from muscle activity (muscle synergies) and with digit motion patterns, finding a greater similarity between neuron activity and single digit movements [62]. The conclusions drawn by the authors were that synergistic control of hand muscles is unlikely and that M1 likely controls single digit kinematics more than the components of hand motion. For this reason, when discussing the strategies adopted by the brain to control the hand, somatotopy-based models and their recent revisions need to be extensively taken into account.

4.9 Techniques for Hand Movement Recordings: Motion Capture and EMG

Human hand movements have been studied so far with different measurement systems. These systems, hereinafter referred to as Hand Pose Reconstruction (HPR) systems (see Chap. 15), can be roughly grouped into remote, or optical-based systems, and wearable, or glove-based, systems [63, 64]. The first type of HPR systems records the 3d positions of active or passive markers attached to the skin through the usage of cameras, such as the Phase Space System (by Phase Space LLC), which uses active infrared markers, that was also adopted for the experiments described in this chapter. The 3d information on markers is then used to estimate joint angles, knowing hand geometric parameters [65], as also discussed in Chap. 14. Glove-based systems directly provide the values of joint angles, after a suitable calibration, and they were used e.g. in [50, 53]. For a review of such systems, the reader is invited to refer to [66] and Chap. 15. It is important to notice that optical-based systems are usually more accurate than their glove-based counterpart [63, 64].

In literature, synergies have been often computed from a set of measured provided by HPR, either glove-based or optical-based [50, 53, 54, 65], see e.g. Chaps. 2, 8, 9 and 12. This topic is widely discussed in Chaps. 14 and 15, where tools and techniques for recording of hand kinematics are described.

Another widely used technique to study hand synergies is EMG: intramuscular recordings are widely used in animals [55] while surface EMG, as discussed in Chaps. 10 and 11, is a non-invasive technique with an easy set-up which has been widely used in studies of gait and locomotion [51] and hand movements [67]. However, the placement of the self-adhesive electrodes for surface EMG can cover only a small fraction of the muscles in hand and forearm. For this reason, only a part of the intrinsic or extrinsic hand muscles can be recorded, making EMG-recorded postures less accurate than the ones recorded with motion capture systems. The data are analyzed by extracting a wide number of time-domain features that describe the EMG signal [68]. Muscle synergies can then be computed applying linear dimensionality reduction methods such as Principal Component Analysis (PCA) [53] or Non-Negative Matrix Factorization [59, 60], see e.g. Chaps. 3 and 9.

4.10 Encoding Techniques: Integrating Behavioral and fMRI Data

In the last years, the advent of machine learning methods has strongly shaped the world of cognitive neuroscience [69]. Data acquired with fMRI or EEG are expressed as time series, representing the signal variation in a brain region across time. These data are usually analyzed with univariate approaches and single fMRI voxels or EEG electrodes are treated as single, mutually independent variables. Multivariate techniques overcome these limitations and consider instead the complex, multiunit populations and have a greater power to detect differences between mental states [70]. In the first report regarding the use of multivariate classifiers on fMRI data [71], pattern analysis was developed to discriminate among object categories (e.g. faces, houses, chairs, cats, etc.) in the temporo-occipital extrastriate region that form the ventral visual stream. That study demonstrated for the first time that cognitive representations of perceived information can arise from distributed and overlapping systems which are based on the integration of many anatomically distinct brain areas instead of relying on a single anatomical region.

The advent of multivariate techniques has therefore moved the explanative power of neuroimaging far beyond the capabilities of classical inference and univariate methods (i.e. a region's function is identified by determining which task activated it most strongly). Among the most interesting new features introduced, there are their ability to discriminate and classify mental states [69, 71, 72], to perform discriminations between classes of stimuli [72, 73] and even to allow for the reconstruction—or decoding—of perceived information from the brain activity associated to it [74, 75]. The decoding of visual content without an a priori inference [74, 76, 77], together with similar approaches on different tasks such as games or mathematical operations [78, 79], have underlined the important putative applications of these techniques to the programming of Brain Machine Interfaces (BMIs) that may be able to "read" brain activity and use that information—for example—to control external devices. These studies, however, relied on supervised machine learning classifiers that needed to be trained on part of the stimuli before being applied on the remaining activity patterns. Important advancements come from later studies in which unsupervised classifiers were used, creating instead a set of models that were tested on the stimulus-evoked activity patterns [80, 81]. The 'successful' method—i.e. the one with the highest prediction accuracy—informed in a totally data driven way about the processing strategy applied by that particular brain region to process perceptual content. The first study using that approach tested a set of different models based on specific perceptual properties of the stimuli to reconstruct (encode) the responses evoked by those stimuli in primary visual areas [80]. Encoding models have therefore confirmed the criteria provided by physiologists and computational neuroscientists to explain the processing in visual cortex, establishing a strong link between these disciplines and cognitive neuroimaging. Encoding and decoding do represent therefore complementary and mutually reinforcing techniques [82] that can be used in association to increase the predictive power of neuroimaging and its potential future application for the development and control of BMIs.

However, even if these techniques are potentially interesting since they test the consistency of many different models with brain activity, they could demonstrate even more applications when integrating fMRI data with external measurements, such as behavioral performance data or peripheral biomarkers. The study that introduced this method for the first time was the seminal paper by Mitchell and colleagues on the semantic representations of concrete nouns [83]. In this experiment, an encoding algorithm was developed to predict the functional activity patterns associated to sixty concrete concepts that were provided to subjects using their associated image and word (e.g. "dog", "apple", "tomato" etc.). A model was created associating the concept-related words with their co-occurrence with 25 different verbs on the basis of a large corpus (e.g. "apple" is strongly associated with "eat" or "pick" and weakly associated with "walk"). Each noun was therefore described as a set of verbal features, with their associated weights. Then, after training a classifier to discriminate between the sixty nouns on the basis of these descriptions, it was possible to "predict" the functional activity patterns associated with the nouns, measured during an fMRI experiment in which subjects freely draw associations between the nouns and their properties.

While these methods are able to integrate external data sources (e.g. semantic features) and brain functional patterns, they rely on a previously defined "model" which is used to predict brain activations. This point is crucial: a single model will not be able to achieve a complete success (i.e. to achieve a full description of activity patterns), first because of the internal noise of fMRI data, second because the human brain surely applies multiple processing strategies and it is unlikely that a single description could totally explain the activity of a specific region or network. Further studies have therefore tested multiple semantic models on brain activity [84, 85] and a combined approach testing more than a model would draw more conclusions than the evaluation of a single description.

4.11 Combining Techniques to "Decode" Hand Posture

The strategies adopted by the brain to control the hand are a theme of extreme interest for neuroscience as well as engineering and robotics. These strategies allow achieving flexible and adaptable configurations that can be quickly modified without losing their stability. Demonstrating the synergistic control of the hand can therefore be very useful to explain both the neural bases of this effective control—thus adding significant information to the neuroscience of movement—and to achieve a model that can be used to design and develop new prosthetic devices based on efficient control strategies. Additionally, the advent of machine learning techniques applied to brain imaging allows both to increase its explanatory power—which was somehow limited by the use of univariate methods—and to integrate brain functional data with behavioral or electrophysiological measurements.

Within the more general aims of the 'The Hand Embodied' program (supported by the European Commission, under the 7th Framework Programme, call FP7-ICT-2009-4, website: http://thehandembodied.eu/), a multimodal experimental paradigm has been adopted to integrate fMRI and kinematic data, in order to study the cortical encoding of hand movements. This multimodal study was therefore aimed at testing whether hand posture, expressed through synergies, has a correlate in brain activity as measured by fMRI. The use of multivariate encoding techniques gave us new instruments to verify this hypothesis, integrating postural and neural functional data obtained from two different experimental sessions.

4.12 Description and Preliminary Results

Nine healthy right-handed volunteers were recruited for two different experimental sessions: session 1 was a kinematic experiment based on an optical tracking of the hand joints during the execution of grasp-to-use acts towards twenty different common-use objects [65], and session 2 was a functional MRI acquisition during the execution of grasping gestures towards the same objects. Twenty target objects were chosen from an earlier report [53] and were not physically present during the sessions. The two sessions were was organized in randomized trials in which participants first watched a picture of the object then a cue sound—heard after an inter-trial interval—prompted the movement execution. Participants had to grasp objects as if they wanted to use them and to hold the resting position once the movement was over. Each trial was repeated five times, in randomized order.

Hand posture was recorded with an optical motion capture system and optical markers were placed on selected bones and joints of the fingers, to derive the joint angles of the hand. In addition, a local frame of reference was obtained adding a bracelet with two additional markers.

The videos from each of the stereocameras were used for movement reconstruction, estimating all the joint angles for each end posture. Postural synergies were then computed using Principal Component Analysis (PCA) [53], see e.g. Chaps. 2, 9 and 11.

In Session 2 the same paradigm was replicated inside the fMRI scanner. Data underwent standard fMRI preprocessing and the BOLD responses to the twenty postures were then used in a multiple linear regression procedure [83] using, as encoding model, the matrix of postural coefficients (i.e. principal component coefficients) which was obtained from the data acquired in Session 1. The analysis resulted in an accuracy value that describes the goodness of the model performance, i.e. the ratio of activity patterns that could be predicted on the basis of the principal component coefficients. Moreover, a map of the voxels whose activity was predictable with the postural synergy model could be derived for each participant. A probability map was then computed, to retain voxels in which the encoding procedure was successful in a greater number of subjects. All the single-subject accuracy values were tested for significance.

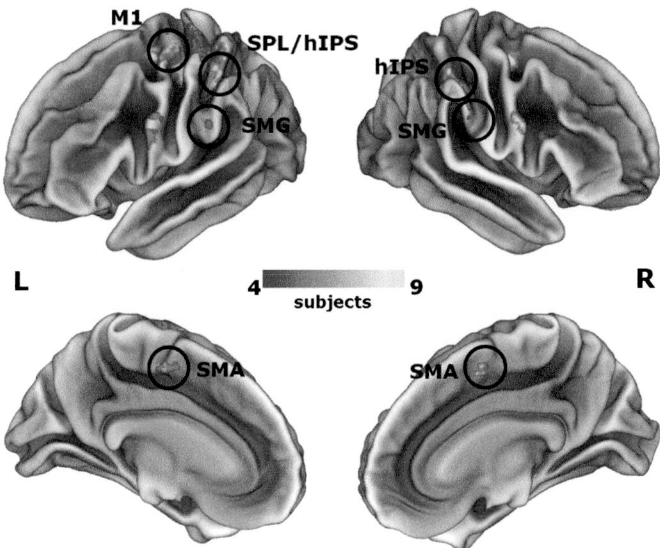

Fig. 4.2 Group map representing the regions with the best encoding performance. The single subject maps of the voxels that effectively predicted neural activity across motor acts were merged, applying a threshold of $p > 0.33$, retaining only the voxels used in at least four subjects. *M1* Primary Motor Cortex; *SMG* Supramarginal Gyrus; *SPL* Superior Parietal Lobule; *IPS* Intraparietal Sulcus; *SMA* Supplementary Motor Area

The encoding procedure resulted in an above-chance level of accuracy in all subjects. The group probability map, which retained the voxels that achieved a successful encoding performance across subjects, identified a well-defined network of regions related to hand control [2, 61]. The network comprised bilateral precentral, supplementary motor and supramarginal areas and left inferior parietal and postcentral cortex (Fig. 4.2). Interestingly, brain activity in these regions was modulated in a posture-specific way, consistently across subjects (Fig. 4.2). This network, therefore, encodes the information for hand motor control relatively to all postures and target objects.

4.13 Conclusions and Future Directions

These preliminary data show for the first time that the network of brain regions that are devoted to the control of hand movement may encode specific 'high-level' representations of single postures through kinematic synergies. Specifically, these motor cortical areas are involved in controlling the end-postures of grasping, and specifically modulate the pattern of hand muscle and joint movements associated to the target objects. These results complement the observations reported in Chap. 5 on the role of the subcortical circuitry in synergistic organization: the extent to which such organization is demanded to different areas of CNS still represents an open challenging issue.

The high accuracy obtained with the encoding approaches described here also opens innovative in-sights on the ability to decode task-specific patterns of neural responses from mo-tor control and action representation networks.

Recently, some interesting Brain Computer Interface (BCI) approaches have surfaced, using direct recording of electrical cortical signals with electrocorticography (ECoG) to extract motor information, obtaining successful control of prosthetic devices [86] and accurate decoding of grasp type (precision vs. whole-hand) [87]. Additional approaches have been obtained from portable devices such as Near-Infrared Spectroscopy (NIRS) and EEG, obtaining a reliable classification of executed movements and motor imagery [88]. These approaches may benefit to a great deal from synergistic control, since the regulation of a reduced number of synergies instead of a full set of joints can increase the reliability of prosthetic devices.

From a theoretical point of view, testing this synergy-based account against alternate hypotheses, such as individual digit control [62] may add some interesting information regarding the models—synergistic or somatotopic—that fit better the neural activity in the network that controls hand posture.

Acknowledgments This work was supported in part by the European Research Council under the Advanced Grant SoftHands "A Theory of Soft Synergies for a New Generation of Artificial Hands" (no. ERC-291166), and by the EU FP7/2007-2013 project (no. 248587) "The Hand Embodied (THE)". We thank Dr. Luca Cecchetti and Ing. Andrea Guidi for their help with fMRI and EMG acquisitions.

References

1. Bernstein NA (1967) The co-ordination and regulation of movements. 1st English edn. Pergamon Press, Oxford
2. Castiello U (2005) The neuroscience of grasping. Nat Rev Neurosci 6(9):726–736. doi:10.1038/nrn1744
3. Grezes J, Decety J (2001) Functional anatomy of execution, mental simulation, observation, and verb generation of actions: a meta-analysis. Hum Brain Mapp 12(1):1–19
4. Bonini L, Ferrari PF, Fogassi L (2013) Neurophysiological bases underlying the organization of intentional actions and the understanding of others' intention. Conscious Cogn 22(3):1095–1104. doi:10.1016/j.concog.2013.03.001
5. Macuga KL, Frey SH (2012) Neural representations involved in observed, imagined, and imitated actions are dissociable and hierarchically organized. Neuroimage 59(3):2798–2807. doi:10.1016/j.neuroimage.2011.09.083
6. Caspers S, Zilles K, Laird AR, Eickhoff SB (2010) ALE meta-analysis of action observation and imitation in the human brain. Neuroimage 50(3):1148–1167. doi:10.1016/j.neuroimage.2009.12.112
7. Muir RB, Lemon RN (1983) Corticospinal neurons with a special role in precision grip. Brain Res 261(2):312–316
8. Kravitz DJ, Saleem KS, Baker CI, Mishkin M (2011) A new neural framework for visuospatial processing. Nat Rev Neurosci 12(4):217–230. doi:10.1038/nrn3008

9. Tunik E, Rice NJ, Hamilton A, Grafton ST (2007) Beyond grasping: representation of action in human anterior intraparietal sulcus. Neuroimage 36(Suppl 2):T77–86. doi:10.1016/j.neuroimage.2007.03.026
10. Gallivan JP, McLean DA, Valyear KF, Culham JC (2013) Decoding the neural mechanisms of human tool use. eLife 2:e00425. doi:10.7554/eLife.00425
11. Fagg AH, Arbib MA (1998) Modeling parietal-premotor interactions in primate control of grasping. Neural Netw: Off J Int Neural Netw Soc 11(7–8):1277–1303
12. Gallese V, Murata A, Kaseda M, Niki N, Sakata H (1994) Deficit of hand preshaping after muscimol injection in monkey parietal cortex. Neuroreport 5(12):1525–1529
13. Logothetis NK, Pauls J, Augath M, Trinath T, Oeltermann A (2001) Neurophysiological investigation of the basis of the fMRI signal. Nature 412(6843):150–157. doi:10.1038/35084005
14. Logothetis NK, Wandell BA (2004) Interpreting the BOLD signal. Annu Rev Physiol 66:735–769. doi:10.1146/annurev.physiol.66.082602.092845
15. Hlustik P, Solodkin A, Gullapalli RP, Noll DC, Small SL (2001) Somatotopy in human primary motor and somatosensory hand representations revisited. Cereb Cortex 11(4):312–321
16. Indovina I, Sanes JN (2001) On somatotopic representation centers for finger movements in human primary motor cortex and supplementary motor area. Neuroimage 13(6 Pt 1):1027–1034. doi:10.1006/nimg.2001.0776
17. Dechent P, Frahm J (2003) Functional somatotopy of finger representations in human primary motor cortex. Hum Brain Mapp 18(4):272–283. doi:10.1002/hbm.10084
18. Moll J, de Oliveira-Souza R, Passman LJ, Cunha FC, Souza-Lima F, Andreiuolo PA (2000) Functional MRI correlates of real and imagined tool-use pantomimes. Neurology 54(6):1331–1336
19. Napier JR (1956) The prehensile movements of the human hand. J Bone Joint Surg British volume 38-B (4):902-913
20. Landsmeer JM (1962) Power grip and precision handling. Ann Rheum Dis 21:164–170
21. Grefkes C, Weiss PH, Zilles K, Fink GR (2002) Crossmodal processing of object features in human anterior intraparietal cortex: an fMRI study implies equivalencies between humans and monkeys. Neuron 35(1):173–184
22. Orban GA, Claeys K, Nelissen K, Smans R, Sunaert S, Todd JT, Wardak C, Durand JB, Vanduffel W (2006) Mapping the parietal cortex of human and non-human primates. Neuropsychologia 44(13):2647–2667. doi:10.1016/j.neuropsychologia.2005.11.001
23. Frey SH, Vinton D, Norlund R, Grafton ST (2005) Cortical topography of human anterior intraparietal cortex active during visually guided grasping. Brain Res Cogn Brain Res 23(2–3):397–405. doi:10.1016/j.cogbrainres.2004.11.010
24. Gallivan JP, McLean DA, Valyear KF, Pettypiece CE, Culham JC (2011) Decoding action intentions from preparatory brain activity in human parieto-frontal networks. J Neurosci: Off J Soc Neurosci 31(26):9599–9610. doi:10.1523/JNEUROSCI.0080-11.2011
25. Ehrsson HH, Fagergren A, Jonsson T, Westling G, Johansson RS, Forssberg H (2000) Cortical activity in precision-versus power-grip tasks: an fMRI study. J Neurophysiol 83(1):528–536
26. Begliomini C, Wall MB, Smith AT, Castiello U (2007) Differential cortical activity for precision and whole-hand visually guided grasping in humans. Eur J Neurosci 25(4):1245–1252. doi:10.1111/j.1460-9568.2007.05365.x
27. Johnson-Frey SH (2004) The neural bases of complex tool use in humans. Trends Cogn Sci 8(2):71–78. doi:10.1016/j.tics.2003.12.002
28. Castiello U, Begliomini C (2008) The cortical control of visually guided grasping. Neurosci: Rev J Bring Neurobiol Neurol Psychiatry 14(2):157–170. doi:10.1177/1073858407312080
29. Tarantino V, De Sanctis T, Straulino E, Begliomini C, Castiello U (2014) Object size modulates fronto-parietal activity during reaching movements. Eur J Neurosci 39(9):1528–1537. doi:10.1111/ejn.12512
30. King M, Rauch HG, Stein DJ, Brooks SJ (2014) The handyman's brain: a neuroimaging meta-analysis describing the similarities and differences between grip type and pattern in humans. Neuroimage 102(Pt 2):923–937. doi:10.1016/j.neuroimage.2014.05.064

31. Bleichner MG, Freudenburg ZV, Jansma JM, Aarnoutse EJ, Vansteensel MJ, Ramsey NF (2014) Give me a sign: decoding four complex hand gestures based on high-density ECoG. Brain Struct Funct doi:10.1007/s00429-014-0902-x
32. Bleichner MG, Jansma JM, Sellmeijer J, Raemaekers M, Ramsey NF (2014) Give me a sign: decoding complex coordinated hand movements using high-field fMRI. Brain Topogr 27(2):248–257. doi:10.1007/s10548-013-0322-x
33. Fridman EA, Immisch I, Hanakawa T, Bohlhalter S, Waldvogel D, Kansaku K, Wheaton L, Wu T, Hallett M (2006) The role of the dorsal stream for gesture production. Neuroimage 29(2):417–428. doi:10.1016/j.neuroimage.2005.07.026
34. Kroliczak G, Frey SH (2009) A common network in the left cerebral hemisphere represents planning of tool use pantomimes and familiar intransitive gestures at the hand-independent level. Cereb Cortex 19(10):2396–2410. doi:10.1093/cercor/bhn261
35. Penfield W, Boldrey E (1937) Somatic motor and sensory representation in the cerebral cortex of man as studied by electrical stimulation. Brain: J Neurol 60(4):389–443. doi:10.1093/brain/60.4.389
36. Woolsey CN, Settlage PH, Meyer DR, Sencer W, Pinto Hamuy T, Travis AM (1952) Patterns of localization in precentral and "supplementary" motor areas and their relation to the concept of a premotor area. Research publications–Association for Research in Nervous and Mental Disease 30:238-264
37. Penfield W, Rasmussen T (1950) The cerebral cortex of man: a clinical study of localization of function. Macmillan, New York
38. Foerster O (1936) The motor cortex in man in the light of Hughlings Jackson's doctrines. 59(2):135-159. doi:10.1093/brain/59.2.135
39. Schieber MH (2001) Constraints on somatotopic organization in the primary motor cortex. J Neurophysiol 86(5):2125–2143
40. Graziano MS, Taylor CS, Moore T (2002) Complex movements evoked by microstimulation of precentral cortex. Neuron 34(5):841–851
41. Sanes JN, Schieber MH (2001) Orderly somatotopy in primary motor cortex: does it exist? Neuroimage 13(6 Pt 1):968–974. doi:10.1006/nimg.2000.0733
42. Aflalo TN, Graziano MS (2006) Partial tuning of motor cortex neurons to final posture in a free-moving paradigm. Proc Natl Acad Sci USA 103(8):2909–2914. doi:10.1073/pnas.0511139103
43. Griffin DM, Hudson HM, Belhaj-Saif A, Cheney PD (2014) EMG activation patterns associated with high frequency, long-duration intracortical microstimulation of primary motor cortex. J Neurosci: Off J Soc Neurosci 34(5):1647–1656. doi:10.1523/JNEUROSCI.3643-13.2014
44. Desmurget M, Richard N, Harquel S, Baraduc P, Szathmari A, Mottolese C, Sirigu A (2014) Neural representations of ethologically relevant hand/mouth synergies in the human precentral gyrus. Proc Natl Acad Sci USA 111(15):5718–5722. doi:10.1073/pnas.1321909111
45. Kaas JH, Gharbawie OA, Stepniewska I (2013) Cortical networks for ethologically relevant behaviors in primates. Am J Primatol 75(5):407–414. doi:10.1002/ajp.22065
46. Schaffelhofer S, Agudelo-Toro A, Scherberger H (2015) Decoding a wide range of hand configurations from macaque motor, premotor, and parietal cortices. J Neurosci: Off J Soc Neurosci 35(3):1068–1081. doi:10.1523/JNEUROSCI.3594-14.2015
47. Latash ML (2010) Motor synergies and the equilibrium-point hypothesis. Motor Control 14(3):294–322
48. Santello M, Baud-Bovy G, Jorntell H (2013) Neural bases of hand synergies. Front Comput Neurosci 7:23. doi:10.3389/fncom.2013.00023
49. Turvey MT (2007) Action and perception at the level of synergies. Hum Mov Sci 26(4):657–697. doi:10.1016/j.humov.2007.04.002
50. Tessitore G, Sinigaglia C, Prevete R (2013) Hierarchical and multiple hand action representation using temporal postural synergies. Exp Brain Res 225(1):11–36. doi:10.1007/s00221-012-3344-9
51. Torres-Oviedo G, Ting LH (2010) Subject-specific muscle synergies in human balance control are consistent across different biomechanical contexts. J Neurophysiol 103(6):3084–3098. doi:10.1152/jn.00960.2009

52. Chiovetto E, Berret B, Delis I, Panzeri S, Pozzo T (2013) Investigating reduction of dimensionality during single-joint elbow movements: a case study on muscle synergies. Front Comput Neurosci 7:11. doi:10.3389/fncom.2013.00011
53. Santello M, Flanders M, Soechting JF (1998) Postural hand synergies for tool use. J Neurosci: Off J Soc Neurosci 18(23):10105–10115
54. Santello M, Flanders M, Soechting JF (2002) Patterns of hand motion during grasping and the influence of sensory guidance. J Neurosci: Off J Soc Neurosci 22(4):1426–1435
55. Overduin SA, d'Avella A, Carmena JM, Bizzi E (2012) Microstimulation activates a handful of muscle synergies. Neuron 76(6):1071–1077. doi:10.1016/j.neuron.2012.10.018
56. Overduin SA, d'Avella A, Carmena JM, Bizzi E (2014) Muscle synergies evoked by microstimulation are preferentially encoded during behavior. Front Comput Neurosci 8:20. doi:10.3389/fncom.2014.00020
57. Saltiel P, Wyler-Duda K, D'Avella A, Tresch MC, Bizzi E (2001) Muscle synergies encoded within the spinal cord: evidence from focal intraspinal NMDA iontophoresis in the frog. J Neurophysiol 85(2):605–619
58. Gentner R, Classen J (2006) Modular organization of finger movements by the human central nervous system. Neuron 52(4):731–742. doi:10.1016/j.neuron.2006.09.038
59. Cheung VC, Piron L, Agostini M, Silvoni S, Turolla A, Bizzi E (2009) Stability of muscle synergies for voluntary actions after cortical stroke in humans. Proc Natl Acad Sci USA 106(46):19563–19568. doi:10.1073/pnas.0910114106
60. Cheung VC, Turolla A, Agostini M, Silvoni S, Bennis C, Kasi P, Paganoni S, Bonato P, Bizzi E (2012) Muscle synergy patterns as physiological markers of motor cortical damage. Proc Natl Acad Sci USA 109(36):14652–14656. doi:10.1073/pnas.1212056109
61. Ehrsson HH, Kuhtz-Buschbeck JP, Forssberg H (2002) Brain regions controlling nonsynergistic versus synergistic movement of the digits: a functional magnetic resonance imaging study. J Neurosci: Off J Soc Neurosci 22(12):5074–5080
62. Kirsch E, Rivlis G, Schieber MH (2014) Primary motor cortex neurons during individuated finger and wrist movements: correlation of spike firing rates with the motion of individual digits versus their principal components. Front Neurol 5:70. doi:10.3389/fneur.2014.00070
63. Bianchi M, Salaris P, Bicchi A (2013) Synergy-based hand pose sensing: optimal glove design. Int J Robot Res 32(4):407–424. doi:10.1177/0278364912474079
64. Bianchi M, Salaris P, Bicchi A (2013) Synergy-based hand pose sensing: reconstruction enhancement. Int J Robot Res 32(4):396–406. doi:10.1177/0278364912474078
65. Gabiccini M, Stillfried G, Marino H, Bianchi M (2013) A data-driven kinematic model of the human hand with soft-tissue artifact compensation mechanism for grasp synergy analysis. IEEE Int C Int Robot 3738-3745
66. Dipietro L, Sabatini AM, Dario P (2008) A survey of glove-based systems and their applications. IEEE T Syst Man Cy C 38(4):461–482. doi:10.1109/Tsmcc.2008.923862
67. Weiss EJ, Flanders M (2004) Muscular and postural synergies of the human hand. J Neurophysiol 92(1):523–535. doi:10.1152/jn.01265.2003
68. Mathiesen JR, Bøg MF, Erkocevic E, Niemeier MJ, Smidstrup A, Kamavuako EN Prediction of grasping force based on features of surface and intramuscular EMG
69. Varoquaux G, Thirion B (2014) How machine learning is shaping cognitive neuroimaging. GigaSci 3:28. doi:10.1186/2047-217X-3-28
70. Pereira F, Mitchell T, Botvinick M (2009) Machine learning classifiers and fMRI: a tutorial overview. Neuroimage 45(1 Suppl):S199–209. doi:10.1016/j.neuroimage.2008.11.007
71. Haxby JV, Gobbini MI, Furey ML, Ishai A, Schouten JL, Pietrini P (2001) Distributed and overlapping representations of faces and objects in ventral temporal cortex. Science 293(5539):2425–2430. doi:10.1126/science.1063736
72. Formisano E, De Martino F, Bonte M, Goebel R (2008) "Who" is saying "what"? Brain-based decoding of human voice and speech. Science 322(5903):970–973. doi:10.1126/science.1164318
73. Mitchell TM, Hutchinson R, Niculescu RS, Pereira F, Wang XR, Just M, Newman S (2004) Learning to decode cognitive states from brain images. Mach Learn 57(1–2):145–175. doi:10.1023/B:Mach.0000035475.85309.1b

74. Thirion B, Duchesnay E, Hubbard E, Dubois J, Poline JB, Lebihan D, Dehaene S (2006) Inverse retinotopy: inferring the visual content of images from brain activation patterns. Neuroimage 33(4):1104–1116. doi:10.1016/j.neuroimage.2006.06.062
75. Miyawaki Y, Uchida H, Yamashita O, Sato MA, Morito Y, Tanabe HC, Sadato N, Kamitani Y (2008) Visual image reconstruction from human brain activity using a combination of multiscale local image decoders. Neuron 60(5):915–929. doi:10.1016/j.neuron.2008.11.004
76. Kamitani Y, Tong F (2005) Decoding the visual and subjective contents of the human brain. Nat Neurosci 8(5):679–685. doi:10.1038/nn1444
77. Haynes JD, Rees G (2005) Predicting the orientation of invisible stimuli from activity in human primary visual cortex. Nat Neurosci 8(5):686–691. doi:10.1038/nn1445
78. Gourtzelidis P, Tzagarakis C, Lewis SM, Crowe DA, Auerbach E, Jerde TA, Ugurbil K, Georgopoulos AP (2005) Mental maze solving: directional fMRI tuning and population coding in the superior parietal lobule. Exp Brain Res 165(3):273–282. doi:10.1007/s00221-005-2298-6
79. Haynes JD, Sakai K, Rees G, Gilbert S, Frith C, Passingham RE (2007) Reading hidden intentions in the human brain. Curr Biol 17(4):323–328. doi:10.1016/j.cub.2006.11.072
80. Kay KN, Naselaris T, Prenger RJ, Gallant JL (2008) Identifying natural images from human brain activity. Nature 452(7185):352–355. doi:10.1038/nature06713
81. Naselaris T, Prenger RJ, Kay KN, Oliver M, Gallant JL (2009) Bayesian reconstruction of natural images from human brain activity. Neuron 63(6):902–915. doi:10.1016/j.neuron.2009.09.006
82. Naselaris T, Kay KN, Nishimoto S, Gallant JL (2011) Encoding and decoding in fMRI. Neuroimage 56(2):400–410. doi:10.1016/j.neuroimage.2010.07.073
83. Mitchell TM, Shinkareva SV, Carlson A, Chang KM, Malave VL, Mason RA, Just MA (2008) Predicting human brain activity associated with the meanings of nouns. Science 320(5880):1191–1195. doi:10.1126/science.1152876
84. Chang KM, Mitchell T, Just MA (2011) Quantitative modeling of the neural representation of objects: how semantic feature norms can account for fMRI activation. Neuroimage 56(2):716–727. doi:10.1016/j.neuroimage.2010.04.271
85. Pereira F, Botvinick M, Detre G (2013) Using Wikipedia to learn semantic feature representations of concrete concepts in neuroimaging experiments. Artif Intell 194:240–252. doi:10.1016/j.artint.2012.06.005
86. Collinger JL, Wodlinger B, Downey JE, Wang W, Tyler-Kabara EC, Weber DJ, McMorland AJ, Velliste M, Boninger ML, Schwartz AB (2013) High-performance neuroprosthetic control by an individual with tetraplegia. Lancet 381(9866):557-564. doi:10.1016/S0140-6736(12)61816-9
87. Pistohl T, Schulze-Bonhage A, Aertsen A, Mehring C, Ball T (2012) Decoding natural grasp types from human ECoG. Neuroimage 59(1):248–260. doi:10.1016/j.neuroimage.2011.06.084
88. Fazli S, Mehnert J, Steinbrink J, Curio G, Villringer A, Muller KR, Blankertz B (2012) Enhanced performance by a hybrid NIRS-EEG brain computer interface. Neuroimage 59(1):519–529. doi:10.1016/j.neuroimage.2011.07.084

Chapter 5
Synergy Control in Subcortical Circuitry: Insights from Neurophysiology

Henrik Jörntell

Abstract Synergy control in the brain is likely to a large extent delegated to subcortical circuitry, with a focus on spinal circuitry and add-on capability provided by the spinocerebellar system for complex synergies. The advantage with this organization is that there is a tight connection, in the sense of shorter delays, between sensor feedback and the continuously updated motor command. By involving the sensory feedback in the motor command, the brain can make sure that the relevant biomechanical properties are properly compensated for. A consequence of this arrangement is that the neocortex, from which all voluntary motor commands originates, needs to learn the properties of the subcortical circuitry rather than the full details of the high-dimensional biomechanical plant. As the subcortical circuitry appears to have primarily linear properties, this arrangement makes it possible for the voluntary system to add synergy components linearly.

5.1 Introduction

The link between Neuroscience and Robotics is both natural and logical as widely discussed in this book. The evolution of the nervous system is driven by the need to attain more effective ways to move, and robotics is naturally first and foremost about motor control. Roboticists have also long taken inspiration from humans and biology (see for example Chap. 9) as biological systems for example can combine tremendous versatility, smooth performance, speed and compliance to interact with other individuals. These factors to a large extent depend on the capabilities of the nervous system. But how does the human brain solve these tasks?

H. Jörntell (✉)
Neural Basis of Sensorimotor Control, Department of Experimental Medical Science, Lund University, BMC F10 Tornavägen 10, SE221 64 Lund, Sweden
e-mail: Henrik.jorntell@med.lu.se

M. Bianchi and A. Moscatelli (eds.), *Human and Robot Hands*,
Springer Series on Touch and Haptic Systems, DOI 10.1007/978-3-319-26706-7_5

The answer to this question requires a profound insight into the organization and function of the brain. Such insights can be obtained from the combination of two main lines of neuroscience research, external observations of the functions characterizing motor behaviour and analysis of the infrastructure at the level of the neuronal circuitry supporting these functions, by means of electrophysiology or imaging techniques (see also Chap. 4). Unfortunately, there has so far been only limited work focusing on the link between neuronal microcircuitry and high level aspects of motor behavior (i.e. overt movement), primarily due to difficulties in understanding how neuronal circuits work. But in this chapter the focus will be on the information that does exist, and on the additional type of information that need to be gained before the mechanisms underlying the versatility of the biological motor control have been understood to a sufficient extent to allow all its advantages to be implemented in robotic systems.

5.2 State-of-the-Art

The concept of synergy-based robots is relatively new, but several cases of synergy control implementation in robotics, inspired by principles observed from biology and behaviour, exist [3] (see for e.g. Chaps. 8–10 and 15). There is sufficient amount of proof-of-concept to show that the concept is useful in terms of simplifying control and for example making robotic hands more human-like in their movement patterns. But there is much more to be learnt from biological synergy control than this—the most striking one being well-trained complex movements performed at high speed. In order to reach such a goal, we need to learn more about the design of the control in biological systems. Although behavioural observation is important, and a main reason the concept of synergies arose, knowledge of the supporting control system (Chap. 4) should lead to deeper insights that could make it possible to reach further towards biological versatility and simplified control. There is today relatively rudimentary information about this issue.

The knowledge about the circuitry and its components in the nervous system is high in the level of detail but poor at the level of the integrated system. However, an integrated neuroscience-robotics approach, where the target is to produce a functional system rather than reproducing it at high level of biological detail, can help making advances in this important field.

5.3 Problem Framing

There are four major issues that need to be understood in order to implement neuroscience control principles in robotics. The first is the basic function of the cellular units, or the neurons, of which the brain is composed. This is a huge subject area in its own right and will not be considered here. Basically, recent findings indicate that neurons are basically linear integrators as long as they are operating in the range of activity they normally display under behaviour. This means that brain function per se is largely determined by the pattern of connectivity between the neurons, plus some delays in information transfer associated with the biological limitations of the neurons. The second issue to be understood is the early developmental localization of various structures in the brain. The various structures will be connected to each other in predetermined, genetically programmed, patterns so this is part of the connectivity issue. The third issue is how the nervous system is wired up against the biomechanical reality it will be acting in—that is, our bodies. Our bodies contain thousands of muscle fibers and hundred thousands of sensors. It is only by connecting to these actuators and sensors that the brain will develop into a working system. The ability to elicit movement, and to receive sensory feedback from that movement, is what brain development is all about the first years of life and cognitive development, that start later, cannot work properly unless there is first established a relationship to our bodies. The fourth issue is the function of the brain in movement control as approximated by external observations. Can the movements somehow be decomposed into basic building blocks, which may be the functions subserved by specific subcomponents of the brain circuitry?

5.4 Synergy Control in Subcortical Circuitry

An important initial fact to note is that neuronal circuitry of the mammalian brain is naturally designed for implementing synergy control—in fact, it is hard to see how the brain could not operate in this fashion. The main reason for this statement is that the individual neurons, which are the building blocks of brain circuitry, have widely divergent outputs, in the sense that each neuron issues its output information to a wide set of receiving neurons [9]. At the level of the final stage of the motor command, the spinal alpha-motorneurons, which directly innervates the muscle cells in the periphery, there is a muscle specific innervation. However, the neurons which activate the alpha-motorneurons, and which thereby mediate the voluntary control of their output, are all heavily divergent. The individual neurons of the neocortex that innervates the spinal cord, and which thereby project the voluntary aspect of the motor command, have terminations that are reaching multiple motor nuclei (the definition of a motor nucleus being that it contains all the alpha-motorneurons that innervate a particular muscle) [10]. In addition, the vast majority of the terminations of the corticospinal axons are made outside the motor nuclei but in the midst of the large population of spinal premotor neurons, or spinal interneurons [2]. The individual spinal interneurons, in turn, individually also target more than one motor nucleus, even though the extent of the divergence of the spinal interneurons in some cases may be more limited than that of the individual corticospinal axons [6]. Hence, there is in the mammalian central nervous system possibly not a single case of a neuron that target the output neurons of a single muscle only, all the neuronal terminations can be expected to be divergent with respect to the muscles/alpha-motorneurons targeted.

It follows that brain circuitry is naturally designed to deal with muscle activation in terms of synergies rather than on a muscle-per-muscle basis.

But what is the structure of the motor controlling circuitry and how can it deal with the organization of motor control in the face of this naturally divergent connectivity? A first and perhaps surprising insight in this respect is that the same blueprint for the central nervous system is used across the vast variety of species in the vertebrate part of the animal series, ranging from non-limbed mammals like the whale to the highest limb controlling capacity found in humans [5, 8]. This means that the same types of basic structures and neuronal subsystems, including the basic pattern of interconnections between these structures, are present in all these species. This is essentially completely true for subcortical circuitry. For the neocortex and the cerebellum, there exist in humans, as opposed to lower mammals, a great expansion of the cortical volumes and additional subsystems are added on top of the hierarchically lower cortical areas that is generally believed to add additional cognitive capacity. However, for the core motor control capacity, the available structures are essentially identical, although the total number of neurons per structure may vary between species. This is an observation that is typically very hard to understand, given the wide differences in movement repertoire that these species display. The reason that these similarities prevail is that the genetic programs that generate the central nervous system are extraordinarily complex and the smallest change in the core of these programs would almost certainly always result in disaster or death of the animal.

The reason that the same basic genetic program for the generation of the central nervous system can work across such a vast array of species, with very different anatomical constraints, is likely to be spelt 'learning'. Learning at the neuronal level can be subdivided into many different categories, but here we are considering learning in the sense of fundamental circuitry structuring in the immature circuitry. The main pathways of connections between different main structures of the brain, and also between the multitude of sensors in the skin, muscles and other internal components and the central nervous system, are generated by genetically preprogrammed signals. However, the local connectivity, or the connectivity between specific types of afferents and specific neurons, is subject to extensive modification during development, to the point that they can presumably be considered as entirely established through learning. Hence, the activation profiles of the different specific sensors that are generated through erratic or organized movements can provide the central nervous system with the signals that are needed for self-organization of the circuitry. The sensor signals that 'belong' to each other, and later also the association between specific motor commands and sensor signals, in the specific setting defined by the biomechanics of the animal species, can be found by practice. Since the anatomy of the periphery and thereby the conditions of movement differs so much between these species, it follows that there is a lot of adaptation or learning that needs to take place in these brains before any purposeful movement control can be exerted. In all mammals including humans, this type of learning appear to first and foremost occur in spinal circuitry, and it is only after the basic connectivity rules have been established here that the neocortex can start interact with the subcortical circuitry in a meaningful manner. It follows that the brain is extremely versatile in terms of range

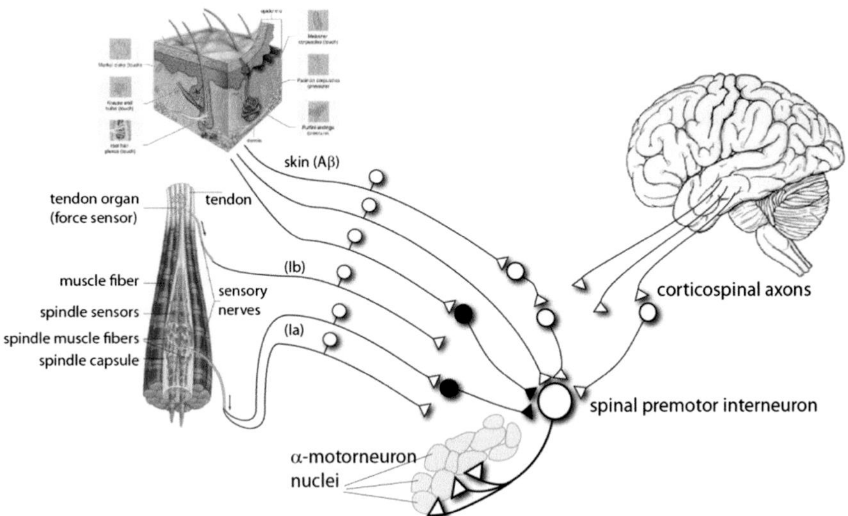

Fig. 5.1 Schematic illustrating the tight link between motor command and sensory feedback in the mammalian nervous system. Corticospinal tract axons target primarily spinal interneurons, which in turn have divergent connections to multiple alpha-motorneuron nuclei. When muscles are activated, they inevitably generate a massive sensory feedback from muscle spindles (*Ia*), from tendons (*Ib*) and from various categories of skin sensors ($A\beta$). This sensory feedback directly or indirectly (through additional spinal interneurons) is provided back to the spinal interneuron. This results in that the drive on the spinal interneuron will vary during the course of the movement, regardless of if there is any variation in the corticospinal tract activity. The sensory feedback can be used by the central nervous system to find the muscles that are naturally co-activated, as a consequence of the biomechanical setting of the body, under a given context (*motor task*). Hence, synergy components are most effectively built at this level of the neuronal circuitry

of the control it can perform, in a sense that no artificial control system currently can match. However, the principles could naturally be exploited and implemented for robotics control as well (Chap. 9).

A second and again perhaps surprising observation is that the same parts of the nervous system are engaged across a variety of movement conditions. This is in contrast to the classical ideas of functional localization within the brain. But consider for example the spinal circuitry, through which all motor commands have to pass. Here it is indisputably the case that the circuitry must either learn to contribute to all kinds of movements, or at the very least make sure that it will mediate the motor commands without destroying them. The second case is not very different from the former, because all motor commands inevitably elicit sensory feedback that in turn alter the state of the spinal circuitry over time as the movement develops (Fig. 5.1), so in principle it can be assumed that the spinal circuitry must learn to contribute to all movements.

But if it is possible in one type of neuronal circuitry, there is nothing that prevents this from being the case also for the cerebellar neuronal circuitry and the neocortical circuitry. Indeed, in all cases where the contribution of a particular type of cerebellar or neocortical neuron to various movements has been tested, they have been found to be correlated with all movement types tested [4, 7, 13]. It follows that neuronal and circuitry functions have to be widely generalizable. In addition, it also follows that the performance of different types of movements may more be a matter of putting the neuronal circuitries globally into different states than to generate activation of specific circuitry components that hypothetically could have driven specific motor programs.

What would be the basis for this portrayed extreme capacity for generalization in neurons? One important element could be linearity in the neuronal elements. Linear properties could make it easier to use the same circuitry set for a vast set of functions, in particular if at least part of the functions are linear in character. It is at any rate difficult to see how a particular type of non-linearity could be made use of in all motor control functions. In a recent study, the properties of the spike generation in both cerebellar and spinal neurons were investigated [11]. Across a wide set of stationary states for each neuron included, it was found that the neurons are almost perfectly linear, at least in the range of firing frequencies these neurons have been found to display in vivo under behaviour. Hence, it is likely that, at least subcortically, all populations of neurons that contribute to movements are linear. This may not have to mean that the circuitry always behave strictly linearly, but may at least facilitate an extreme generalization in terms of the array of motor functions that these neurons can participate in.

Assuming that all neuronal elements are linear, the coordination of activation of synergy components can be relatively easily explained by subcortical mechanisms, at least at a superficial level. As recently described, the spinocerebellar systems, i.e. the connection from last order interneurons to the cerebellum as mossy fibers, play a crucial role in this view [1, 12]. Synergy control and the coordination of synergy components over time then basically becomes a function that resides in the spinal cord and the cerebellum, and the neocortex can learn to utilize the functions that are resident in this subcortical circuitry. The spinal circuitry, based on movement statistics as defined by the biomechanical properties of the body, becomes entrained to the most commonly used muscle combinations, or synergies, and will support the activation of particular muscle combinations depending on the context given by the pattern of descending activation from the corticospinal tract. The link of synergy components over time, as a movement develops, can then be handled by the cerebellum. The cerebellum receives rapid information about the ongoing synergy activation pattern by the spinocerebellar neurons and can use cues, both from these systems and other sources of information providing context, to decide in which phase of the movement that it becomes time to initiate the activation of new synergy components. Again, as the neuronal systems are linear, any synergy component can in principle be added at any time, and it is likely that their summation is primarily linear, although threshold effects and non-linearities in this step cannot be excluded, as for example many of the spinal interneurons are inhibitory and would tend to quench the activation of certain motor nuclei while driving others.

5.5 Conclusions

While the integrated function of neural control systems has not been extensively addressed in the neuroscience literature, recent advances suggest that it is possible to decipher the underlying circuitry structure and its functions. Key factors to understand the low-level circuitry structure include, first; how the learning of how sensors and muscles are related works; secondly, how the brain builds the lowest level models of these relationships. Once obtained, the brain can utilize these models to combine and recruit synergies in a context-dependent fashion. To understand the global solution of the brain to synergy control, we then next need to understand how the neocortex represents and recruits these low level models—this is likely done again by learning and building models, this time at a higher level of abstraction than is done subcortically. A related open question is how many different levels of models that exist in the brain, where the subcortical circuitry can be expected to constitute the innermost and most direct models. An integrated neuroscience-robotics approach, where the target is to produce a functional system rather than reproducing it at high level of biological detail, can help further advances in this important field. Robotics has the major advantage that there are clear consequences when the synthetized control system has flaws, at the same time as it often comes with non-trivial mechanical properties which needs to be taken care of by the control system. Therefore, integrated neuroscience-robotics approaches not only have the advantage that they can help improving robotic systems in terms of versatility and ease-of-control, they also force neuroscience to focus on important outstanding questions for understanding the organization of the brain at a functional level. In addition, for the growing field of robotic implementations in human environments, such as assistive devices, the property of having biologically-driven systems is likely to be advantageous both in terms of safety and in terms of intuitiveness and ease of use.

References

1. Bengtsson F, Jorntell H (2014) Specific relationship between excitatory inputs and climbing fiber receptive fields in deep cerebellar nuclear neurons. PLoS ONE 9(1):e84616. doi:10.1371/journal.pone.0084616
2. Bortoff GA, Strick PL (1993) Corticospinal terminations in two new-world primates: further evidence that corticomotoneuronal connections provide part of the neural substrate for manual dexterity. J Neurosci 13(12):5105–5118
3. Catalano MG, Grioli G, Serio A, Farnioli E, Piazza C, Bicchi A (2012) Adaptive synergies for a humanoid robot hand. In: 2012 12th IEEE-RAS international conference on humanoid robots (Humanoids), pp 7–14, IEEE
4. Georgopoulos AP, Merchant H, Naselaris T, Amirikian B (2007) Mapping of the preferred direction in the motor cortex. Proc Natl Acad Sci USA 104(26):11068–11072. doi:10.1073/pnas.0611597104(0611597104[pii])
5. Grillner S (2003) The motor infrastructure: from ion channels to neuronal networks. Nat Rev Neurosci 4(7):573–586. doi:10.1038/nrn1137

6. Jankowska E (1992) Interneuronal relay in spinal pathways from proprioceptors. Prog Neurobiol 38(4):335–378
7. Naselaris T, Merchant H, Amirikian B, Georgopoulos AP (2006) Large-scale organization of preferred directions in the motor cortex. II. Analysis of local distributions. J Neurophysiol 96(6):3237–3247. doi:10.1152/jn.00488.2006(00488.2006[pii])
8. Ocana FM, Suryanarayana SM, Saitoh K, Kardamakis AA, Capantini L, Robertson B, Grillner S (2015) The lamprey pallium provides a blueprint of the Mammalian motor projections from cortex. Curr Biol 25(4):413–423. doi:10.1016/j.cub.2014.12.013
9. Santello M, Baud-Bovy G, Jorntell H (2013) Neural bases of hand synergies. Front Comput Neurosci 7:23. doi:10.3389/fncom.2013.00023
10. Shinoda Y, Yokota J, Futami T (1981) Divergent projection of individual corticospinal axons to motoneurons of multiple muscles in the monkey. Neurosci Lett 23(1):7–12
11. Spanne A, Geborek P, Bengtsson F, Jorntell H (2014) Spike generation estimated from stationary spike trains in a variety of neurons in vivo. Front Cell Neurosci 8:199. doi:10.3389/fncel.2014.00199
12. Spanne A, Jorntell H (2013) Processing of multi-dimensional sensorimotor information in the spinal and cerebellar neuronal circuitry: a new hypothesis. PLoS Comput Biol 9(3):e1002979. doi:10.1371/journal.pcbi.1002979
13. van Kan PL, Horn KM, Gibson AR (1994) The importance of hand use to discharge of interpositus neurones of the monkey. J Physiol 480(Pt 1):171–190

Chapter 6
Neuronal "Op-amps" Implement Adaptive Control in Biology and Robotics

Martin Nilsson

Abstract Animals control their limbs very efficiently using interconnected neuronal populations. We propose that these populations can be seen as general-purpose *neuronal operational amplifiers*, or neuronal "op-amps", forming adaptive feedback networks. The neuronal op-amp is an interdisciplinary concept offering tentative explanations of animal behaviour as well as approaches to biologically inspired high-dimensional robot control. For instance, in biology, the concept indicates the origin of synergies and saliency in the mammalian central nervous system; in robotics, it presents a design of simple but robust adaptive controllers that identify unknown sensors online. Here, we introduce the neuronal op-amp concept and its biological basis. We explore its biological plausibility, its application, and its performance in adaptive control both theoretically and experimentally.

6.1 Introduction

As long as several hundred million years ago, evolution developed efficient solutions to many difficult control problems that puzzle today's biologists and engineers. Spurred on by curiosity about how biological systems work and whether the same principles can be applied in robotics, we have focused on two particularly interesting problems: the *correspondence* (or sensorimotor association) problem and the *adaptive servo* (or adaptive reference signal tracking) problem.

6.1.1 Two Central (Nervous System) Problems

The correspondence problem concerns how the brain "knows" which strings to pull to move a limb (Fig. 6.1). How does the brain associate sensory feedback signals with the corresponding motor commands? The ability to bootstrap this coupling

M. Nilsson (✉)
SICS (Swedish Institute of Computer Science), POB 1263, 164 29 Kista, Sweden
e-mail: neuronalopamps@drnil.com

M. Bianchi and A. Moscatelli (eds.), *Human and Robot Hands*,
Springer Series on Touch and Haptic Systems, DOI 10.1007/978-3-319-26706-7_6

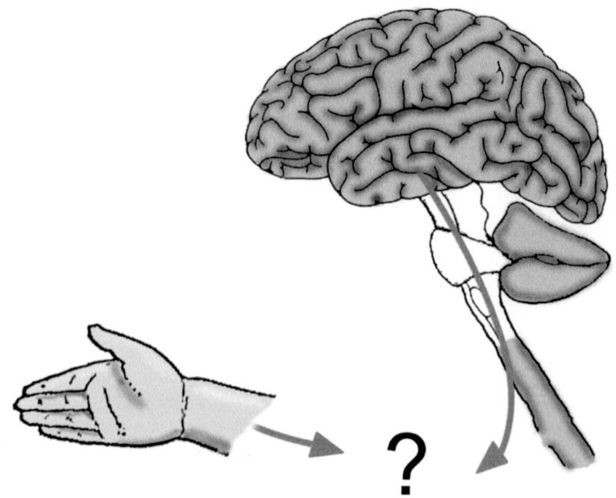

Fig. 6.1 The correspondence problem: How does the brain "know" which sensors and actuators belong together? (Figure partially based on [13])

from scratch is perplexing if sensors are unreliable and uncalibrated and if models are not pre-specified. Nevertheless, biological systems are robust and tolerate noise and fault. These properties are also highly desirable in modern mechatronic systems, such as robot arms and prosthetic devices. There are two main putative approaches to answering the question posed by the correspondence problem. One possibility is that sensorimotor association is genetically preprogrammed (phylogenetic); however, this appears unlikely due to the evolutionary fragility and the large amount of configuration information that must be processed. The alternative is that the association is bootstrapped during development (ontogenetic)—but how?

The adaptive servo problem focuses on how a controller can track a time-varying reference signal, despite continuous changes in system properties. This behaviour is built into biological systems performing such tasks as reaching and grasping. A complicating factor for biological systems is their huge numbers of sensors and actuators/motors, which we will refer to collectively as *transducers*. For instance, a single human hand has more than 30 muscles, and its glabrous skin alone contains approximately 20,000 tactile sensors [16], to which a significant number of muscle spindles and Golgi tendon organs should be added. How can such high-dimensional systems be controlled efficiently?

The answers to these questions could help us understand how the mammalian central nervous system (CNS) operates in addition to helping us build biomimetic or biologically inspired robots with many degrees of freedom, which allow for efficient, robust, reliable, and inexpensive movement.

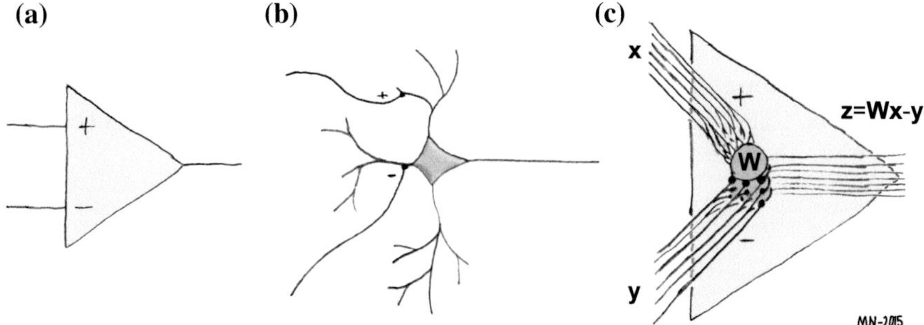

Fig. 6.2 **a** Op-amp, **b** Neuron, **c** Neuronal op-amp model of a population, having vector inputs and outputs

6.1.2 Main Objective

We propose the concept of the *neuronal operational amplifier*, or "neuronal op-amp", which is inspired by the ubiquitous electronic operational amplifier, "op-amp" (Fig. 6.2). Emphasis is placed on the op-amp as a linear amplifier with non-inverting and inverting inputs that deliver an output signal proportional to their difference. By placing an op-amp in a feedback circuit, a wide variety of functions can be obtained [28]. An electronic op-amp resembles a neuron with its excitatory and inhibitory inputs. Our main objective is to demonstrate the expressional power of neuronal op-amps integrated into feedback circuits. We show through theory and experiments that such control circuits do not require *a priori* models, but can acquire internal models in real time; can implement biologically feasible, multi-variable adaptive feedback control; show stable and robust reference signal tracking; and are scalable to high dimensions.

6.1.3 Scope and Assumptions

An important consideration in both biological and robotic systems is the type of transducers involved. Transducers can be *tonic* or *phasic*, where "tonic" means that for an input x, the output $y = f(x)$ is independent of time. This is a special case of a phasic transducer, in which the output can additionally be a function of time, $y = f(x;t)$. The primary advantage of tonic models over phasic models is a substantial mathematical simplification. The tonic model is still sufficiently powerful to be effectively applied in robotics applications. In robotics, sensors are usually tonic, as are servo motors. Explaining the function of tonic systems is possible using straightforward mathematical techniques, and this provides a valuable stepping stone for clarifying and applying a comprehensive phasic theory. Therefore, we assume that transducers are tonic, or can be considered tonic as a first approximation. The system to be controlled is $M \times N$ multiple-input-multiple-output (MIMO), where M is the number of actuators and $N \geq M$ (typically $N \gg M$) is the number of sensors.

Upper and lower case boldface letters in mathematical notation denote matrices and vectors, respectively. Unless specified otherwise, noise is assumed to be zero-mean white Gaussian.

6.1.4 Outline

After this introduction, in Sect. 6.2, we define the concept of a neuronal op-amp, including its plasticity and use in the internal model control structure, in the next section. We then describe two experiments illustrating the concept and its performance in Sects. 6.3 and 6.4, respectively. We discuss the results, related work, and conclude the text in Sect. 6.5.

6.2 The Neuronal Op-amp

Neurons are similar to electronic op-amps in the sense that both have excitatory (non-inverting) and inhibitory (inverting) inputs (Fig. 6.2). In addition, it has been shown experimentally that the output is a linear function of the inputs for several types of neurons, including CA1 pyramidal neurons [3], cerebellar Purkinje neurons [30], and spinal interneurons [27]. Although this suggests a direct analogy between single neurons and op-amps, neurons in the nervous system of mammals notably tend to group into *populations* or *ensembles*, i.e., homogeneous groups or layers of cytoarchitectonically similar neurons, all receiving input from and delivering output (projecting) to the same groups of source and target neurons, respectively. Therefore, we define a neuronal op-amp as an entire population of neurons having two *vector* inputs, an excitatory input $\mathbf{x}$ and an inhibitory input $\mathbf{y}$, not necessarily of the same dimension, and a *vector* output $\mathbf{z} = \mathbf{W}\mathbf{x} - \mathbf{V}\mathbf{y}$, where $\mathbf{W}$ and $\mathbf{V}$ are synaptic weight matrices. Furthermore, because primarily excitatory synapses have been linked to neuronal plasticity [4], we assume that only $\mathbf{W}$ is affected by learning. Without restriction, $\mathbf{V}$ can be assumed to be the identity matrix because the choice of coordinate system is arbitrary. In summary, a neuronal op-amp is a linear mapping from $(\mathbf{x}, \mathbf{y})$ to $\mathbf{z}$ such that

$$\mathbf{z} = \mathbf{W}\mathbf{x} - \mathbf{y} \; . \tag{6.1}$$

6.2.1 Plasticity of Neuronal Op-amps

Suppose that during actual operation, $\mathbf{x}$ and $\mathbf{y}$ are not completely uncorrelated, but are related by the equation $\mathbf{x} = \mathbf{A}\mathbf{y} + \mathbf{n}$ where $\mathbf{A}$ is a fixed non-zero matrix and $\mathbf{n}$

is a noise vector. By updating $\mathbf{W}$ as described below when new inputs $\mathbf{x}, \mathbf{y}$ arrive, the output $\mathbf{z}$ can be forced to converge to $\mathbf{A}^{+}\mathbf{x} - \mathbf{y}$, where $\mathbf{A}^{+}$ denotes the Moore-Penrose matrix pseudoinverse [2]. In a control loop, $\mathbf{A}^{+}$ may represent, for example, an acquired inverse plant model.

More specifically, in biological systems, we assume that learning operates as follows: When new inputs arrive and a new output is generated, the synaptic weights $\mathbf{W}$ are updated according to the Hebbian learning rule

$$\mathbf{W} \leftarrow \mathbf{W} + \mathbf{g}(\mathbf{z})\ \mathbf{h}(\mathbf{x})^{T}\ , \tag{6.2}$$

where $\mathbf{g}$ and $\mathbf{h}$ are vector-valued gain functions. The best-known update rule is perhaps the LMS-rule [14], a linearization where $\mathbf{g}(\mathbf{z}) = \mathbf{z}$, $\mathbf{h}(\mathbf{x}) = \mu\mathbf{x}$ and μ is a suitable scalar constant. However, any gain functions $\mathbf{g}$ and $\mathbf{h}$ can be used as long as persistent excitation of the input—sufficient random motion for the system to allow exploration of the local environment—guarantees convergence of $\mathbf{z}$ to $\mathbf{A}^{+}\mathbf{x}-\mathbf{y}$. Many such functions are available [22], allowing great flexibility and biological plausibility in the choice of $\mathbf{g}$ and $\mathbf{h}$.

In engineering, we are not limited to biologically plausible update rules of the form (6.2), but can use any update procedure for which $\mathbf{z} \to \mathbf{A}^{+}\mathbf{x} - \mathbf{y}$. We explore this possibility in Sect. 6.4, where we use Kalman filtering to achieve an optimal adaptation rate.

6.2.2 *Internal Model Control Using Neuronal Op-amps*

In the field of automatic control, *Internal Model Control* (IMC) [12, 20] refers to a particular class of control structures that continuously compare an internal process model with the actual process (Fig. 6.3 shows an adaptive version). The difference between process output and model output is fed back to the input stage as an error signal. Frank [8] appears to have published the first systematic investigation of IMC design, including a section on the early history of model feedback ("Modellrückkopplung").

IMC may superficially appear similar to observer schemes, such as Linear-Quadratic-Gaussian (LQG) control [12], but although IMC does contain an internal model, it uses neither state representations nor state variable feedback. IMC controllers have many attractive properties in spite of their simplicity; practice has shown them to be robust, reliable, and easy to generalize to high dimensions.

Control experience has shown that there is a tradeoff between optimality and robustness; it is often necessary to detune a nearly-optimal controller to achieve a better stability margin [7]. In this respect, IMC controllers can be said to prioritize robustness over optimality [20].

The combination of neurophysiology and control theory was pioneered by Wiener [31], and elaborated by Ito [15], in particular for the cerebellum. The general idea of internal models in CNS feedback control spread rapidly and has been reviewed

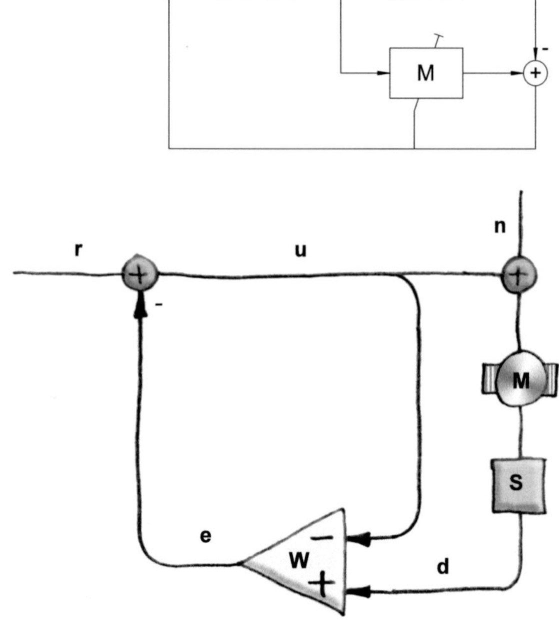

Fig. 6.3 An adaptive IMC system, where C is the controller, P the process to be controlled (the "plant"), and M an adjustable model

Fig. 6.4 A neuronal op-amp in adaptive IMC configuration (cf. Fig. 6.3), where the motors are **M**, the sensors are **S**, the neuronal op-amp is **W**, the reference signal is **r**, the motor control signal is **u**, the load disturbance is **n**, the sensor output is **d**, and the error signal is **e**

by Kawato [17]. These internal models are usually viewed as black boxes and their implementation left open. However, we propose that internal models and IMC controllers can be constructed directly from neuronal op-amps (Fig. 6.4). Motor centra send a motor command vector **r** to the motors. On the way, this command combines with the feedback error vector **e** into the motor control signal vector $\mathbf{u} = \mathbf{r} - \mathbf{e}$ which arrives at the motor unit **M**, potentially also affected by an additive load disturbance vector **n**. The actions of the motors are sensed by a large number of sensors **S**. The sensor output vector $\mathbf{d} = \mathbf{SM}\,(\mathbf{u} + \mathbf{n})$ excites a neuronal population/op-amp **W**, which is simultaneously inhibited by the original motor control signal **u**. The output error $\mathbf{e} = \mathbf{Wd} - \mathbf{u}$ is fed back as a correction of the motor command **r**. Here, the sensor output **d** is high-dimensional, whereas the other vectors are relatively low-dimensional. The matrix **W** typically has a large number of columns, matching the dimension of the measurements **d**, but a smaller number of rows, matching the dimension of the motor control signal **u**.

We do assume that **S** is sufficiently "dense" such that all states of **M** can be uniquely observed. Mathematically, this can be expressed by requiring the null space of **SM** to equal the null space of **M**, $\mathbf{M}_\perp = (\mathbf{SM})_\perp$. This provides a stronger sense of observability than the classical concept in control theory, but is perhaps more appropriate in the context of a large number of sensors, such as in a biological system.

To understand the operation of the model, we trace the loop from the reference signal, first assuming an equilibrium situation in which $\mathbf{e} = \mathbf{0}$ and $\mathbf{n} = \mathbf{0}$. Suppose that the motor is suddenly affected by a load disturbance $\mathbf{n}$. For readability, we introduce the abbreviation $\mathbf{A}$ for the matrix product $\mathbf{SM}$.

The excitatory input to $\mathbf{W}$ becomes $\mathbf{d} \leftarrow \mathbf{Au}+\mathbf{An} = \mathbf{Ar}+\mathbf{An}$. The error becomes $\mathbf{e} = \mathbf{Wd} - \mathbf{u} = \mathbf{A}^{+}\mathbf{An} - \left(\mathbf{I} - \mathbf{A}^{+}\mathbf{A}\right)\mathbf{r}$. The term $\left(\mathbf{I} - \mathbf{A}^{+}\mathbf{A}\right)\mathbf{r}$ belongs to the null space of $\mathbf{A}$, and components of u in this space will not affect the motors $\mathbf{M}$ because $\mathbf{M}_{\perp} = (\mathbf{SM})_{\perp} = \mathbf{A}_{\perp}$, according to our earlier assumption. Once in the null space of $\mathbf{A}$, these components will remain there after passing into $\mathbf{e}$, so they can safely be ignored. Consequently, on the next turn of the loop, we have $\mathbf{u} \leftarrow \mathbf{r} - \mathbf{A}^{+}\mathbf{An}$. Incidentally, this operation can be performed by a second neuronal op-amp, forming a disynaptic loop. The updated sensor output becomes $\mathbf{d} \leftarrow \mathbf{A}\left(\mathbf{r} - \mathbf{A}^{+}\mathbf{An}\right) + \mathbf{An} = \mathbf{Ar}$, demonstrating that the loop now has neutralized the disturbance $\mathbf{n}$. The error is $\mathbf{e} \leftarrow \mathbf{Ar} - (\mathbf{r} - \mathbf{A}^{+}\mathbf{An}) = \mathbf{A}^{+}\mathbf{An} - (\mathbf{I} - \mathbf{A}^{+}\mathbf{A})\mathbf{r}$, showing that the feedback has stabilized.

In practice, low-pass filtering $\mathbf{u}$ is required for stability [20]. A biological neuron always includes an implicit low-pass filter via its membrane capacitance and resistance.

6.3 Experiment: Neuronal Op-amps in Biology

To explore how biological systems solve the correspondence problem, a simple two-arm motion apparatus was built to mimic biological limbs by incorporating viscoelastic transmissions and using a reconstruction of biological adaptive multivariate feedback control. The goal of the experiment was for the device to learn how to associate motor commands from the joystick with proprioceptive sensor feedback produced by encoders, and then maintain good tracking of the joystick reference signal.

In ordinary engineering applications, such tasks would typically involve Kalman filtering, calibrated models, and floating-point hardware, even when omitting adaptivity. Here, using biological principles, it was possible to implement the adaptive controller on a tiny 8-bit microprocessor. Below, we recount the experiment in detail, demonstrating the automatic bootstrapping ability and effective adaptive control of the system.

6.3.1 Setup

The experimental setup shown in Fig. 6.5 consists of a human-operated reference input device (J). This input symbolizes a volitional motor command from the brain's prefrontal cortex. The signal is fed to a microprocessor board, which represents the motor centra and the spinal cord. The processed motor command is sent to two servo motors (M1-2), connected by viscoelastic transmissions ("tendons", T1-2) to a pair

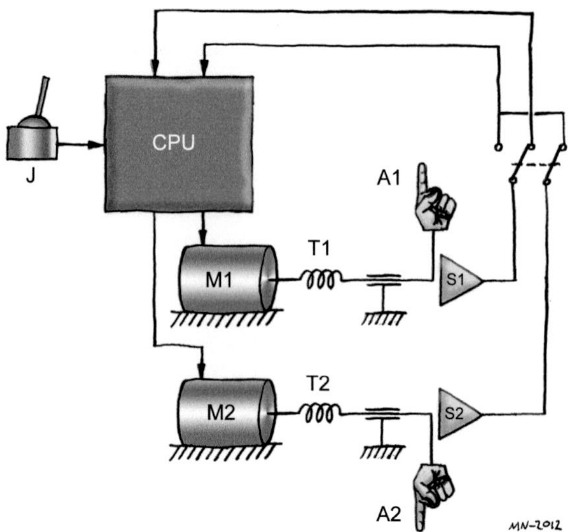

Fig. 6.5 Schematic of experiment, where a 2D motor command is input via the joystick J, M1-2 are servo motors, T1-2 are flexible transmissions, A1-2 are end effectors, S1-2 are angular encoders, and SW is a switch exchanging S1 and S2

of arms (A1-2). For each arm, an angle sensor (S1-2) feeds its position back to the CPU board, but *which sensor is connected to which motor is not preprogrammed in the processor*. Technically, this is implemented in the setup by connecting the sensor outputs to the board's A/D-converter inputs via a double-pole double-throw (DPDT) switch, which allows exchanging the connections manually.

The input device of the physical apparatus (Fig. 6.6) is a joystick, the CPU board is an 8-bit-microprocessor Arduino board [1] without floating-point hardware, and the actuators are standard R/C servo motors. Pieces of PVC tubing constitute the viscoelastic transmissions, and rotary encoders [19] are angle sensors. The board has an additional switch for disconnecting feedback, useful for comparing closed- and open-loop feedback behaviour. Also included is a potentiometer, through which the learning rate (plasticity) can be varied.

The adaptation algorithm employs the LMS-rule [14] of the IMC controller described in Sect. 6.2.2.

6.3.2 Execution

The principal experiment shows that the system is able to develop and maintain stable feedback control, regardless of sensor connectivity. It was conducted as follows:

1. The internal model was automatically initialized to zero on startup.
2. Arm motion was observed not to compensate well for disturbances. Here, disturbances can be internal, such as measurement noise and model error, or external, such as manual arm pushing.

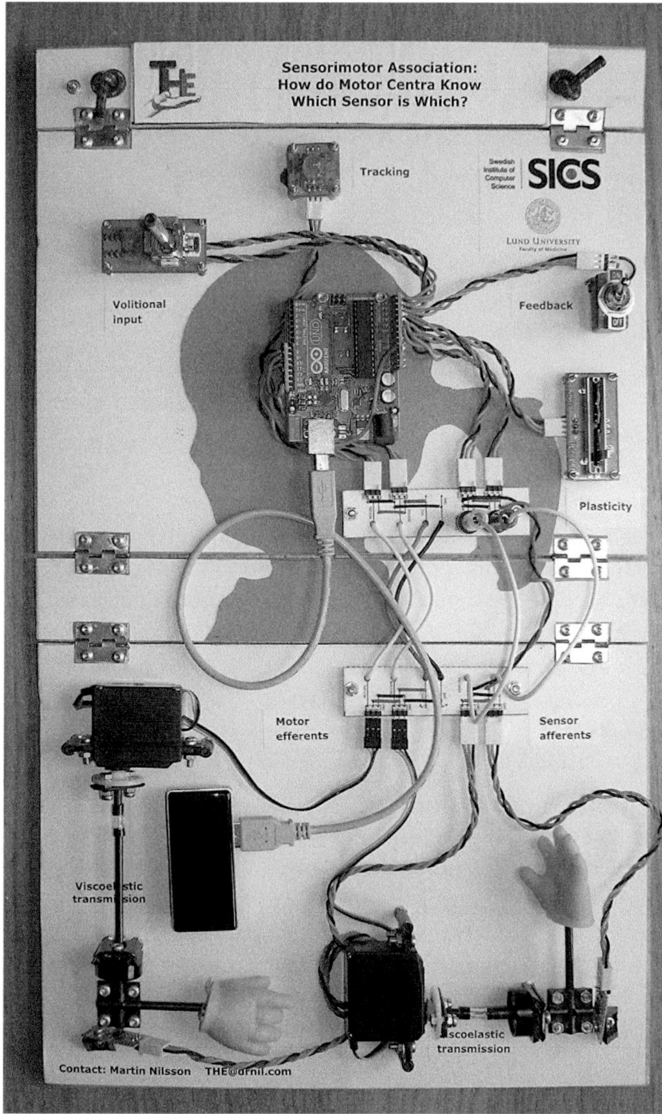

Fig. 6.6 Experiment board physically realizing the schematic in Fig. 6.5, where the DPDT switch is implemented with two banana plugs (the *black box* between the servo motors is a battery)

3. Gradually, an internal model emerged, although the motion of the system was still erratic.
4. As the model improved, the system gradually responded better and better to disturbances, with the DPDT switch/plugs remaining in fixed positions.

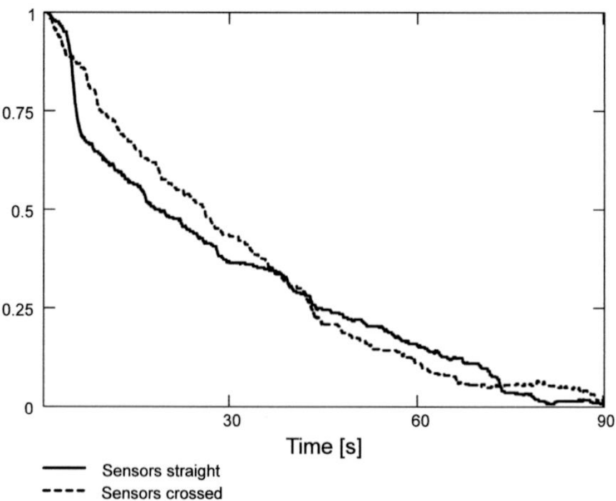

Fig. 6.7 The convergence rate of model acquisition depends little on the motor-sensor connection

5. The sensor connections were changed by switching the plugs, the system was reset, and the experimental protocol restarted from the beginning.
6. The system bootstrapped and re-learned the new sensor connectivity.

6.3.3 Results

No substantial difference was observed between the runs (Fig. 6.7). This experiment demonstrates that the ability to bootstrap is independent of sensor configuration. The figure shows a comparison of the convergence rates of an acquired model for straight-connected (solid line) versus cross-connected (dashed line) sensors for 90-second trial runs. The model was acquired automatically during normal feedback operation, while the system was excited by manual random joystick operation. The adaptation error is measured as the expression $||\mathbf{W}(t) - \mathbf{W}(90\,\mathrm{s})||_F/||\mathbf{W}(90\,\mathrm{s})||_F$, where $\mathbf{W}(t)$ is the acquired model at time t, and $||\cdot||_F$ denotes the Frobenius norm.

6.4 Experiment: Neuronal Op-amps in Engineering

In engineering, we are not limited to biologically plausible methods for acquiring the pseudoinverse, such as Hebbian learning, but free to use any technically feasible method. We can take advantage of the fact that the number of transducers is usually much smaller in non-biological systems than in biological systems. Thus, assuming a relatively small number of transducers (on the order of 10), the controller can employ linear Kalman filters (LKF) to achieve optimal noise filtering properties, a feature unavailable to biological systems. Although we lose scalability due to the

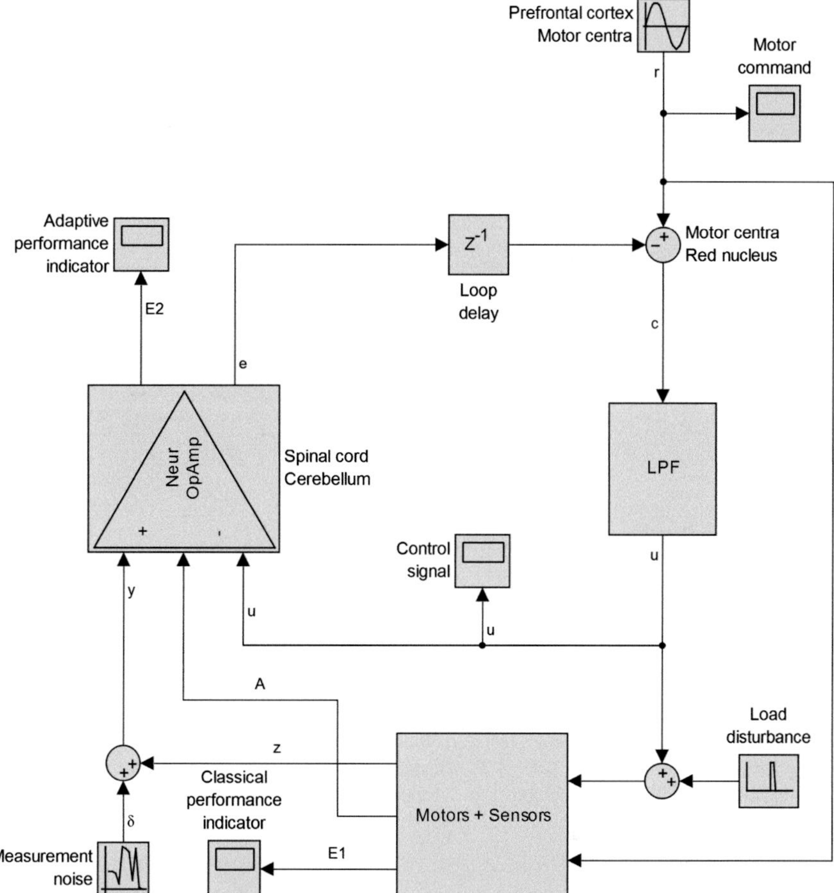

Fig. 6.8 Neuronal op-amp in a multivariate IMC configuration, Simulink diagram; the internal structure of the LKF neuronal op-amp block is given in Fig. 6.9

computational complexity of the Kalman filter, we accelerate the adaptation rate considerably.

Below, we consider another adaptive controller inspired by the mammalian hand-arm control system. However, this time we do not constrain ourselves to biologically plausible implementations of the neuronal op-amp. The new controller quickly determines the optimal use of an unmodelled sensorimotor configuration for adaptive feedback control. The experiment is implemented as an executable *Simulink* [26] graph (Fig. 6.8).

The goal of this experiment was to maximize the online adaptation rate while simultaneously compensating for disturbances.

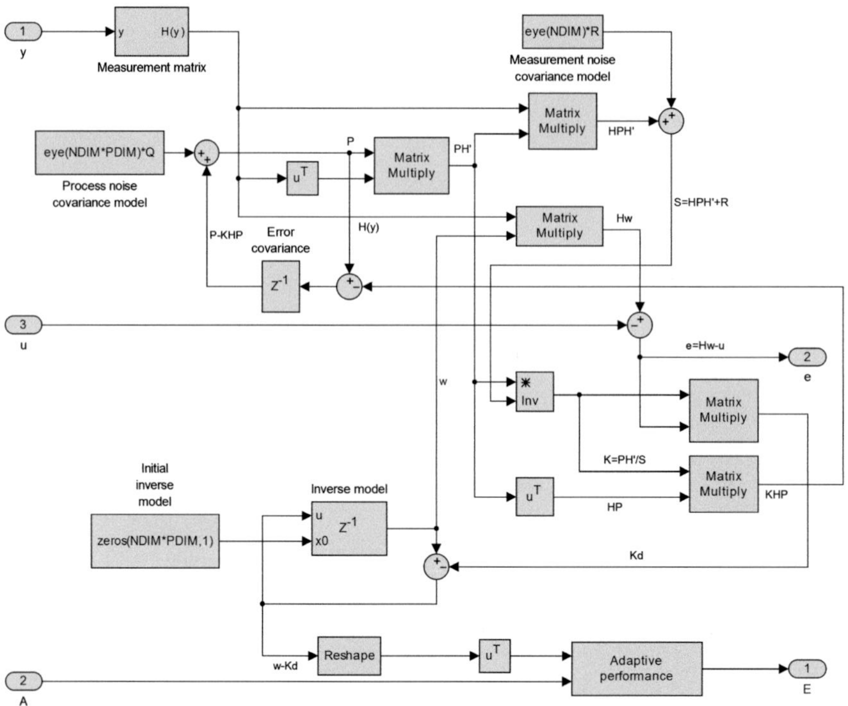

Fig. 6.9 Neuronal op-amp "cheat" using Kalman filtering for computing the pseudoinverse, Simulink diagram; the Kalman gain is $\mathbf{K}$, the measurement matrix is $\mathbf{H}(\mathbf{y})$, the process noise covariance is $\mathbf{P}$, and the measurement noise covariance is $\mathbf{R}$

6.4.1 Setup

The setup consists of a number of Simulink modules. Below, details of the controller are furnished together with brief descriptions of the main constituent modules.

The LKF module: The heart of the controller is the LKF-module (Fig. 6.9), which is functionally equivalent to a neuronal op-amp but is implemented using linear Kalman filtering rather than Hebbian learning. The Kalman filter is optimal for minimizing the mean square error (MSE) and therefore converges quickly. This approach works well because the task of finding the pseudoinverse in this application can be solved via a linear model.

In addition to the LKF module's two differential inputs, the sensor feedback signal $\mathbf{y}$ and the motor control signal $\mathbf{u}$, there is a third input of a motor-sensor model matrix, only used for performance evaluation in this module. The outputs are the adaptive performance index E_2 (see Sect. 6.4.3) and the error estimate $\mathbf{e} = \mathbf{W}\mathbf{y} - \mathbf{u}$, where $\mathbf{W}$ is expected to converge to the inverse model $\mathbf{A}^+$. The E_2 output does not exist in biological systems, but is provided in this device for debugging and performance analysis. The module has two Kalman-filter derived tuning parameters: Q, indicating the process noise covariance, and R, the measurement noise covariance.

Motor-sensor module: This module inputs the control signal $\mathbf{u}$, and outputs the product $\mathbf{Au}$, where $\mathbf{A}$ defines the motor-sensor function, here chosen as a matrix of random numbers uniformly drawn from [0, 1]. This module also receives the motor command reference signal as input to compute conventional control performance as the diagnostics output E_1 (see Sect. 6.4.3), together with the motor-sensor matrix $\mathbf{A}$.

Low-pass filter module: Stability of the IMC controller requires a low-pass filter. The filter has one settable constant, which determines the cut-off frequency. Rules of thumb for choosing this constant are identical to those for IMC controllers [12, 20].

Classical control performance measurement module: This module measures the distance between the sensor outputs and the product $\mathbf{Ar}$ of the motor-sensor model matrix $\mathbf{A}$ and the motor command reference $\mathbf{r}$ (see Sect. 6.4.3).

Adaptive performance measurement module: This module measures the distance between the identity matrix and the product $\mathbf{WA}$ of the motor-sensor model matrix $\mathbf{A}$ and the inverse model $\mathbf{W}$ (see Sect. 6.4.3). If adaptation is successful, this performance index approaches zero.

Measurement matrix construction module: As part of generating the internal model, each component of $\mathbf{W}$ is recursively estimated by the Kalman filter. For this purpose, a measurement matrix $\mathbf{H}(\mathbf{y})$ needs to be constructed from the $\mathbf{y}$ input.

6.4.2 Execution

The reference signal input is shown in Fig. 6.10 for a six-motor by twelve-sensor system. The motor command is composed of six sine waves of varying phase and frequency, one of which is shown in panel **a**. The corresponding control signal is shown in panel **b**. Here, the Q parameter (process noise covariance) was set to 10^{-8} in each dimension; all noise sources were assumed to be independent; the R parameter (measurement noise covariance) was set to 1 in each dimension; the low-pass filter time constant was approximately 3.3 ms; the sampling frequency was 100 kHz; the load disturbance was a pulse of duration 5 ms and amplitude 0.2; and the actual measurement noise was zero-mean white Gaussian noise of standard deviation 0.1.

6.4.3 Results

There are two performance aspects to the controller: first, the steady-state control performance of the controller as a “classical” IMC controller once the model has

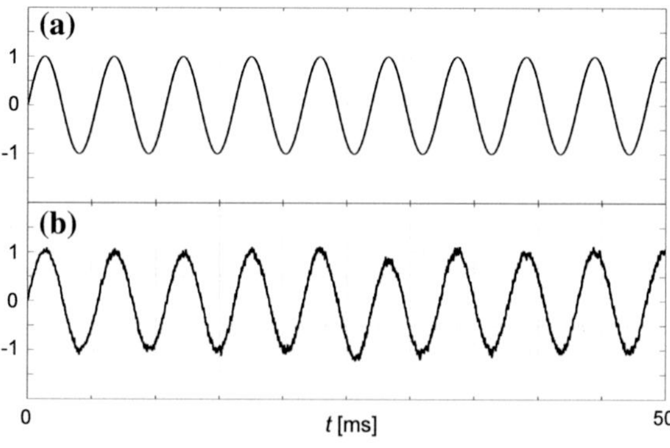

Fig. 6.10 **a** Reference signal r, **b** control signal u (one of six)

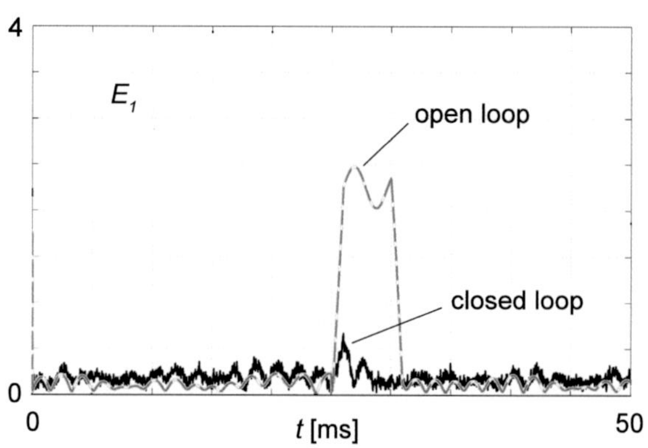

Fig. 6.11 Control performance E_1 for closed (*solid line*) and open (*dashed line*) loops

converged, and second, the convergence to an inverse model (adaptive performance). The former aspect has been studied extensively for IMC [12, 20], and can be illustrated e.g. by disconnecting the feedback $\mathbf{e}$ and watching the actual performance under load disturbances $\mathbf{n}$ (Fig. 6.11). We compute it as $E_1 = |\mathbf{Ar} - \mathbf{z}|$, where $\mathbf{A}$ is the motor-sensor model matrix, $\mathbf{r}$ is the motor command reference, and $\mathbf{z}$ are the sensor outputs.

The importance of feedback for correcting the load disturbance $\mathbf{n}$ can be appreciated by comparing the open- and closed-loop traces in Fig. 6.11.

The second aspect is perhaps less intuitive: Does the inverse model converge at all, and if so, how quickly? One test is a comparison with the generalized inverse of the motor-sensor model $\mathbf{A}$: The convergence of the product $\mathbf{WA}$ to the identity matrix $\mathbf{I}$ indicated by the Frobenius norm $E_2 = ||\mathbf{I} - \mathbf{WA}||_F$ characterizes the adaptive performance of the neuronal op-amp. The performance index E_2 (Fig. 6.12) reveals stable and rapid adaptation, regardless of whether the loop is closed or open. Since adaptation is based on Kalman filtering, there is a trade-off between adaptation rate and smoothness.

Yet another aspect of the controller is its stability. Proving the stability of adaptive IMC controllers is difficult, even in the single-input-single-output (SISO) case [6]. Here, the controller is multiple-input-multiple-output (MIMO), and in a cascaded

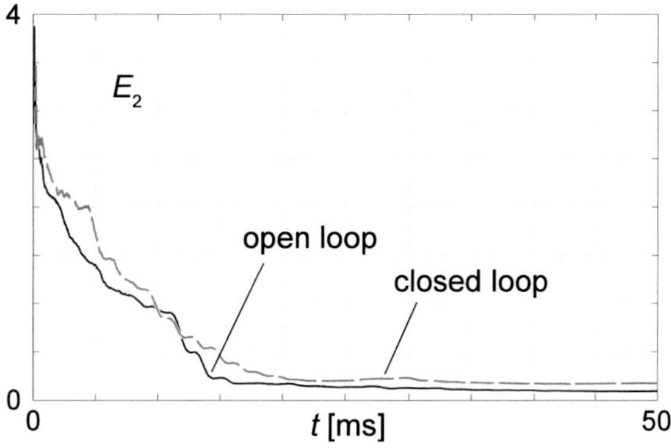

Fig. 6.12 Adaptive performance E_2 for closed (*solid line*) and open (*dashed line*) loops

configuration, there are several adaptive processes executing simultaneously. Proving stability for such systems is formidable, suggesting that testing the stability of this type of controller is currently limited to simulation.

6.5 Discussion and Conclusions

We have addressed two problems related to adaptive control, both in a biological and an engineering context. For biology, our interest is analytical, whereas for engineering, we would like to synthesize new solutions based on our understanding of biological function.

We performed both physical and simulation experiments for the proposed MIMO control structures. The experiments indicate that the neuronal op-amp concept is a feasible approach, both for understanding biological function and for using this knowledge as a basis for new designs in robotics and engineering. However, the best implementation of neuronal op-amp internals depends on the application. Because Hebbian learning is scalable, it is suitable for high-dimensional systems. For lower-dimensional systems, a Kalman filter approach similar to that described in Sect. 6.4 may be more appropriate.

A disadvantage of adaptive controllers in engineering applications is the difficulty of theoretically guaranteeing stability. It should be noted that persistent excitation is necessary for stability, as is the case for all adaptive controllers. In biological systems, this is not a problem due to the large amounts of naturally occurring noise, but in artificial systems, random perturbations or other provisions may be necessary.

Nonetheless, in our experiments, the neuronal op-amp IMC configuration operated robustly, both in steady-state operation and identifying (bootstrapping) inverse process models online. The simplicity of the structure is striking; the loop consists of only two reciprocally connected neuronal op-amps, a pervasive pattern in the nervous system [11].

The simplicity of the neuronal op-amp adaptive IMC model, its robustness, high performance, and biological "hi-fidelity", positions it as a biologically plausible candidate for CNS motion control. However, the acceptance of this hypothesis is conditional upon the input-output linearity of neurons. Such linearity has been demonstrated experimentally [3, 27, 30], but a mechanistic explanation [5] has not yet been reported. Although there is no linearity requirement for the update rule (6.2), the linearity of the feedforward operation (6.1) is essential.

The proposed controller should be considered only one representative of a family of controllers; many variations of the IMC scheme are possible (such as the introduction of a Smith regulator, or transformation to allow stabilization of unstable plants without risking internal instability) [20].

Neuronal op-amps resemble adaptive filters [14], which have been used to model the CNS [10, 21]. An important difference between adaptive filters and neuronal op-amps is that the feedback path for adaptation is external for adaptive filters, and often used for modifying the system properties. For neuronal op-amps, the feedback path is strictly internal and cannot be accessed externally.

The neuronal op-amp model helps explaining the appearance of *synergies* [23]: In principle, synergies express the fact that the output range dimension is lower than the output space dimension. Mathematically, synergies have been characterized as manifolds [25]. Although a human hand has many degrees of freedom, a low-dimensional manifold in the configuration space dominates the representation [24]. For the neuronal op-amp, the existence of synergies is detected by a low-rank acquired inverse model $\mathbf{W} = \mathbf{A}^+$, implying that the original system $\mathbf{A}$ must also be low-rank. From a geometrical perspective, the neuronal op-amp linearity predicts that synergy manifolds are linear (i.e., linear subspaces).

Another consequence of the neuronal op-amp model is that in a well-adapted steady state, its output is approximately zero. In other words, its basic functionality is to remember "normal" situations and to deliver a signal out only when the situation deviates from the learned (accustomed) situation. The output represents the *innovation* as introduced in signal processing [22], reminiscent of concepts previously discussed in the literature, but at a higher, cognitive level, including *novelty* [18], *free energy* or *surprise* [9], and *saliency* [29]. Thus, the neuronal op-amp provides a potential link between neuronal substrate and cognition in the CNS.

Acknowledgments The author thanks the THE project partners and reviewers for many inspirational interdisciplinary discussions across the robotics and neuroscience camps, and is especially grateful for support by Dr. Henrik Jörntell at the Department of Experimental Medical Science, Lund University. The author is grateful to Prof. Rolf Johansson at the Department of Automatic Control, Lund University, who introduced him to the fascinating field of adaptive control.

This research was funded by the European Union FP7 research project THE, "The Hand Embodied", under grant agreement 248587.

References

1. Arduino Organization http://www.arduino.cc: Arduino Uno Data Sheet (2012). http://arduino.cc/en/Main/ArduinoBoardUno. Accessed 5 March 2012
2. Ben-Israel A, Greville TNE (2003) Generalized Inverses: theory and applications, 2nd edn. Springer, New York
3. Cash S, Yuste R (1999) Linear summation of excitatory inputs by CA1 pyramidal neurons. Neuron 22:383–394. doi:10.1016/S0896-6273(00)81098-3
4. Chater TE, Goda Y (2014) The role of AMPA receptors in postsynaptic mechanisms of synaptic plasticity. Front Cell Neurosci 8:1–14. doi:10.3389/fncel.2014.00401
5. Craver CF (2006) When mechanistic models explain. Synthese 153:355–376. doi:10.1007/s11229-006-9097-x
6. Datta A (1998) Adaptive internal model control. Springer, London
7. Doyle JC (1978) Guaranteed margins for LQG regulators. IEEE Trans autom control 23(4):756–757. doi:10.1109/TAC.1978.1101812
8. Frank, PM (1974) Entwurf von Regelkreisen mit vorgeschriebenem Verhalten. G Braun
9. Friston K, Kilner J, Harrison L (2006) Free energy principle for the brain. J Physiol Paris 100:70–87. doi:10.1016/j.jphysparis.2006.10.001
10. Fujita M (1982) Adaptive filter model of the cerebellum. Biol Cybern 45(3):195–206. doi:10.1007/BF00336192
11. Gilbert CD, Li W (2013) Top-down influences on visual processing. Nature Rev Neurosci 14:350–363. doi:10.1038/nrn3476
12. Glad T, Ljung L (2000) Control theory: multivariable and nonlinear methods. Taylor & Francis, London
13. Gray H (1918) Anatomy of the human body. Lea and Febiger, Philadelphia, USA
14. Haykin SS, Widrow B (eds) (2003) Least-mean-square adaptive filters. John Wiley, Hoboken, NJ
15. Ito M (1984) The cerebellum and neural control. Raven Press, New York
16. Johansson RS, Vallbo ÅB (1979) Tactile sensibility in the human hand: relative and absolute densities of four types of mechanoreceptive units in glabrous skin. J Physiol 286:283–300. doi:10.1113/jphysiol.1979.sp012619
17. Kawato M (1999) Internal models for motor control and trajectory planning. Curr Opin Neurobiol 9(6):718–727. doi:10.1016/S0959-4388(99)00028-8
18. Kohonen T (1977) Associative Memory: a system-theoretical approach. Springer, Berlin
19. Melexis microelectronic systems, Ieper, Belgium: Melexis MLX 90316 Data sheet (2012). http://www.melexis.com/Assets/MLX90316-DataSheet-4834.aspx
20. Morari M, Zafiriou E (1989) Robust process control. Prentice Hall, Englewood Cliffs
21. Porrill J, Dean P, Anderson SR (2013) Adaptive filters and internal models:multilevel description of cerebellar function. Neural Networks 47:134–149. doi:10.1016/j.neunet.2012.12.005
22. Proakis J, Manolakis D (2007) Digital signal processing, 4th edn. Prentice Hall, Pearson Education
23. Santello M, Baud-Bovy G, Jörntell H (2013) Neural bases of hand synergies. Front Comput Neurosci 7(23):1–15. doi:10.3389/fncom.2013.00023
24. Santello M, Flanders M, Soechting JF (2002) Patterns of hand motion during grasping and the influence of sensory guidance. J Neurosci 2(4):1426-1435 http://www.jneurosci.org/content/22/4/1426.long
25. Scholz JP, Schöner G (1999) The uncontrolled manifold concept: identifying control variables for a functional task. Exp Brain Res 126(3):289–306. doi:10.1007/s002210050738
26. SIMULINK manual (2012) Mathworks Inc, Natick MA, USA. http://www.mathworks.se/help/toolbox/simulink/
27. Spanne A, Geborek P, Bengtsson F, Jörntell H (2014) Spike generation estimated from stationary spike trains in a variety of neurons in vivo. Front Cell Neurosci 8:1–15. doi:10.3389/fncel.2014.00199

28. Stout DF, Kaufman M (1976) In: Handbook of operational amplifier circuit design. McGraw-Hill, New York
29. van Polanen V, Tiest WMB, Kappers AML (2012) Haptic search for hard and soft spheres. PLOS One 7(10):1–10. doi:10.1371/journal.pone.0045298
30. Walter JT, Khodakhah K (2006) The linear computational algorithm of cerebellar Purkinje cells. J Neurosci 26:12861–12872. doi:10.1523/JNEUROSCI.4507-05.2006
31. Wiener N (1948) Cybernetics, or control and communication in the animal and the machine. MIT Press, Cambridge, MA

Chapter 7
Sensorymotor Synergies: Fusion of Cutaneous Touch and Proprioception in the Perceived Hand Kinematics

Alessandro Moscatelli, Matteo Bianchi, Alessandro Serio, Antonio Bicchi and Marc O. Ernst

Abstract According to classical studies in physiology, muscle spindles and other receptors from joints and tendons provide crucial information on the position of our body and our limbs. Cutaneous cues also provide an important contribution to our sense of position. For example, it is possible to induce a vivid sensation of movement in the anesthetized finger, by stretching the skin around the proximal interphalangeal joint. However, much of proprioceptive literature did not consider the role of tactile interaction with external objects as position and motion cues. Whenever we touch an external, stationary object, the contact forces produce a mechanical deformation of the skin which changes with the hand posture and movement. Therefore, these cutaneous contact cues might also provide proprioceptive information. In this paragraph, evaluated this hypothesis based on recently published experimental data.

A. Moscatelli (✉) · M.O. Ernst
Department of Cognitive Neuroscience and Cognitive Interaction Technology Centre of Excellence, University of Bielefeld, Bielefeld, Germany
e-mail: alessandro.moscatelli@uni-bielefeld.de

M.O. Ernst
e-mail: marc.ernst@uni-bielefeld.de

A. Moscatelli
Department of Systems Medicine and Centre of Space Bio-Medicine, Università di Roma"Tor Vergata", Rome, Italy

M. Bianchi · A. Serio · A. Bicchi
Advanced Robotics Department, Istituto Italiano di Tecnologia, Genova, Italy

M. Bianchi
e-mail: matteo.bianchi@iit.it

A. Serio
e-mail: alessandro.serio@centropiaggio.unipi.it

A. Bicchi
e-mail: bicchi@centropiaggio.unipi.it

M. Bianchi · A. Serio · A. Bicchi
Centro di Ricerca E. Piaggio, Università di Pisa, Pisa, Italy

M. Bianchi and A. Moscatelli (eds.), *Human and Robot Hands*,
Springer Series on Touch and Haptic Systems, DOI 10.1007/978-3-319-26706-7_7

7.1 Introduction

According to classical studies in physiology [12, 18], muscle spindle, Golgi tendon organ and receptors from joints provide crucial information on the static position and movement of our body and our limbs. Edin and colleagues [6, 7] showed that information from cutaneous mechanoreceptors also contribute to our sense of position, that is, to proprioception. Limb and hand movements produces a stereotyped strain pattern on the skin. Since mechanoreceptors respond consistently to these cutaneous stimuli, they provide the central nervous system with detailed kinematic information [6]. Accordingly, Edin et al. [7] showed that it was possible to induce a vivid sensation of movement in the anesthetized finger by stretching the skin around the proximal interphalangeal joint. Whenever we touch a static object, as for example the surface of a table, the reaction force produces a strain pattern on the skin which changes in relation to our own movement. Here, we evaluated if these types of tactile stimuli produced during contact with external objects would also provide a cue for proprioception.

Hayward et al. [10] classified four fundamental types of cutaneous contact stimuli. When we first touch the surface of the object, the interaction force produces an area of high strain on the skin which expands over time [2, 3, 10]. Hayward et al. [10] defined this stimulus as 'contact on'. On the contrary, releasing the contact produces a shrinking in the high-strain area ('contact off'). Furthermore, the authors defined a 'slip' and a 'roll' stimulus as the displacement, in somatosensory coordinates, of the high and low strain area, respectively. These types of stimuli are supposed to be the fundamental input feature in touch—the same as for example the stimulus orientation in vision. Accordingly, different neurons the cuneate nuclei respond preferentially to one of these tactile stimuli [11]. If the external object is stationary, each of these stimuli maps to a restricted angular displacement of the joints involved in the movement. Consider a scenario as illustrated in Fig. 7.1: The pad of the index finger is in contact with a stationary, stable object. Different movements of the hand or the finger would produce one of the four tactile stimuli on the finger's pad. We used a transparent plane in the figure to show the change in the skin deformation during hand movements. For example, (a) the rotation of the wrist joint produces a roll motion on the skin; (b) the hand displacement tangential to the contact plane produces a slip motion, the direction of the tactile motion having same orientation and opposite direction as the hand movement; (c) the flexo-extension of the metacarpophalangeal or proximal joints, with the finger outstretched, produces a contact-on and contact off stimuli.

Does the somatosensory system use this contact information to estimate the displacement of our hands and our limbs? In this chapter, Sects. 7.2 and 7.3, we present the results of two previous studies addressing this question for two types of tactile stimuli, contact on/off [15] and slip motion [16]. The information provided by cutaneous touch is maximized when (i) the external object is stationary, (ii) the changes in the mechanical properties of the object (e.g. the stiffness) are negligible and (iii) motor synergies limit the degrees of freedom of the hand movement.

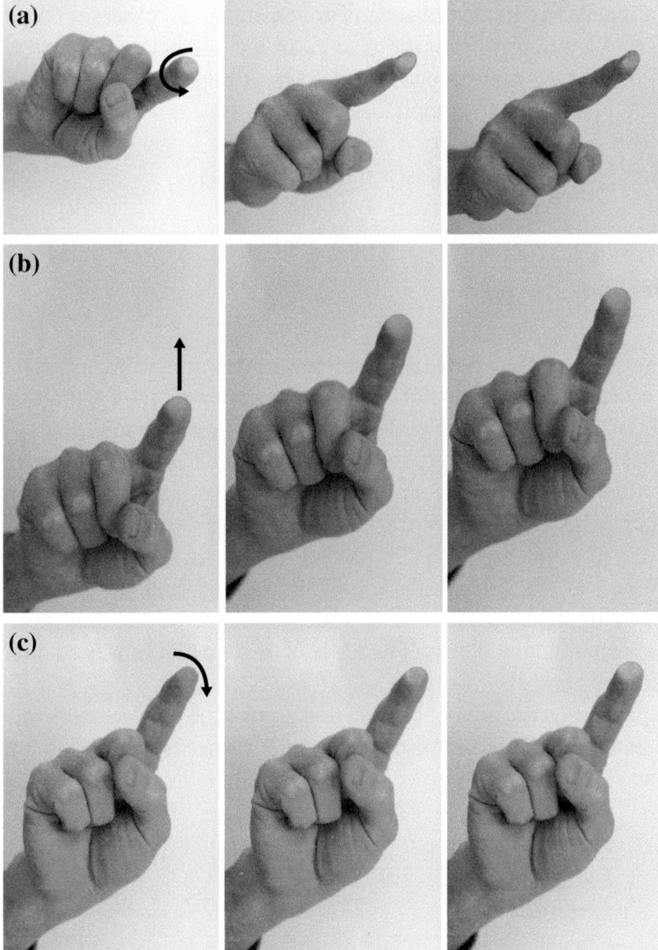

Fig. 7.1 Different hand movement maps to one of the four fundamental tactile stimuli: **a** roll, **b** slip, **c** contact on and contact off

Assumptions underlying sensory fusion between proprioception and touch are discussed in Sect. 7.4.

7.2 Contact Area

If we push the finger against an external surface and then lift the finger to release the contact, we produce first an increase and then a decrease in the area of contact. The relation between the area of contact and the angular position of the finger joint is mostly evident for a compliant contact surface, as illustrated in [15]. If the object is stationary and its compliance is not changing, an observer could infer the displacement of the finger from the change in the contact area, that is, from the recruitment of the mechanoreceptors in the skin. Along the same line of reasoning, an unexpected change in the compliance of the surface, and therefore in the area of contact, should induce an illusory displacement of the finger. Moscatelli et al. [15] run a psychophysical experiment to test this hypothesis.

7.2.1 Methods

Participants compared the passive displacement of their index finger between a reference and a comparison stimulus. In each stimulus interval, the apparatus illustrated in Fig. 7.2a contacted the finger pad and lifted the finger up and down. The compliance of the contacted surface unexpectedly changed between the reference and the comparison stimulus. Therefore, the area of contact in the comparison stimulus the was either wider or narrower than in the reference stimulus (*wide* and *narrow* condition, respectively; Fig. 7.2b). In each stimulus, we set the compliance of the surface by means of the softness display FYD-2 [20], without the participants being aware of this compliance change. The wide and narrow condition were tested in two different experimental blocks. The order of the two blocks was counterbalanced between participants.

If participants used the area of contact as a cue for the displacement of the finger, they would perceive a larger movement extent for those stimuli having a wider area of contact, and *vice versa*. This perceptual bias should induce an opposite shift of the

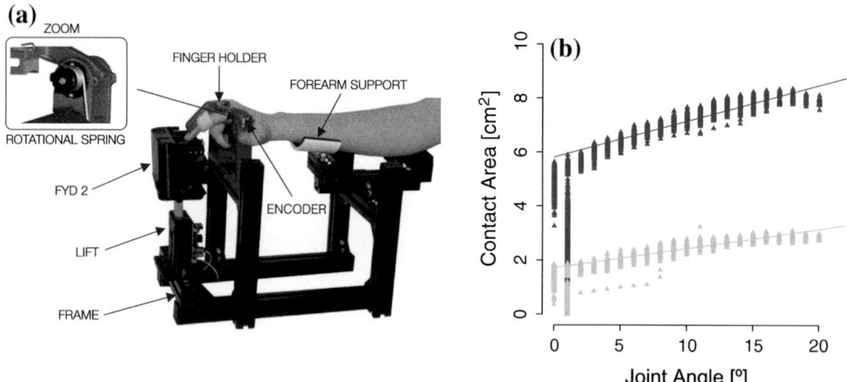

Fig. 7.2 **a** The area of contact between the skin and the sponge (marked in *red* in the figure) increases as the finger moves towards the bottom edge of the object. **b** The setup including the lift, the FYD-2 device and the angle encoder. **c** The area of contact changes as a function of the displacement of the finger and of the compliance of the surface (in *red wide* condition and in *grey narrow* condition; results from a representative participant). Adapted with permission from [15]

response between the two experimental conditions. To test if this was the case, we fit the binary responses of each participant with two psychometric functions, one for each experimental condition. In each condition, the function related the perceived and the physical width of the passive movement. Each of the two psychometric function had the form,

$$\Phi^{-1}\left[P(Y_j = 1)\right] \sim \beta_0 + \beta_1 \gamma_j, \tag{7.1}$$

where $P(Y_j = 1)$ is the probability that, in trial j, the participant reported a larger joint movement in the comparison than in the reference stimulus, $\Phi^{-1}[\cdot]$ is the *probit* transformation of the response probability (i.e., the inverse function of the cumulative normal distribution), and γ_j is the measured rotation of the joint angle in the comparison stimulus. The *point of subjective equality* (PSE),

$$\text{PSE} = -\frac{\beta_0}{\beta_1}, \tag{7.2}$$

is an indicator of the perceptual bias. If participants used the width of the contact area as a cue to solve the task, the PSE should be significantly different between the two experimental conditions, with $\text{PSE}_{\text{wide}} < \text{PSE}_{\text{narrow}}$. For statistical inference at the population level, we analyzed the data of all participants (n = 6) toghether using a *Generalized Linear Mixed Model* (GLMM) and estimated the PSE and the 95 % confidence interval (CI) as explained in [14].

7.2.2 Results

Figure 7.3a, b show the psychometric function in a representative participant and the PSE estimates in the experimental population, respectively. In all participants, the PSE_{wide} was significantly smaller than the $\text{PSE}_{\text{narrow}}$ (paired t-test; $t_5 = 4.5$; $p = 0.006$), in accordance with our predictions. Likewise, the 95 % CI of the two PSEs were not overlapping (Fig. 7.3b). The estimated PSE_{wide} was equal to 10.5° ($95\,\%\,\text{CI} : 9.9 - 11.1°$), and the estimated $\text{PSE}_{\text{narrow}}$ to 12.8° ($95\,\%\,\text{CI} : 12.1 - 13.6°$). That is, the wider the contact area was in the comparison stimulus, the larger was the perceived extension of the joint angle. This is in accordance with our hypothesis.

In conclusion, results of the experiment showed that participants took the change of the area of contact into account to estimate the angular displacement of the finger. In the next section, we evaluated the contribution of a different type of tactile stimulus, the slip motion, to the hand displacement.

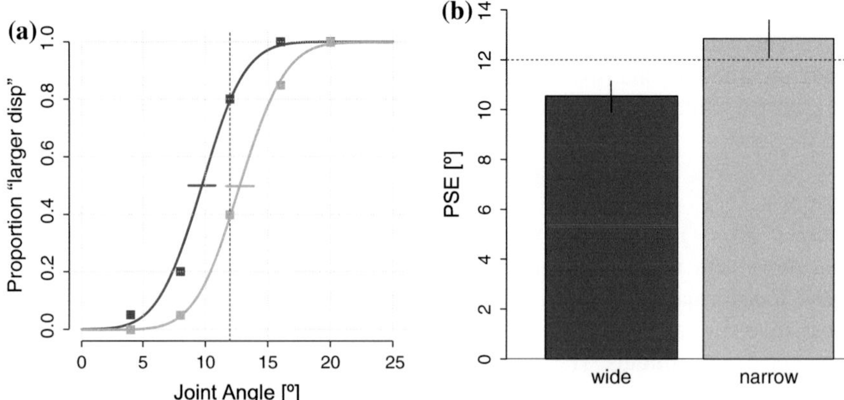

Fig. 7.3 **a** The psychometric functions for a representative participant, in the two experimental conditions (in *red wide* condition and in *grey narrow* condition). The reference finger displacement (12°) is indicated with a *dashed blue line*. **b** The *point-of-subjective-equality* (PSE) in the two experimental conditions (n = 6). Adapted with permission from [15]

7.3 Slip Motion

Whenever we move the hand across the surface of an object, tactile slip provide information about the relative motion between the skin and the surface. If the observer were able to integrate the velocity of tactile slip over time, this would provide him or her with an estimate of the displacement of the hand relative to the object. In [16], we investigated whether humans are able to form a reliable representation of hand displacement from tactile cues only, integrating motion information over time.

7.3.1 Methods

Using the Slip Force Device [8] we rendered the displacement of a surface along different paths. The participant touched the moving surface with the tip of the right index finger. We asked participants to keep their hand world-stationary during the presentation of the stimulus. Therefore proprioceptive inputs were not informative about the path of motion. The surface moved along a right-angle-triangular path, counterclockwise from the x-cathetus to the hypotenuse (Fig. 7.4a). The displacement of the simulated surface always formed a closed triangular figure, whose perimeter and angles were unknown to the participant. During the presentation of the stimulus, participants had to imagine the movement of the finger on a stationary surface (e.g. the plane of a table) that matched the tactile sensation produced by the device (Fig. 7.4a). After the presentation of the stimulus, participants reproduced the movement of the finger on a squared sheet of paper (21 × 21 cm) and then drew it with a

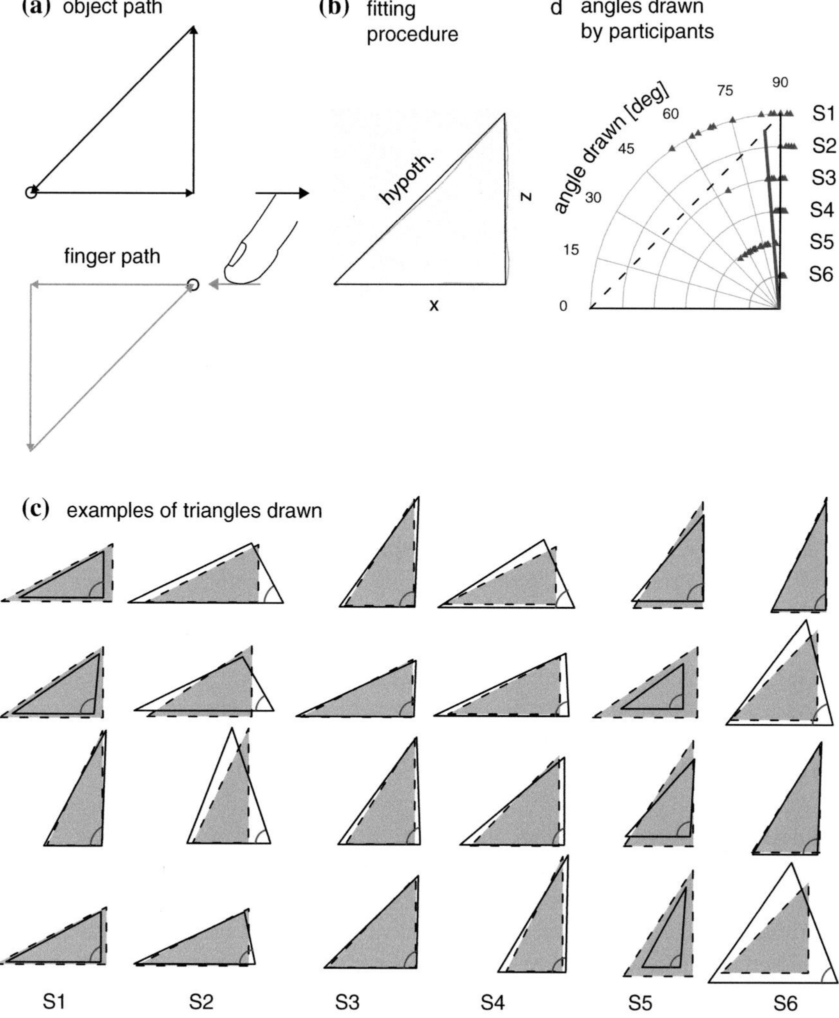

Fig. 7.4 **a** During the administration of the stimulus, the participant imagined a movement of the finger (finger path) that would match with the tactile stimulus produced by the device (object path). After that, they drew the imagined movement on a squared paper sheet. The home position is indicated as a *black dot*. **b** For the analysis and the plot, we digitized the participants' drawings by connecting the vertices of the drawn triangle by straight lines. **c** The physical path (filled triangles), and the drawings from the participants (blank triangles). Examples for 6 different participants are illustrated in different columns. **d** The polar plot represents the perceived size of the bottom-right angle (i.e. the right angle of the stimulus, in *red* in the figure). For each trial, the size of the angle drawn by participants is represented as a triangle-shaped data point. The two *solid lines* are for illustrative purposes and show the 90° angle of the stimulus. The *red line* represents a fit to the participant's data and shows the average angle drawn by participants. Adapted with permission from [16]

pencil (Fig. 7.4b). We asked participants to accurately reproduce the size, the shape, and the orientation of the simulated triangular path.

7.3.2 *Results*

Figure 7.4c provides some examples of the stimuli together with the figures drawn by the participants (the stimuli are represented by grey shapes and the responses of the participants by the solid lines). We independently analyzed the perceived length of the displacement, summarized by the perimeter of the triangles, and the perceived direction of motion, summarized by the first angle. The estimate of the length of the displacement was very accurate: the average of the extent of the simulated displacement was 22.4 cm and the grand mean of the perimeters drawn by participants was 21.9 cm. Thus, the accuracy was $21.9/22.4 = 0.98$. This value was confirmed by fitting the data with a Linear Mixed Model (LMM) which takes both the variability between and within participants into account (see [16] for details). The statistical model confirmed the high accuracy of the response (LMM estimated accuracy: 0.98).

Next, we focused on the direction errors. As explained before, the surface moved first along the x-cathetus, then along the z-cathetus, and finally along the hypotenuse, with the right angle always occurring first. Therefore, the perception of the right angle was not affected by the geometrical constrains of the figure and the drawn angle provides a fair estimate of the direction error. Figure 7.4d shows the angle drawn by each of the 6 observers for each of the 15 trials. The grand mean of the angle was 84.7°. The angle was clearly underestimated in 3 out of 6 participants. There are two possible explanations for this underestimation of the right angle. The participants might have misestimated the direction per se, irrespective of the motion along the first cathetus. Alternatively, participants might have misperceived the direction of motion of the second cathetus, due to possibly a motion aftereffect. In order to evaluate the Aftereffect hypothesis, in a second experiment we reverted the direction of motion of the surface. This way, the motion path was the mirror image along the sagittal axis of the stimulus used in the previous experiment. If the Aftereffect hypothesis were true, the inner angle between the two catheti would be underestimated in a similar way in the two experiments. Instead, if the angle bias was the consequence of a direction anisotropy in world- or skin-framed reference, the angle would be overestimated in the second experiment and not underestimated as in the first one. However, four out of six participants underestimated the inner angle between the x- and the z-cathetus in the second experiment. The average angle bias was -5.4 ± 2 degrees (Mean $\pm$ SE). These results are only consistent with the Motion Aftereffect hypothesis. The bias was larger (i.e., the angle was more underestimation) for short Inter Stimulus Intervals (ISI); the effect was statistically significant ($p < 0.05$). Note that the angle bias was negligible ($\sim$3°) for ISI = 0.8 s, which is a plausible delay in haptic exploration. The reproduced path length was accurate in 4 participants and slightly overestimated in 2 of them. The average drawn perimeter was equal to 26.3 ± 2.0 cm (Mean $\pm$ SE), corresponding to an accuracy of 22.4/26.3 = 0.85.

In summary, we showed that participants are able to integrate tactile slip to reproduce the displacement of the hand, quite accurately. Tactile slip is only informative on the relative motion between the skin and the external object. That is, infinite combinations of tactile and object motion would correspond to the same hand motion,

$$\mathbf{v}_{ha} = \mathbf{v}_{ob} - \mathbf{v}_{ts}. \tag{7.3}$$

where $\mathbf{v}_{ha}$, $\mathbf{v}_{ob}$, $\mathbf{v}_{ts}$ are the hand, object and tactile slip velocity, respectively. In the current experiment, we required participants to match the passive tactile slip with the active movement on a stationary surface, therefore the experimental procedure disambiguated whether the tactile movement depended on the object (passive stimulus) or the hand (active reproduction). In a real life scenario the uncertainty might be much larger as both our limbs and, occasionally, the external object can move. In order to reduce the uncertainty, the observer may assumes a priori that objects are world-stationary [9, 17], so that $\mathbf{v}_{ha} = -\mathbf{v}_{ts}$. We will further discuss this assumption in the next section.

7.4 Discussion

In Sects. 7.2 and 7.3 we showed that the somatosensory system fuses the information from muskuloskeletal system and cutaneous touch to produce a unified percept of the finger or hand displacement. Other studies support this hypothesis. For instance, Bicchi et al. [3] showed that the change in the strain pattern produced during tactile slip conveys information on self-motion similar to the optic flow in vision. The authors introduced the term "tactile flow" to stress the analogy in motion encoding between vision and touch. Roll motion can be also informative on the hand and finger displacement. Accordingly, Dostmohamed and Hayward [5] induced an illusory hand displacement along a curved object by modulating the roll stimulus on the finger pad.

In accordance with Eq. (7.3), the velocity of a tactile stimulus would be informative on the displacement of the finger only if the external object remains stationary. Most of the inanimate objects around us are either at rest or in slow motion—at least, objects in a scale relevant for perception, from millimeters to meters. Therefore, it might be convenient for the observer assuming a priori that objects are world-stationary. Previous studies showed that this is indeed the case and observers perceive the movement of the objects consistently with a Bayesian model combining the sensory measurement with a stationarity prior. This Bayesian model can explain several motion illusions, both in vision [22, 23] and in touch [9, 17]. The second, similar assumption is that the mechanical properties of the object—as for example its stiffness—do not change over time. As illustrated in Sect. 7.3, violating this mechanical-constancy assumption produces an illusory displacement of the finger.

All the examples discussed above involve dynamic stimuli, as in most of daily-life scenarios, static touch does not provide information on our limb position. However, if the surface of the object is convex and its distance to the observer is known, the strain pattern produced by static touch may also provide a cue on the position of the finger. Due to contact mechanics, the force vector produced by a probe on a convex object is unique for each point on the surface of the object [1]. Serio et al. [21] used this principle to localize the contact point on the object from the force vector. The tactile system might take advantage of this mechanical property in order to localize the position of the finger with respect to the object. Static touch would provide the observer with a different 'snapshot' for each different point of contact. To our knowledge, this hypothesis has not been tested yet in human studies.

The discussion above focused on the properties of the external object, which has to remain stationary and maintain a constant stiffness. We also suggested that, in specific scenarios, the shape of the object might convey information on the layout of the finger. Are there assumptions concerning our own body, specifically, the motor control of our hand and our limbs, which may simplify the sensory fusion between proprioception and touch? As showed in the Chaps. 2–5 of this book, a specific co-activation of skeletal muscles occurs when performing a given action. This clustered activity of different muscles in relation to a specific goal is referred to as *motor synergy* [19] and is supposed to reflect a structural of functional organization of the central nervous system aiming to reduce the degrees of freedom of the end effector, e.g. of the hand. The advantages of synergies are not only in simplifying the motor planning, but also in reducing the number of ambiguous interpretation of the sensory feedback. Human hand has roughly 30 degrees of freedom [13]. If the motor system could control independently the angular position of each different joint, several movement would produce identical tactile stimuli, posing a problem of correspondence between the joint configuration and the cutaneous strain pattern. Instead, motor synergy limits the actual number of hand postures, reducing the dimensionality of the system. This suggest the possibility of a mapping or correspondence between the strain pattern on the skin and the coordinated muscle activity, peculiar for a given action. In other words, a sensorimotor synergy. As suggested in [4], the sensorimotor synergy could simplify the interpretation of the complex sensory information from touch. The strength of the correspondence between hand posture and skin deformation could be evaluated experimentally.

Acknowledgments This work was supported by the EU Collaborative Project no. 248587: "THE Hand Embodied", Project no. 601165: "WearHap" and by the European Research Council under the Advanced Grant SoftHands "A Theory of Soft Synergies for a New Generation of Artificial Hands" no. ERC-291166. We thanks Simon Jetzschke for providing us with the photographs used in Fig. 7.1.

References

1. Bicchi A, Salisbury JK, Brock DL (1993) Contact sensing from force measurements. Int J Robot Res 12(3):249–262
2. Bicchi A, Scilingo E, De Rossi D (2000) Haptic discrimination of softness in teleoperation: the role of the contact area spread rate. IEEE Trans Robot Autom 16(5):496–504. doi:10.1109/70.880800
3. Bicchi A, Scilingo EP, Ricciardi E, Pietrini P (2008) Tactile flow explains haptic counterparts of common visual illusions. Brain Res Bull 75(6):737–741. doi:10.1016/j.brainresbull.2008.01.011
4. Bicchi A, Gabiccini M, Santello M (2011) Modelling natural and artificial hands with synergies. Philos Trans R Soc B Biol Sci 366:3153–3161. doi:10.1098/rstb.2011.0152
5. Dostmohamed H, Hayward V (2005) Trajectory of contact region on the fingerpad gives the illusion of haptic shape. Exp Brain Res Exp Hirnforsch Expérimentation Cérébrale 164(3):387–394
6. Edin BB, Abbs JH (1991) Finger movement responses of cutaneous mechanoreceptors in the dorsal skin of the human hand. J Neurophysiol 65(3):657–670
7. Edin BB, Johansson N (1995) Skin strain patterns provide kinaesthetic information to the human central nervous system. J Physiol 487(Pt 1):243–251
8. Fritschi M, Ernst MO, Buss M (2006) Integration of kinesthetic and tactile display—a modular design concept. In: Proceedings of the eurohaptics conference, pp 607–612
9. Hayward V (2008) Haptic shape cues, invariants, priors and interface design. In: Human haptic perception: basics and applications, Birkhäuser Basel
10. Hayward V, Terekhov AV, Wong SC, Geborek P, Bengtsson F, Jörntell H (2014) Spatio-temporal skin strain distributions evoke low variability spike responses in cuneate neurons. J R Soc Interface 11(93):20131,015
11. Jörntell H, Bengtsson F, Geborek P, Spanne A, Terekhov AV, Hayward V (2014) Segregation of tactile input features in neurons of the cuneate nucleus. Neuron 83(6):1444–1452. doi:10.1016/j.neuron.2014.07.038
12. Kandel ER, Schwartz JH, Jessell TM (eds) (2000) Principles of neural science. McGraw-Hill, New York
13. Lin J, Wu Y, Huang TS (2000) Modeling the constraints of human hand motion. In: Proceedings of the workshop on human motion. IEEE, pp 121–126
14. Moscatelli A, Mezzetti M, Lacquaniti F (2012) Modeling psychophysical data at the population-level: the generalized linear mixed model. J Vis 12(26):1–17
15. Moscatelli A, Bianchi M, Serio A, Atassi OA, Fani S, Terekhov A, Hayward V, Ernst MO, Bicchi A (2014a) A change in the fingertip contact area induces an illusory displacement of the finger. In: Auvray M, Duriez C (eds) Haptics: neuroscience, devices, modeling, and applications. Springer, Berlin, pp 72–79
16. Moscatelli A, Naceri A, Ernst MO (2014) Path integration in tactile perception of shapes. Behav Brain Res 274:355–364. doi:10.1016/j.bbr.2014.08.025
17. Moscatelli A, Scheller M, Kowalski GJ, Ernst MO (2014c) The haptic analog of the visual aubert-fleischl phenomenon. In: Auvray M, Duriez C (eds) Haptics: neuroscience, devices, modeling, and applications. Springer, Berlin. http://link.springer.com/chapter/10.1007/978-3-662-44196-1_5
18. Proske U, Gandevia SC (2012) The proprioceptive senses: their roles in signaling body shape, body position and movement, and muscle force. Physiol Rev 92(4):1651–1697
19. Santello M, Baud-Bovy G, Jörntell H (2013) Neural bases of hand synergies. Front Comput Neurosci 7:23
20. Serio A, Bianchi M, Bicchi A (2013) A device for mimicking the contact force/contact area relationship of different materials with applications to softness rendering. IEEE/RSJ international conference on intelligent robots and systems, IROS 2013. Tokyo, Japan, pp 4484–4490

21. Serio A, Riccomini E, Tartaglia V, Sarakoglou I, Gabiccini M, Tsagarakis N, Bicchi A (2014) The patched intrinsic tactile object: a tool to investigate human grasps. In: IEEE/RSJ international conference on intelligent robots and systems (IROS 2014), pp 1261–1268
22. Aa Stocker, Simoncelli EP (2006) Noise characteristics and prior expectations in human visual speed perception. Nat Neurosci 9(4):578–585
23. Weiss Y, Simoncelli EP, Adelson EH (2002) Motion illusions as optimal percepts. Nat Neurosci 5(6):598–604

Part II
Robotics, Models and Sensing Tools

Chapter 8
From Soft to Adaptive Synergies: The Pisa/IIT SoftHand

Manuel G. Catalano, Giorgio Grioli, Edoardo Farnioli, Alessandro Serio, Manuel Bonilla, Manolo Garabini, Cristina Piazza, Marco Gabiccini and Antonio Bicchi

Abstract Taking inspiration from the neuroscientific findings on hand synergies discussed in the first part of the book, in this chapter we present the Pisa/IIT SoftHand, a novel robot hand prototype. The design moves under the guidelines of making an hardware robust and easy to control, preserving an high level of grasping capabilities and an aspect as similar as possible to the human counterpart. First, the main theoretical tools used to enable such simplification are presented, as for example the notion of *soft synergies*. A discussion of some possible actuation schemes shows that a straightforward implementation of the soft synergy idea in an effective design

M.G. Catalano (✉) · G. Grioli · E. Farnioli · A. Serio · M. Gabiccini · A. Bicchi
Department of Advanced Robotics, Istituto Italiano di Tecnologia, Via Morego 30, 16163 Genoa, Italy
e-mail: m.catalano@iit.it; manuel.catalano@iit.it

G. Grioli
e-mail: g.grioli@iit.it

E. Farnioli
e-mail: e.farnioli@iit.it

M. Gabiccini
e-mail: m.gabiccini@iit.it

A. Bicchi
e-mail: a.bicchi@iit.it

M.G. Catalano · G. Grioli · E. Farnioli · A. Serio · M. Bonilla · M. Garabini · C. Piazza · M. Gabiccini · A. Bicchi
Research Center E. Piaggio, Università di Pisa, Largo Lucio Lazzarino 1, 56122 Pisa, Italy
e-mail: a.serio@centropiaggio.unipi.it

M. Bonilla
e-mail: m.bonilla@centropiaggio.unipi.it

C. Piazza
e-mail: c.piazza@centropiaggio.unipi.it

M. Garabini
e-mail: m.garabini@centropiaggio.unipi.it

M. Garabini · M. Gabiccini
Department of Civil and Industrial Engineering, Università di Pisa, Largo Lucio Lazzarino 1, 56126 Pisa, Italy

M. Bianchi and A. Moscatelli (eds.), *Human and Robot Hands*,
Springer Series on Touch and Haptic Systems, DOI 10.1007/978-3-319-26706-7_8

is not trivial. The proposed approach, called *adaptive synergy*, rests on ideas coming from underactuated hand design, offering a design method to implement the desired set of soft synergies as demonstrated both with simulations and experiments. As a particular instance of application of the synthesis method of adaptive synergies, the Pisa/IIT SoftHand is described in detail. The hand has 19 joints, but only uses one actuator to activate its adaptive synergy. Of particular relevance in its design is the very soft and safe, yet powerful and extremely robust structure, obtained through the use of innovative articulations and ligaments replacing conventional joint design. Moreover, in this work, summarizing results presented in previous papers, a discussion is presented about how a new set of possibilities is open from paradigm shift in manipulation approaches, moving from manipulation with rigid to soft hands.

8.1 Introduction

In the first part of the book, the neuroscientific concept of hand synergies, i.e. motor primitives or common actuation patterns of neuro-muscular activities for the human hand, has been widely discussed and considered at different levels (neural, muscular, kinematic and sensory, see Chaps. 2–7).

Recently, different approaches in robotics tried to take advantage from the idea of synergies, aiming to reproduce a similar "coordinated and ordered ensemble" of human hand motions. To transfer part of the embodied intelligence, typical of the human hand, into a robotic counterpart, a promising possibility is the re-creation of synergy patterns as a feature of the mechatronic hand system. This approach has already been tried in recent literature (see next section for a short review), although a purely kinematic model of synergies leads to inconsistent grasp force distribution models. To solve such problems, the concept of *soft synergies* was introduced [1, 2], which provides a model of how synergies may generate and control the internal forces needed to hold an object.

In this chapter, we summarize the results discussed in [3–5], presenting how soft synergy idea can be exploited to build robot hands, such as the Pisa/IIT SoftHand, that can grasp a large variety of objects in a stable way, while remaining very simple and robust. Moreover, we show that the Pisa/IIT SoftHand can afford grasping capabilities that are comparable to natural one. Through the observation of human-directed operations of the prototype it appears how fundamental in everyday grasping and manipulation is the role of hand compliance. Indeed, the Pisa/IIT SoftHand can be functionally shaped using both the object to grasp and the environmental constraints, going beyond nominal kinematic limits by suitably exploiting its structural softness.

The approach to the principled simplification of hand design can be summarized as follows. From statistical observations of human grasping, we derive the hand postures most often used in the grasp approach phase (aka synergies, see also Chaps. 2–5) and a mathematical description on the basis of the soft synergy model (see also

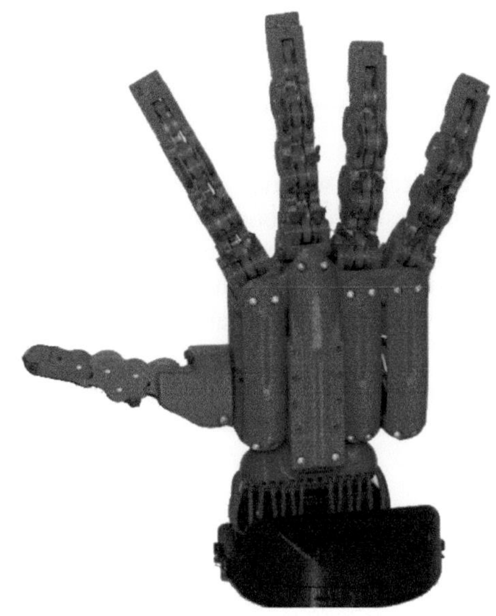

Fig. 8.1 Skeleton of the Pisa/IIT SoftHand advanced anthropomorphic hand prototype implementing one adaptive synergy. The prototype dimensions are comparable to those of the hand of an adult human

Chaps. 12 and 13). Indeed, as also discussed in Chap. 9, human–like hand movement has great influence in the possibility of successfully achieving a large number of grasps belonging to the sphere of activities of daily living (ADL). The actual realization of the hand mechanics is not however a straightforward implementation of the soft synergy model. Indeed, to achieve a simple and compact design and better robustness, we recur to the technology of underactuated hands [6], complementing it with innovative joint and ligament design. A relevant point discussed in this work regards how the design parameters of an underactuated hand can be chosen so that its motion replicates a given set of synergies, in a sense allowing the translation of the concept of *soft synergies* into *adaptive synergies*.

The result of our design method is the Pisa/IIT SoftHand (see Fig. 8.1), a 19-joint hand with anthropomorphic features, which grasps objects of rather general shape by using only a single actuator, and employing an innovative design of articulations and ligaments, which provides a high degree of compliance to external solicitations.

The chapter is organized as follows: Sect. 8.2 briefly presents the analytical model of grasping problem for fully actuated hand and underactuated hands, both via soft and adaptive synergies. Section 8.3 describes in detail the architecture of the Pisa/IIT SoftHand and Sect. 8.4 presents the grasping results experimentally obtained. Section 8.5 addresses the issue of the change of paradigm in the problem of grasping with SoftHands, i.e. compliant yet robust robotic hands. Finally, conclusions are drawn in Sect. 8.6.

8.2 Hand Actuation, Synergies and Adaptation

8.2.1 Fully Actuated Hands

In this section, we briefly present a description of the principal actuation paradigms for the design of robotic systems. The nomenclature and notation are synthesized in Table 8.1. Finally, we present a map between the human inspired *soft synergy* model (see also Chap. 13) and the *adaptive synergy* design used in the Pisa/IIT SoftHand. More theoretical details about grasp analysis are presented in Chaps. 12, 13 and in [7–9].

Starting from fully actuated hands, a quasi-static description of the problem of object grasping can be formalized through a system of three equations as

$$\delta w + G\delta f_c = 0, \tag{8.1}$$

$$\delta\tau = Q\delta q + U\delta u + J^T\delta f_c, \tag{8.2}$$

$$\delta f_c = K_c(J\delta q - G^T\delta u). \tag{8.3}$$

More specifically, the object equilibrium equation, in (8.1), establishes a relationship between external disturbances acting on the object and contact forces that the hand exerts on the object; Eq. (8.2) describes the joint torque variation required to

Table 8.1 Notation for grasp analysis

Notation	Definition
δx	Variation of variable x
$\bar{x}$	Value of x in the reference configuration
$\sharp x$	Dimension of vector x
$w \in \mathbb{R}^6$	External wrench acting on the object
$u \in \mathbb{R}^6$	Pose of the object frame
$f_c \in \mathbb{R}^c$	Contact forces exerted by the hand on the object
c	Number of contact constraints
$\tau \in \mathbb{R}^{\sharp q}$	Joint torque
$q \in \mathbb{R}^{\sharp q}$	Joint configuration
$q_r \in \mathbb{R}^{\sharp q}$	Reference joint configuration
$\sigma \in \mathbb{R}^{\sharp\sigma}$	Soft synergy configuration
$\varepsilon \in \mathbb{R}^{\sharp\sigma}$	Soft synergy forces
$z \in \mathbb{R}^{\sharp z}$	Adaptive synergy configuration
$\eta \in \mathbb{R}^{\sharp z}$	Adaptive synergy forces
$G \in \mathbb{R}^{6\times c}$	Grasp matrix in object frame
$J \in \mathbb{R}^{c\times\sharp q}$	Hand Jacobian matrix in object frame
$S \in \mathbb{R}^{\sharp q\times\sharp\sigma}$	Soft synergy matrix
$A \in \mathbb{R}^{\sharp z\times\sharp q}$	Adaptive synergy matrix

compensate contact force variation and/or kinematic displacement of the system, and the contact constitutive equation (8.3) relates contact force variation with the mutual displacements of the hand and the object contact points.

More in detail, in Eq. (8.1), the symbol $w \in \mathbb{R}^6$ indicates the external wrench acting on the object, described in a local frame $\{O\}$, while $f_c \in \mathbb{R}^c$ are the forces that the hand exerts on the object, described in local contact frames, fixed to the object. The value of c, for the contact force vector, depends on the number and type of contact constraints. For example, a *hard finger* contact, allows the presence of three components of forces, thus it contributes three. The *soft finger* contact, with respect to the hard one, adds the possibility to exert a moment around the normal vector to the contact surface, thus it contributes four. Through the introduction of the *grasp matrix* $G \in \mathbb{R}^{6\times c}$, the object equilibrium condition is written as

$$w + Gf_c = 0. \tag{8.4}$$

Because the equation is written in a reference frame attached to the object, the grasp matrix is constant, hence by differentiating (8.4), (8.1) follows.

The hand equilibrium equation relates contact forces with joint torques, $\tau \in \mathbb{R}^{\sharp q}$, through the transpose of the hand Jacobian matrix $J^T \in \mathbb{R}^{\sharp q\times c}$, as

$$\tau = J^T f_c. \tag{8.5}$$

It is worth observing that the Jacobian matrix is here a function both of the hand configuration q and of the object configuration $u \in \mathbb{R}^6$. This is a consequence of the choice to describe the contact interaction in a local frame attached to the grasped object. From this fact it follows that, differentiating (8.5), (8.2) is obtained, where the terms $Q = \frac{\partial J^T f_c}{\partial q} \in \mathbb{R}^{\sharp q\times \sharp q}$ and $U = \frac{\partial J^T f_c}{\partial u} \in \mathbb{R}^{\sharp q\times 6}$ have to be considered in order to properly take into account the initial contact force preload.

As described in [10], a rigid model of hand/object interaction does not allow the computation of the contact force distribution. The problem can be simply solved by introducing a *virtual spring* at the contact points. One extreme of each virtual spring is attached to the hand and the other to the object, both in the nominal contact location. The virtual spring model generates a force variation corresponding to the local interpenetration of the hand and object parts. Correspondingly, a contact force variation is described in (8.3) through the introduction of the *contact stiffness* matrix $K_c \in \mathbb{R}^{c\times c}$.

The basic grasp equations (8.1)–(8.3) can be rearranged in matrix form as

$$\begin{bmatrix} I & 0 & G & 0 & 0 \\ 0 & I & -J^T & -Q & -U \\ 0 & 0 & I & -K_c J & K_c G^T \end{bmatrix} \begin{bmatrix} \delta w \\ \delta \tau \\ \delta f_c \\ \delta q \\ \delta u \end{bmatrix} = 0. \tag{8.6}$$

This is a linear homogeneous system of equations in the form $\Phi\delta\varphi = 0$, where $\Phi \in \mathbb{R}^{r_\Phi \times c_\Phi}$ is the coefficient matrix, and $\delta\varphi \in \mathbb{R}^{c_\Phi}$ is the vector containing all system variables. From (8.6), we easily obtain that Φ is always full row rank, and its dimensions are

$$\begin{aligned} r_\Phi &= \sharp w + \sharp q + \sharp f, \\ c_\Phi &= 2\sharp w + 2\sharp q + \sharp f. \end{aligned} \tag{8.7}$$

These facts imply that a basis for the solution space of the system has dimension $c_\Phi - r_\Phi = \sharp w + \sharp q$. Thus, a perturbed configuration of the system can be completely described knowing the values of the external wrench variation, δw, and the displacements of the joint configuration,[1] δq. We will refer to these as the *independent variables* of the system. The *dependent variables* will be indicated as $\delta\varphi_d = \left[\delta\tau^T, \delta f_c^T, \delta u^T\right]^T$.

Acting on the coefficient matrix of the system, it is possible to obtain a formal method to get an explicit expression of the dependent variables of the system, as a function of the independent ones. This result is achievable extending the *elementary Gauss operations*, defined for typical linear systems of equations, in order to act on a block partitioned matrix. A general algorithm to obtain the desired form starting from (8.6), called GEROME-B, as completely described in Chap. 13. The final result of the procedure is a set of equations of the type

$$\delta\varphi_d = W_d\,\delta w_c + R_d\delta q, \tag{8.8}$$

where W_d and R_d are matrices of suitable dimensions. In the rest, we will mostly focus on the study of the controllability of grasping with different hand actuation systems, hence considering a null external wrench variation in (8.8). For the sake of completeness, we report here on the structure of this matrix, which can be partitioned as $R_d = \left[\, R_\tau^T \; R_f^T \; R_u^T \,\right]^T$, with the following explicit formulae

$$\begin{aligned} R_\tau &= Q + J^T K_c J + (U - J^T K_c G^T)\left(G K_c G^T\right)^{-1} G K_c J, \\ R_f &= K_c J - K_c G^T \left(G K_c G^T\right)^{-1} G K_c J, \\ R_u &= \left(G K_c G^T\right)^{-1} G K_c J. \end{aligned} \tag{8.9}$$

8.2.2 Approaches to Simplification

Full independent actuation of the joints, in principle, offers the widest range of grasping and manipulation possibilities, limited only by the hand kinematics. As a counterpart, the large number of actuators needed causes complication in the design and a growth of the costs. Even disregarding the hardware aspects, however, the

[1] From the previous considerations, it follows that other choices are possible. However, a complete discussion about these cases is out of the scope of this work.

exploitation of the potential of full independent actuation requires sophisticated programming and control of the hand. Programming complexity turns often out to represent a major obstacle to usability and efficiency in real-world applications of robot hands.

Recently, researchers tried to find a trade off between the full utilization of the hand capabilities and the simplicity in control. Neuroscience studies, as widely described in Chaps. 2–6, 10, 11, 15 and e.g. in [11–13], showed that humans control their hand by organized motion patterns or primitives. Particular muscular activation patterns produce correlated movements of the hand joints, which form a base set [14]. As extensively discussed in this book, especially in the first part (e.g. Chap. 2), such base is referred to as the space of the *postural synergies*, or *eigengrasp space* [15–17]. What makes the bio–aware synergy basis stand out among other possible choices for the basis to describe the hand configuration is the fact that most of the hand grasp posture variance, actually the 80 %, is explained just by the first two synergies, and the 87 % by the first three [18]. This renders the synergy space a credible candidate as a basis for simplification.

In order to transfer this concept in robotics, the *synergy matrix* $S \in \mathbb{R}^{\sharp q \times \sharp \sigma}$ is introduced, describing the principal components of a dataset of grasping postures, where $\sharp\sigma \leq \sharp q$ is the number of used synergies. With this actuation scheme, a hand configuration can be represented in the synergy space by the coordinate vector $\sigma \in \mathbb{R}^{\sharp\sigma}$ as

$$q = S\sigma. \tag{8.10}$$

A similar approach was used in [15], where *software synergies* were used to simulate a correlation pattern between joints of a fully actuated robotic hand. Software synergies can substantially simplify the design phase of a grasp, by reducing the number of control variables (see also [19]). However, software synergies clearly do not impact the simplification of the design of physical hands. The synergy concept was also applied via hardware, as in the design proposed in [20], where the authors used a train of pulleys of different radii to simultaneously transmit different motions to each joint.

Both the aforementioned software and hardware implementation of hand synergies assumed a model of the hand with a number of independent actuators (or Degrees of Actuation, DoAs) smaller than the number of joints (or Degrees of Freedom, DoFs). This causes the hand to move in a way that does not necessarily comply with the shape of an object to be grasped, hence resulting in few contacts being established between the hand and the object. To face this problem, some fixes can be considered, such as e.g. stopping the motion of each finger when it comes in contact with the grasped object, while prosecuting motion of others, or introducing a complementary actuation system for modifying the shape of synergies. While these techniques can be considered to simplify the grasp approaching phase design, they do not benefit from synergy concept to control grasping forces.

8.2.3 Soft Synergies

To fully take advantage of the synergistic approach, avoiding the previously explained limitations, the idea of *soft synergies* was introduced and discussed in [2]. In this model, synergy coordinates define the configuration of a *virtual hand*, toward which the real one is attracted by an elastic field. To describe this situation, we introduce a *reference configuration* vector $q_r \in \mathbb{R}^{\sharp q}$, describing the configuration of the virtual hand. In this model, the motion of the virtual hand is directly controlled in the synergy space as

$$\delta q_r = S\delta\sigma. \tag{8.11}$$

The difference between the real position of the hand and its reference configuration generates the joint torques, which, at equilibrium, balance the interaction forces between the real hand and the grasped object. In formulae, defining a *joint stiffness* matrix $K_q^s \in \mathbb{R}^{\sharp q \times \sharp q}$, the joint torques in the soft synergy model are given by

$$\delta\tau = K_q^s(\delta q_r - \delta q). \tag{8.12}$$

By kineto-static duality, introducing the generalized force in the synergy space $\delta\varepsilon \in \mathbb{R}^{\sharp\sigma}$, we immediately get that

$$\delta\varepsilon = S^T\delta\tau. \tag{8.13}$$

One important aspect of the soft synergy model, is that it enables to reduce the number of degrees of actuation, while retaining all the kinematic degrees of freedom leaving the fine adjustment of the $\sharp q - \sharp\sigma$ remaining movements to the hand compliance. A conceptual hardware implementation of this idea is shown in Fig. 8.2b. This kind of underactuation scheme can be easily modeled by considering (8.11)–(8.13), along with the grasp equation (8.8). In particular, for zero external wrenches on the object, for the joint torques it holds

$$\delta\tau = R_\tau\delta q, \tag{8.14}$$

where $R_\tau \in \mathbb{R}^{\sharp q \times \sharp q}$ was described in (8.9). Substituting (8.14) in (8.12), taking into account (8.11), we obtain

$$\delta q = \left(K_q^s + R_\tau\right)^{-1} K_q^s S\delta\sigma := H^s K_q^s S\delta\sigma. \tag{8.15}$$

Although the idea of soft synergy actuation sketched in Fig. 8.2b appears to provide an elegant solution to the problem of simple hand design, merging the natural motion inherited from the postural synergy approach with adaptivity due to compliance, its implementation in a mechanical design unfortunately turned out not to be very easy or practical, at least in our attempts.

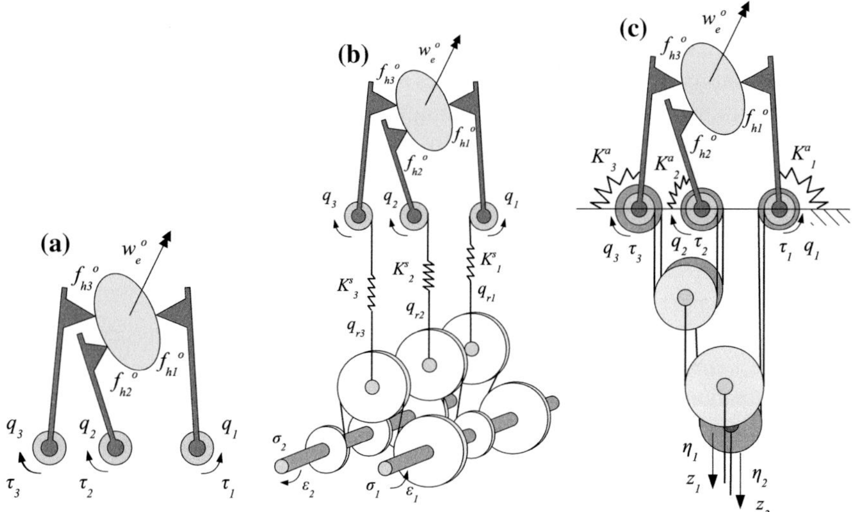

Fig. 8.2 A simple hand grasping an object with different types of (under)actuation. More in detail, panel **a** shows a conceptual schematics of a simple bi-dimensional fully actuated hand grasping an object. Panel **b** shows the Soft Synergy Actuation scheme applied to the same hand. Panel **c** shows two adaptive synergies used for the underactuation mechanism

8.2.4 Adaptive Synergies

A distinct thread of research work has addressed the design of simple robot hands via the use of a small number of actuators without decreasing the number of DoF. This approach, authoritatively described in [6], is referred to as *underactuation* and has produced a number of interesting hands since the earliest times of robotics. For further details the reader can refer to [21–23], and also to [24–27].

The basic idea enabling shape adaptation in underactuated hands is that of a differential transmission, the well-known mechanism used to distribute motion of a prime mover to two or more DOFs. Differentials can be realized in various forms, e.g. with gears [28], closed-chain mechanisms [28], or tendons and pulleys [29], and concatenated so as to distribute motion of a small number of motors to all finger joints q. Letting the vector $z \in \mathbb{R}^{\sharp z}$, with $\sharp z \leq \sharp q$, denote the position of the prime movers, a general differential mechanism is described by the kinematic equation

$$A\delta q = \delta z, \tag{8.16}$$

where $A \in \mathbb{R}^{\sharp z \times \sharp q}$ is the transmission matrix, whose element A_{ij} is the transmission ratio between the ith actuator to the jth joint. By kineto-static duality, the relationship between the actuation force vector $\eta \in \mathbb{R}^{\sharp z}$ and the joint torques is

$$\delta\tau = A^T\delta\eta. \tag{8.17}$$

The kinematic model (8.16) highlights the non-uniqueness of the position attained by an underactuated hand. Indeed, being the transmission matrix A a rectangular fat matrix, an infinity of possible hand postures δq exist which satisfy (8.16) for a given actuator position δz, their difference belonging to the kernel of A.

While these kernel motions are exactly those that provide underactuated hands with the desirable feature of *shape adaptivity*, in practice these hands associate to differential mechanisms the usage of passive elements such as mechanical limits, clutches, and springs [6]. Reasons for adding passive elements are various, including avoiding tendon slackness and ensuring the uniqueness of the position of the hand when not in contact with the object.

Let us define the model of an underactuated hand with elastic springs depicted in Fig. 8.2c as *adaptive synergy* actuation. Notice that springs are arranged in parallel with the actuation and transmission mechanism, as opposed to the soft synergy model in Fig. 8.2b where they are in series. Defining a joint stiffness matrix as $K_q^a \in \mathbb{R}^{\sharp q \times \sharp q}$, the balance equation (8.17) is rewritten as

$$\delta\tau = A^T \delta\eta - K_q^a \delta q. \tag{8.18}$$

Considering (8.18) and (8.14), it immediately follows that

$$\delta q = \left(K_q^a + R_\tau\right)^{-1} A^T \delta\eta := H^a A^T \delta\eta. \tag{8.19}$$

Thus, substituting this in (8.8), we obtain a description of the hand/object equilibria caused by the application of given actuator forces.

If instead actuators are modelled as position sources, by substituting equation (8.19) in (8.16) and inverting, we find

$$\delta\eta = \left(A H^a A^T\right)^{-1} \delta z. \tag{8.20}$$

Substituting this result in (8.19), we then obtain

$$\delta q = H^a A^T \left(A H^a A^T\right)^{-1} \delta z. \tag{8.21}$$

Finally, a complete system description in the case of actuator position control is given by substituting (8.21) in (8.8).

8.2.5 From Soft to Adaptive Synergies

Summarizing the discussion so far, we have seen that two design techniques, i.e. *soft* and *adaptive* synergies, for multiarticulated hands with simple mechanics stand out for different reasons. Soft synergies provide a robust theoretical basis for the design of anthropomorphic hands but without an effective technological implementation. On the contrary, under-actuated hands (adaptive synergies) can be developed exploiting simple elements, such as differential and elastic ones.

For this reason, assuming that a desired soft synergy model is assigned trough its synergy and stiffness matrices, S and K_q^s respectively, our goal is to find a corresponding adaptive synergy model, identified by a transmission matrix A and a joint stiffness K_q^a, which exhibits the same behavior, at least locally, around an equilibrium configuration.

As shown in the previous sections, the behavior of the hand/object system is slightly different if the hand is position controlled or force controlled. This holds true for both soft and adaptive synergy model. Nevertheless, in all of the cases, the system is described as a linear map from an independent variable $\delta\varphi_i$, that can be $\delta\sigma$ or δz, and the joint displacement as $\delta q = \Phi_i \delta\varphi_i$, where Φ_i is taken from one of (8.15) or (8.21), respectively. By means of (8.8), the joint displacement describes the variation of the dependent variables at the hand/object level.

The map can be also defined in the opposite direction, starting from a given adaptive synergy, to obtain a corresponding soft synergy. The total amount of possible maps is eight, considering both the case of position and force control.

We will describe now the procedure to find one of such mappings, from a given position controlled soft synergy model to the corresponding position controlled adaptive synergy hand. All the other maps can be found with similar procedures.

The hand/object behavior for a position controlled soft synergy hand is defined by (8.15), while the behavior of an adaptive underactuated hand is controlled by (8.21). To match them means to impose

$$H^s K_q^s S\, \delta\sigma = H^a A^T \left(A H^a A^T\right)^{-1} \delta z. \tag{8.22}$$

Looking at the span of the second term of the previous equation, it is possible to see that

$$\operatorname{span}\left\{H^a A^T \left(A H^a A^T\right)^{-1}\right\} = \operatorname{span}\left\{H^a A^T\right\}, \tag{8.23}$$

since the term $\left(A H^a A^T\right)^{-1}$ is a square full rank matrix. As a consequence, the span of the two terms in (8.22) can be matched by imposing

$$H^a A^T = H^s K_q^s S M, \tag{8.24}$$

where matrix M can be any full rank square matrix of suitable dimensions, which can be used as design parameter and accounts also for measurement units harmonization. Given the choice on (8.24), a suitable relationship from $\delta\sigma$ to δz can be found in the form

$$\delta z = \left(A H^a A^T\right) M^{-1} \delta\sigma, \tag{8.25}$$

completing the map between the two actuation systems.

8.3 The Pisa/IIT SoftHand

In this section we apply the adaptive synergy design approach previously described and depicted in Fig. 8.2c to the design of a humanoid hand. The hand was designed according to few specifications. On the functional side, the main requirement is the capability of grasping as wide a variety of objects and tools as possible, among those commonly used by humans in everyday tasks. The hand should be primarily able to execute *whole hand grasp* of tools, properly and strongly enough to operate them under arm and wrist control, but it should also be able to achieve *tip grasps*. No in–hand dexterous manipulation is required for this prototype. The main nonfunctional requirements are resilience against force, overexertion and impacts, and safety in interactions with humans. The hand should be lightweight and self-contained, to avoid encumbering the forearm and wrist with motors, batteries and cabling, along with cost effectiveness.

In order to meet the first functional requirement, the hand was designed anthropomorphically, with 19 DOFs arranged in four fingers and an opposable thumb (Fig. 8.3a). To maximize simplicity and usability, however, the hand uses only one actuator. According to our design approach, the motor actuates the adaptive synergy as derived from a human postural database, see also Chaps. 2–5 and 15. The mechanical implementation of the first soft synergy through shape–adaptive underactuation was obtained via the numerical evaluation of the corresponding transmission matrix R and joint stiffness matrix K_q^a appearing in (8.16) and (8.18).

The hand assembly design is shown in Fig. 8.3b. Each finger has four phalanges, while the thumb has three. The hand palm is connected to a flange, to be fixed at the forearm, through a compliant wrist allowing for three passively compliant DOFs.

The wrist of the SoftHand is composed by two curved surfaces, able to roll one on the other. The contact between them is guaranteed by the use of elastic ligaments,

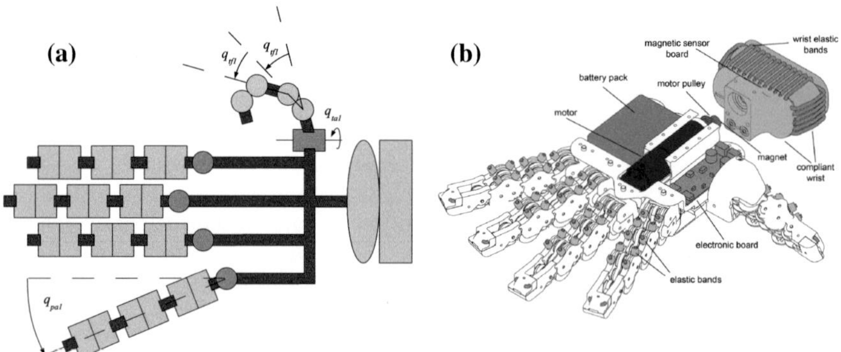

Fig. 8.3 Panel **a** shows the kinematics of the Pisa/IIT SoftHand. Revolute joints are in *dark gray*, while rolling–contact joints are in *light gray*. In panel **b**, a three–dimensional view of the Pisa/IIT SoftHand. Main components (the motor, the battery pack and the electronic control board), joints and wrist architecture are highlighted

arranged along the perimeter of the wrist. When relative motion of the surfaces arises, for example caused by an external load, a set of elastic forces appears. The wrist comes back to the original configuration when the external load is removed.

In rest position, with fingers stretched out and at a relative angle of about 15° in the dorsal plane, the hand spans approximately 230 mm from thumb to little finger tip, is 235 mm long from the wrist basis to the middle finger tip and has 40 mm maximum thickness at the palm. The weight of the hand is approximately 0.5 kg. The requirement on power grasp implies that the hand is able to generate a high enough grasping force, and to distribute it evenly through all contacts, be them at the fingertips, the inner phalanges, or the palm. These goals are naturally facilitated by the shape adaptivity of the soft synergy approach, yet they also require strong enough actuation and, very importantly, low friction in the joints and transmission mechanisms.

The requirement on resilience and safety was one of the most exacting demands we set out for our design, as we believe these to be crucial features that robots must possess to be of real use in interaction with, and assistance to, humans. This is only more true for hands, the body part primarily devoted to physical interaction with the environment for exploration and manipulation. To achieve this goal, we adopted a non-conventional "soft robotics" design of the mechanics of the hand, that fully exploits the potential of modern material deposition techniques to build a rather sophisticated design with rolling joints and elastic ligaments at very low cost. A first departure form conventional design is the use of rolling contact articulations to replace standard revolute joints. Our design takes inspiration from a class of joints known as COmpliant Rolling-contact Elements (CORE) [30, 31], aka "Rolamite" or "XRjoints" joints [32] (see Fig. 8.4a). Among these, Hillberry's design of a rolling joint [33] is particularly interesting to our purposes.

A Hillberry joint consists of a pair of cylinders in rolling contact on each other, held together by metallic bands, which wrap around the cylinders on opposite sides as schematically shown in Fig. 8.4a. In Hillberry joints (see Fig. 8.4b), the band arrangement results in a compliant behavior in flexion but rigid in traction. The joint forms a higher kinematic pair, whose motion is defined by the profile of the cylinders, and exhibits very low friction and abrasive wear. The joint behaves more similarly to the human articulation than simple revolute joint, and for this reason was originally proposed for knee prostheses [33]. Hillberry joints have been used in few robotic applications before, including robot hands [34]. Figure 8.3a shows how we used CORE joints in the design of the Pisa/IIT SoftHand. In particular, we adopted CORE joints for all the interphalangeal, flexion/extension articulations. Conversely, conventional revolute joints were used for metacarpo-phalangeal, abduction/adduction articulations. Our design introduces a few important modifications of existing rolling–contact joints, which are illustrated in Fig. 8.4c, d. Firstly, metallic bands were replaced by elastic ligaments, realized a polyurethane rubber able to withstand large deformations and fatigue, and are fixed across the joint with an offset in the dorsal direction. Suitable pretensioning of the ligaments, together with a carefully designed profile of the two cylinders, introduces a desirable passive stability behavior, with an attractive equilibrium at the rest configuration with fingers

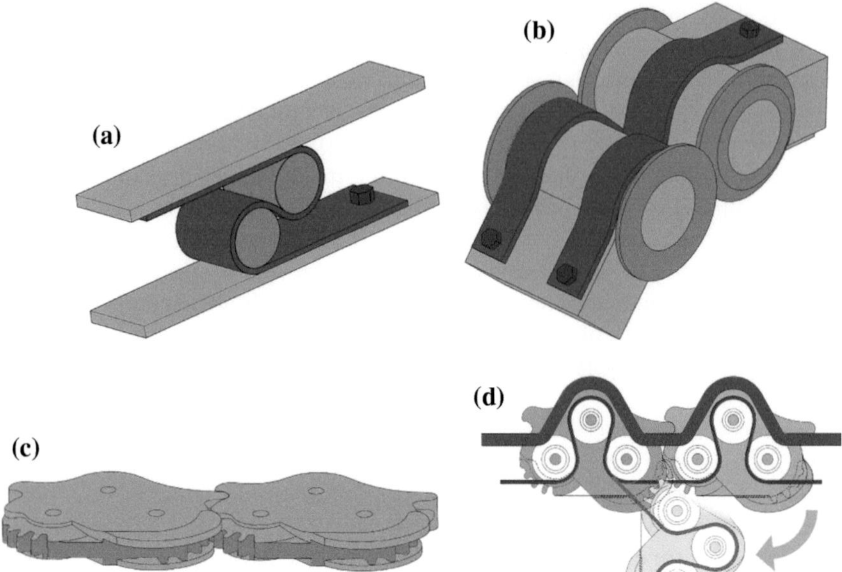

Fig. 8.4 Schematic illustration of two examples of COmpliant Rolling–contacts Elements (CORE): a Rolamite joint (**a**) and a Hillberry joint (**b**). Design of the compliant rolling–contact joint used in the interphalangeal joints of the Pisa/IIT SofHands: **c** perspective view with rolling cylinders with matching multi-stable profile; **d** lateral view, showing the arrangement of ligaments and tendons

stretched. The elastic ligaments are polyurethane rubber segments of 2 mm diameter, characterized by 88 Shore A hardness. The rest length of the ligaments is 10 mm. Some pre-tensioning is applied with a stretch in the range between 2–5 mm. All the long fingers proximal flexion joints have lower values of pre-tensioning, with respect to all the other joints, in order to guarantee a hand motion similar to the first human postural synergy (see Chaps. 2 and 4), as explained in Sect. 8.2.4 (Fig. 8.5).

The coupled rolling cam profiles are designed on a circular primitive with radius 6.5 mm. The actuation tendon is wrapped around pulleys with radius 3.5 mm. All the radii are the same for all the rolling profiles and the pulleys, in order to obtain a modular design. The rolling cam profile is realized on cylinder portions flanked by lateral walls on both sides, whose slope is about 80° (see Figs. 8.4c, d and 8.6). When two phalanges are assembled, such walls are housed in a fitting recess of the matching phalanx. These features of our design are particularly important for the system to behave softly and safely at contact, and to recover from force overexertion, due e.g. to impacts or jamming of the hand, making the hand automatically return to its correct assembly configuration. Indeed, these joint can withstand severe disarticulations (cfr. Fig. 8.6) and violent impacts (Fig. 8.7).

The design of interphalangeal joints does not require the use of screws, shafts, bearings or gears. As it can be seen in Fig. 8.4c, d, a few teeth of an involute gear of

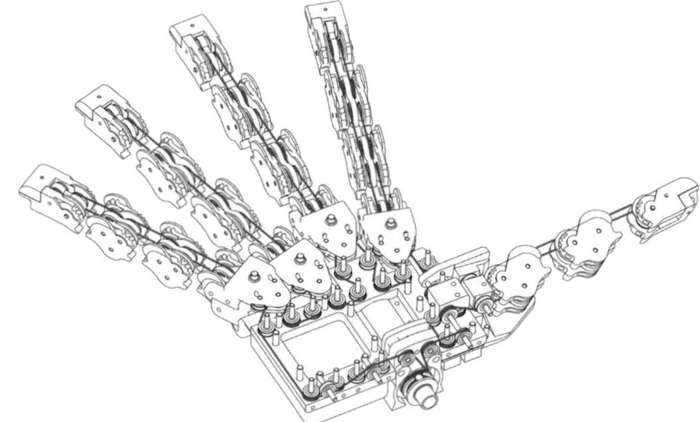

Fig. 8.5 Partially exploded line sketch of the Pisa/IIT SoftHand. The tendon routing distributes the motion to all joints

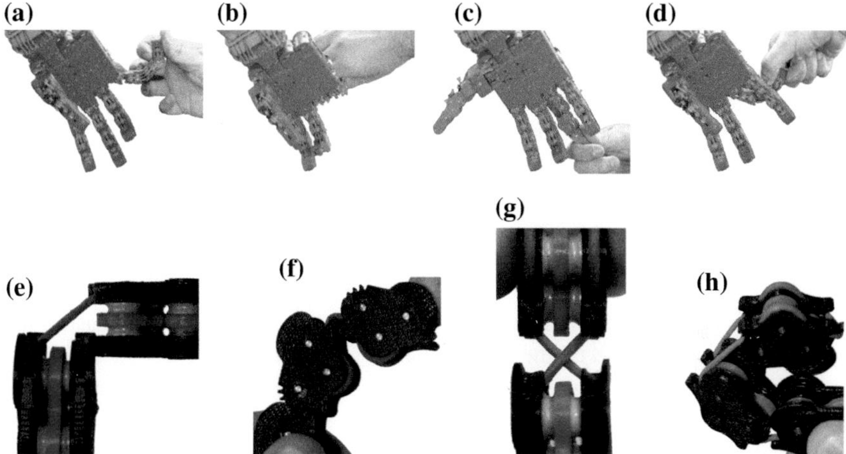

Fig. 8.6 The Pisa/IIT SoftHand joints can withstand severe force overexertion in all directions, automatically returning to the correct assembly configuration. **a** Finger side bend. **b** Finger back bend. **c** Finger twist. **d** Finger skew bend. **e** Side bend. **f** Back bend. **g** Twist. **h** Skew bend

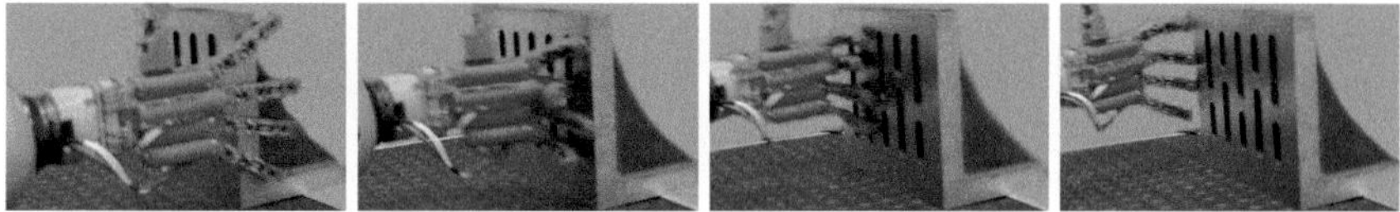

Fig. 8.7 A photo-sequence showing the PISA/IIT SoftHand during a violent impact with a stiff surface

vanishing height are indeed integrated in the cam shape, to better support tangential loads at the joint.

Actuation of the hand is implemented through a single Dyneema tendon routed through all joints using passive anti-derailment pulleys. The tendon action flexes and adducts fingers and thumb, counteracting the elastic force of ligaments, and implementing adaptive underactuation without the need for differential gears (Fig. 8.5).

8.4 Experimental Results

The prototype of the Pisa/IIT SoftHand (Fig. 8.1) was built to perform experimentally tests. The actuator powering the hand is a 6 W Maxon motor RE-max21 with a reduction ratio of 84:1 equipped with a 12 bit magnetic encoder (Austrian Microsystems AS5045) with a resolution of 0.0875°.

The embedded electronic unit hosting sensor processing, motor control and communication is located in the hand back, along with the battery pack. The opening/closing of the hand is controlled via a single set point reference, communicated via one of the available buses (SPI and RS-485).

During experiments the hand worn an off-the-shelf working glove with padded rubber surfaces, supplying contact compliance and grip.

8.4.1 *Force and Torque Measurements*

An interface equipped with an ATI nano 17 F/T sensor was used to measure the holding force and torque of the robotic hand. In the first case, a split cylinders, represented in Fig. 8.8a, was used to measure the grasp force. The cylinder is 120 mm high and has a diameter of 45 mm. The disk to measure maximum holding torque is represented in Fig. 8.8b. The disk is 20 mm high and has a diameter of 95 mm.

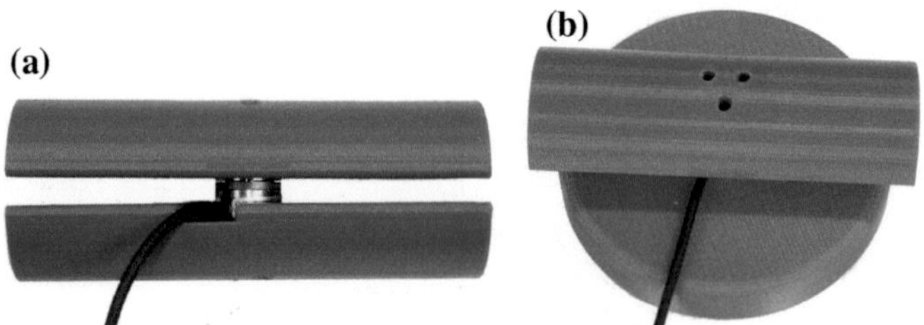

Fig. 8.8 Sensorized object for force measurements (**a**), sensorized object for torque measurements (**b**). **a** Grasp force test object. **b** Holding torque test object

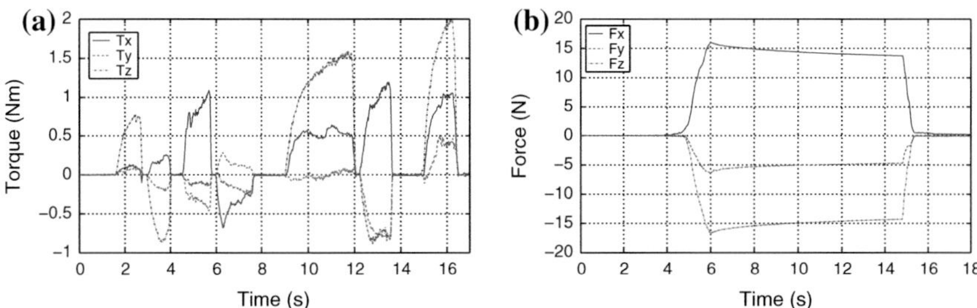

Fig. 8.9 Torques and forces of the robotic hand during grasp task. **a** Torques. **b** Forces

In Fig. 8.9b we report force acquisitions while the hand grasped the sensorized object. It is possible to notice how forces increased when fingers get in contact with the sensorized cylinder (step behavior of the lines in Fig. 8.9b). We achieved a maximum holding torque of 2 Nm and maximum holding force of about 20 N along the z axis. These limits appear to be dictated by the motor size rather than by the hand construction. Although we did not go through an exhaustive analysis, in an occasional experiment with a stronger motor, we obtained holding torque of 3.5 Nm and holding force of 28 N.

8.4.2 Grasp Experiments

To test the adaptiveness of the robotic hand, the grasp of several objects of daily use in a domestic or lab environment were performed. Grasp experiments were performed in three different conditions: (1) the hand wrist fixed on a table and the object placed in the grasp; (2) the object placed on a table and the hand mounted on a robot arm, and (3) the hand wrist fixed to the forearm of a human operator.

Some examples of grasps achieved in the first condition are reported in Fig. 8.10.

To test usage of the Pisa/IIT SoftHand in a robotic scenario, the hand was mounted at the end-effector flange of a KUKA Light–Weight robot arm. Figure 8.11 shows some of the grasps tested. Notice that the robot was manually programmed to reach an area were the object was approximately known to lie, and no grasp planning phase was executed. Rather, the hand was given a closure command by software. The closure time, as well as the robot trajectory, were preprogrammed in the examples shown, and were the same for each object lying roughly in the same area.

Finally, to test the capability of the Pisa/IIT SoftHand to acquire complex grasps of objects randomly placed in the environment, we developed a wearable mechanical interface (see Fig. 8.12) allowing an operator to use our hand as a substitution of his/her own. The interface can be strapped on the operator's forearm and can be controlled by the operator acting on a lever with his/her real hand (see Fig. 8.12a). In Fig. 8.13 we report some grasps executed with the human interface (condition 2 above). In summary, a total of 107 objects of different shape was successfully grasped, with a whole hand or a tip grasp, in all conditions previous considered,

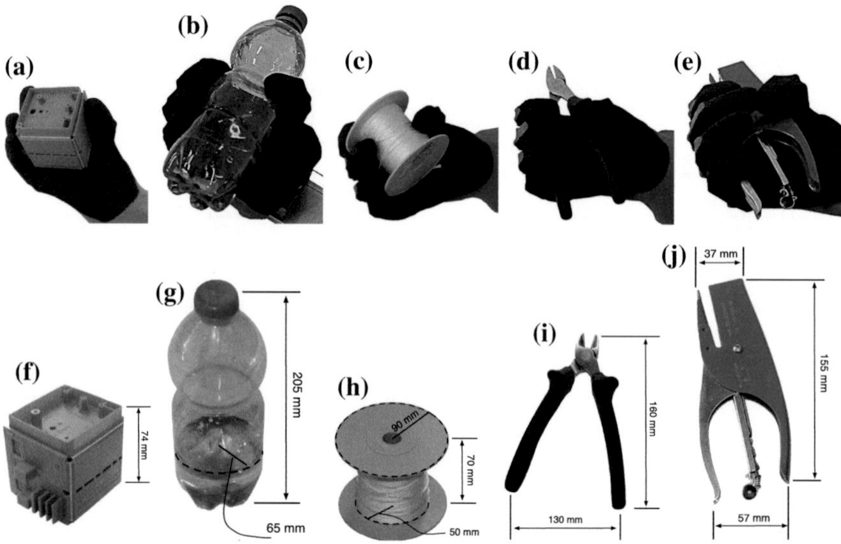

Fig. 8.10 Some experimental grasps performed with the Pisa/IIT SoftHand, with the object placed in the hand by a human operator

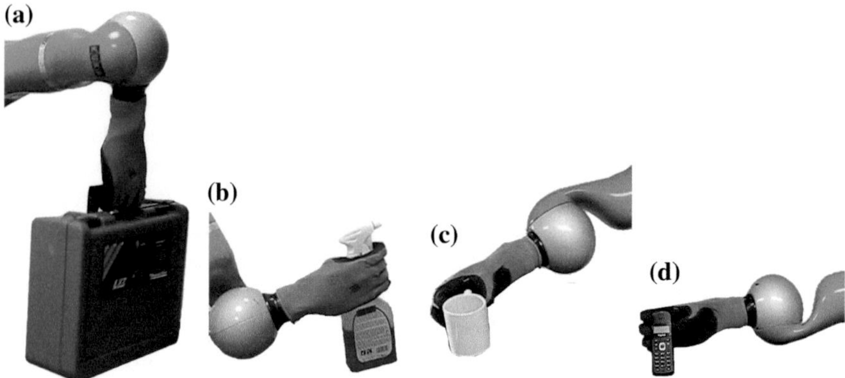

Fig. 8.11 Grasps executed with the Pisa/IIT SoftHand mounted on a Kuka Light Weight Robot: handbag (**a**), spray (**b**), cup (**c**) and telephone (**d**)

during our tests: bottle, reel, pincer, stapler, pen, phone handset, plier, teddy bear, cup, handle, spray, computer mice, hot–glue gun, human hand, cell phone, glass, screw–driver, hammer, file, book, coin, scotch tape holder, ball, tea bag, ketchup bottle, hamburger, camera, tripod stand, cutter, trash can, keyboard, torch, battery container, battery (AA), small cup, measuring tape, caliper, wrench, lighter, eraser, world map (globe), remote control, hex key, AC adapter, keyring, spoon, fork, knife, hand tissue box, liquid soap dispenser, corkscrew, rag, candy, calculator, slice of cake, rubber-stamp, spring, paper, cellphone case, rubber band, bottle top, watch, umbrella, broom, garbage scoop, scarf, chair, schoolbag, USB cable, glue stick,

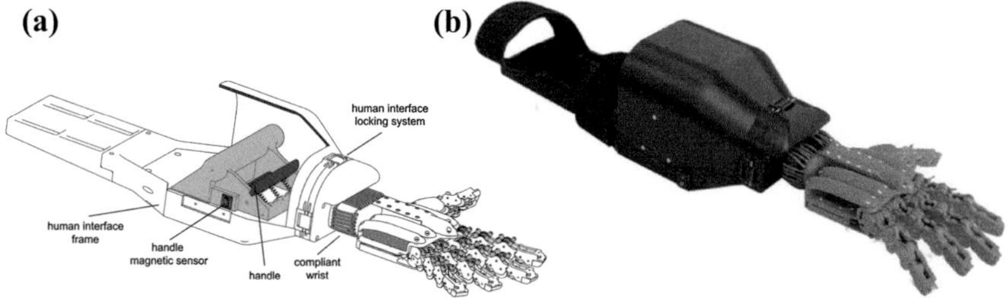

Fig. 8.12 **a** Components of the interface for human use of the Pisa/IIT SoftHand. The angle position of the lever is adopted as reference position to drive the actuator of the hand. **b** Appearance of the assembled human interface prototype

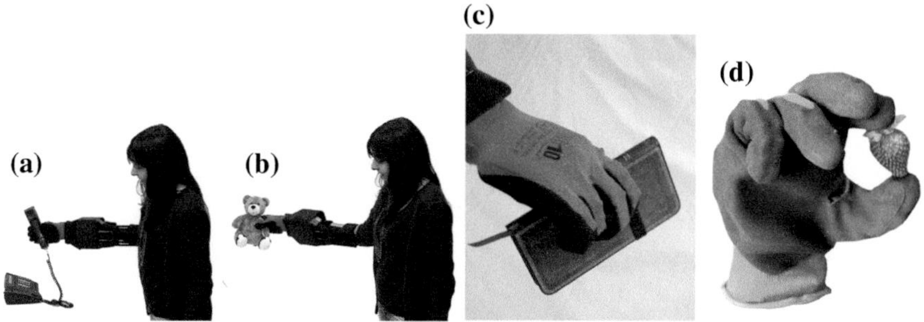

Fig. 8.13 Some examples of grasps executed with the robotic hand mounted on a wearable Human Robot Interface: telephone (**a**), Teddy Bear (**b**), book (**c**), and strawberry (**d**)

wallet, credit card, sponge, pencil sharpener, straight edge, safety lock, mouse pad, hard disk, jacket, drill, chalk, notebook, blackboard eraser, door lock, square ruler, scissors, eyeglasses, deodorant, USB key, hat, headphones, cigarette, helmet, screw (M8), clamp, fridge magnet, drill bit, table calendar, saw, tape cassette, beauty case, bubble gum box, bubble gum, tissue pocket, dish, poster.

The more difficult situation is in grasping very thin objects. However, since grasp limitations of the prototype are also influenced by the operator training, it is not easy to quantify grasp limitations without resorting to further investigations on operator's capabilities, which are out of the scope of this paper.

8.5 A New Set of Possibilities

One of the main lessons learned through these experiments is that, while all grasps could be easily achieved by the hand when operated by a human, programming the robot to achieve the same grasps was in some cases rather complex. One of the main reason for this is, in our understanding, that the human operator quickly learns how to exploit the intrinsic adaptivity of the hand, including the wrist compliance, to shape the hand before and during grasp. This is done with the help of object features and/or environmental constraints, notably the table top and walls. This observation hints that autonomous learning and planning for soft robot grasping might have to be focused on constraint–based motion rather than on free–space, multi-DoF hand shaping. This kind of features allow to use these hands in more "daring" interactions with the objects and the environment, in contrast with the "timid" approach typically adopted with rigid hands. In fact, Soft Hands can use their full surface for enveloping grasps, and exploiting objects and environmental constraints, in order to functionally shape themselves, going beyond their nominal kinematic limits by exploiting structural softness.

The differences between a rigid and soft approach to manipulation are sketched in Fig. 8.14. In the classical paradigm (cfr. Fig. 8.14a), the planner searches for suitable points on the object that generate a nominal grasp of good quality, and for trajectories that can bring there the fingertips while avoiding contacts of the hand with the environment. In the example of Fig. 8.14c, to grasp the green cup while avoiding

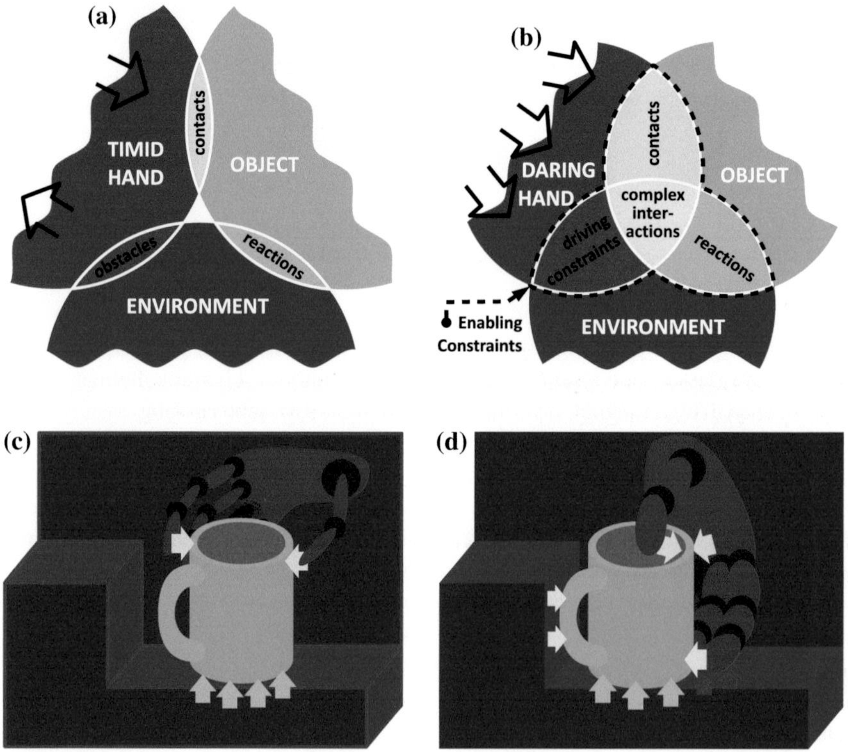

Fig. 8.14 Paradigm shift in manipulation, from rigid manipulation (*left*) to soft manipulation (*right*). Primary colors identify the scenario main actors: *red* for the robotic hand, *blue* for the environment, green for the target object. Secondary colors codify simple interactions between the actors: *yellow* for hand-object, *cyan* for object-environment and *magenta* for environment-hand. Finally, complex interactions, which involve all the three actors at the same time, are *white* colored. **a** Rigid manipulation paradigm. **b** Soft Manipulation Paradigm. **c** Rigid Manipulation example. **d** Soft Manipulation example

the wall on the left the planner has to find a path in a narrow passage. However, soft manipulation subverts this scheme (Fig. 8.14b). In the example of Fig. 8.14d, hand-object, object-environment and hand-environment contacts are not avoided but rather sought after and exploited to shape the hand itself around the object.

The set of all possible physical interactions between the hand, the object and the environment, which define the hand-object functional interaction, will be referred to as the set of *Enabling Constraints*. The analysis of such possibilities constitute a rather new challenge for existing grasping algorithms: adaptation to totally or partially unknown scenes remains a difficult task, toward which only some approaches have been investigated so far. Some of them are model-free and propose geometrical features which indicate good grasps [35, 36], some others evaluate also topological object properties such as holes [37]. Typically, grasps are ranked on a fixed list of suitable hypotheses, do not require supervised learning, but do not adapt over time. Other methods are based on learning the success rate of grasps given some descriptor extracted from sensor data, either evaluated on a real robotic system [38, 39], or on simulated sensor data [40, 41].

Moreover, beside vision-based methods, hand compliance offers the real possibility to use tactile exploration for 3D reconstruction of unknown environments and objects. Indeed, tactile sensing can solve some severe limitations of computer vision, such as sensitivity to illumination and limited perspective. As an example, a combined procedure based on dynamic potential fields, that aims at reconstructing 3D object models, which are then used for grasp planning and execution was presented in [42] and recently extended in [43].

An important observation from the way humans use their real hands is that, in everyday grasping and manipulation tasks, the role of hand compliance is fundamental. In the first place it is used to adapt to the shape of the hand surroundings: both the target object and the rest of the environment. On the other hand, it is important to notice how the objects and the environment constraints are used, in turn, to functionally shape the hand, going beyond its nominal kinematic limits by exploiting its structural softness (Fig. 8.15).

Although one could ascribe such levels of dexterity to the high levels of sensory-motor capabilities of the human hand itself, it is astounding to compare the performance of the human naked hand with that of a person using a simple robot hand, as the Pisa/IIT SoftHand arm-mounted device shown in Fig. 8.16.

Thanks to its under-actuated mechanisms the SoftHand is capable to grasp several number of objects by matching to their shape. These combination of simplicity, adaptivity and robustness lets the person experiment in a very natural way with the robotic hand, and soon achieve a level of performance comparable, often similar, to that obtained with their true hands. This achievement is obtained despite the presence of just one degree of actuation on the mechanism and an almost total lack of tactile feedback, and has encouraged the exploitation of the Pisa/IIT SoftHand as an ideal platform for the development of a novel hand prosthesis, as described in Chap. 10. Figure 8.15, shows three very different ways to grasp a cup, implemented with both the bare human hand and the SoftHand.

Fig. 8.15 A human hand grasping a cup with three different approaches (*top panels*) and the same grasps reproduced with the Pisa/IIT SoftHand (*bottom panels*)

Fig. 8.16 The Pisa/IIT SoftHand mounted on an a human arm

Notice that the SoftHand can substantially match the grasping performance of the human hand thanks to its possibility of exploring and exploiting the *Enabling Constraints* that define, at a very basal level, the problem of grasping and manipulation.

As a further example, consider Fig. 8.17, where the combined action of adaptability and robustness allow the user to manipulate and interact with both the environment and the object at the same time, in a complex way (refer also to Fig. 8.14). Exploiting all the physical constraints that are external with respect to the hand itself: walls, surfaces and edges, force closures of the object between the hand and the environment can be obtained and used to generate simple and effective manipulation tasks, in this case sliding and pivoting a book.

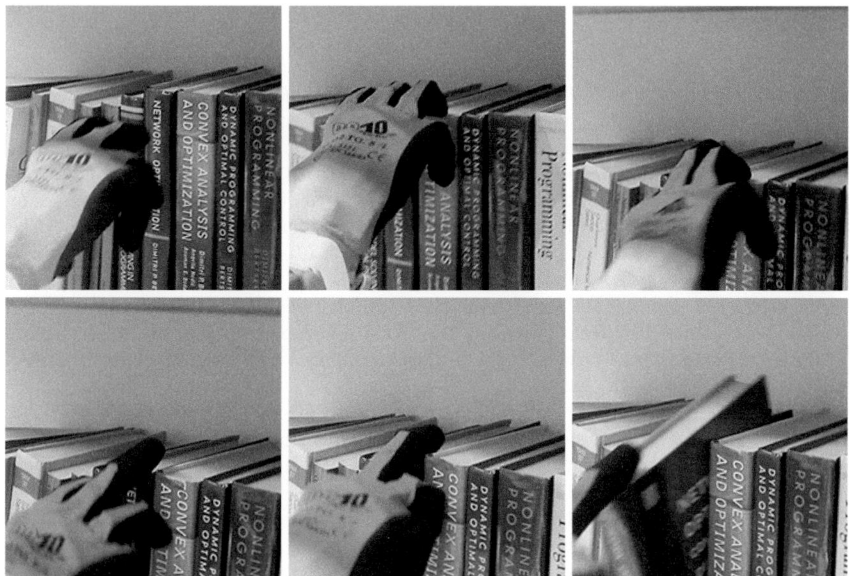

Fig. 8.17 A person with the arm-mounted SoftHand can seamlessly execute also difficult manipulation tasks which involve combined interactions between hand, object and environment

8.6 Conclusion

This chapter presented the design and implementation of the Pisa/IIT SoftHand, along with the theoretical framework behind that justifies the main design choices. The important aspect of the hand actuation pattern is considered first, reviewing various past and recent approaches, and finally considering adaptive synergies as preferred choice.

The first prototype of the Pisa/IIT SoftHand, a highly integrated robot hand characterized by a humanoid shape and good robustness and compliance, is presented and discussed. The hand is validated experimentally through extensive grasp cases and grasp force measurements. Finally considerations on new sets of possibilities were shown, discussing the possibility of a paradigm shift in manipulation approaches, from rigid to soft manipulation, with further developments toward prosthetics, as discussed in Chap. 10.

Acknowledgments The authors would like to thank Andrea Di Basco, Fabrizio Vivaldi, Simone Tono and Emanuele Silvestro for their valuable help in the realization of the prototypes.
This work was supported by the European Commission under the CP-IP grant no. 248587 "THE Hand Embodied", within the FP7-2007-2013 program "Cognitive Systems and Robotics", the grant no. 645599 "SOMA: Soft-bodied Intelligence for Manipulation", funded under H2020-EU-2115, the ERC Advanced Grant no. 291166 "SoftHands: A Theory of Soft Synergies for a New Generation of Artificial Hands".

References

1. Gabiccini M, Bicchi A, Prattichizzo D, Malvezzi M (2011) On the role of hand synergies in the optimal choice of grasping forces. Auton Robots 31(2–3):235–252
2. Bicchi A, Gabiccini M, Santello M (2011) Modelling natural and artificial hands with synergies. Philos Trans R Soc B: Biol Sci 366(1581):3153–3161
3. Catalano MG, Grioli G, Serio A, Farnioli E, Piazza C, Bicchi A (2012) Adaptive synergies for a humanoid robot hand. In: IEEE-RAS international conference on humanoid robots, Osaka, Japan
4. Catalano MG, Grioli G, Farnioli E, Serio A, Piazza C, Bicchi A (2014) Adaptive synergies for the design and control of the pisa/iit softhand. Int J Robot Res 33:768–782
5. Bonilla M, Farnioli E, Piazza C, Catalano MG, Grioli G, Garabini M, Gabiccini M, Bicchi A (in Press) Grasping with soft hands. In International conference on humanoid robots IEEE-RAS 2014, Madrid, Spain, 18–20 Nov
6. Birglen L, Gosselin C, Laliberté T (2008) Underactuated robotic hands, vol 40. Springer
7. Gabiccini M, Farnioli E, Bicchi A (2012) Grasp and manipulation analysis for synergistic underactuated hands under general loading conditions. In: 2012 IEEE international conference on Robotics and Automation (ICRA), IEEE, pp 2836–2842
8. Gabiccini M, Farnioli E (2013) Bicchi A (2013) Grasp analysis tools for synergistic underactuated robotic hands. Int J Robot Res 32:1553–1576
9. Farnioli E, Gabiccini M, Bonilla M, Bicchi A (2013) Grasp compliance regulation in synergistically controlled robotic hands with VSA. In: IEEE/RSJ international conference on intelligent robots and systems, IROS 2013, Tokyo, Japan, pp 3015 –3022, 3–7 Nov 2013
10. Bicchi A (1994) On the problem of decomposing grasp and manipulation forces in multiple whole-limb manipulation. Int J Robot Auton Syst 13:127–147
11. Weiss EJ, Flanders M (2004) Muscular and postural synergies of the human hand. J Neurophysiol 92:523–535
12. Santello M, Baud-Bovy G, Joerntell E (2013) Neural bases of hand synergies. Note: in review
13. Castellini C, van der Smagt P (2013) Evidence of muscle synergies during human grasping. Biol Cybern 107:233–245
14. Easton T (1972) On the normal use of reflexes: the hypothesis that reflexes form the basic language of the motor program permits simple, flexible specifications of voluntary movements and allows fruitful speculation. Am Sci 60(5):591–599
15. Ciocarlie M, Goldfeder C, Allen P (2007) Dexterous grasping via eigengrasps: a low-dimensional approach to a high-complexity problem. In: Proceedings of the robotics: science and systems 2007 workshop-sensing and adapting to the real world, Electronically published, Citeseer
16. Prattichizzo D, Malvezzi M, Bicchi A (2010) On motion and force controllability of grasping hands with postural synergies. In: Proceedings of robotics: science and systems, Zaragoza, Spain
17. Wimboeck T, Reinecke J, Chalon M (2012) Derivation and verification of synergy coordinates for the DLR hand arm system. In: CASE, IEEE, pp 454–460
18. Santello M, Flanders M, Soechting J (1998) Postural hand synergies for tool use. J Neurosci 18(23):10105–10115
19. Ficuciello F, Palli G, Melchiorri C, Siciliano B (2011) Experimental evaluation of postural synergies during reach to grasp with the ub hand iv. In: 2011 IEEE/RSJ international conference on Intelligent Robots and Systems (IROS), IEEE, pp 1775–1780
20. Brown C, Asada H (2007) Inter-finger coordination and postural synergies in robot hands via mechanical implementation of principal components analysis. In: IEEE/RSJ International conference on intelligent robots and systems, IROS 2007, IEEE, pp 2877–2882
21. Tomovic R, Boni G (1962) An adaptive artificial hand. IRE Trans Autom Control 7(3):3–10
22. Hirose S, Umetani Y (1978) The development of soft gripper for the versatile robot hand. Mech Mach Theory 13(3):351–359

23. Rovetta A (1981) On functionality of a new mechanical hand. J Mech Des 103:277
24. Laliberté T, Gosselin C (1998) Simulation and design of underactuated mechanical hands. Mech Mach Theory 33(1):39–57
25. Carrozza M, Suppo C, Sebastiani F, Massa B, Vecchi F, Lazzarini R, Cutkosky M, Dario P (2004) The spring hand: development of a self-adaptive prosthesis for restoring natural grasping. Auton Robots 16(2):125–141
26. Gosselin C, Pelletier F, Laliberte T (2008) An anthropomorphic underactuated robotic hand with 15 dofs and a single actuator. IEEE international conference on robotics and automation, ICRA 2008:749–754
27. Dollar A, Howe R (2010) The highly adaptive sdm hand: design and performance evaluation. Int J Robot Res 29(5):585
28. Laliberte T, Birglen L, Gosselin C (2002) Underactuation in robotic grasping hands. Mach Intell Robot Control 4(3):1–11
29. Hirose S (1985) Connected differential mechanism and its applications. In: Proceedings of the 2nd ICAR, pp 319–326
30. Cannon JR, Howell LL (2005) A compliant contact-aided revolute joint. Mech Mach Theory 40:1273–1293
31. Jeanneau A, Herder J, Laliberté T, Gosselin C (2004) A compliant rolling contact joint and its application in a 3-DoF planar parallel mechanism with kinematic analysis. ASME Conf Proc 46954:689–698
32. Cadman R (1970) Rolamite–geometry and force analysis. Technical Report, Sandia Laboratories, April 1970
33. Hillberry B, Hall A Jr (1976) Rolling contact joint. US Patent 3,932,045, 13 Jan 1976
34. Ruoff C (1985) Rolling contact robot joint. US Patent 4,558,911, 17 Dec 1985
35. Hsiao K, Chitta S, Ciocarlie M, Jones EG (2010) Contact-reactive grasping of objects with partial shape information. In: 2010 IEEE/RSJ international conference on Intelligent Robots and Systems (IROS), pp 1–8
36. Klingbeil E, Rao D, Carpenter B, Ganapathi V, Ng AY, Khatib O (2011) Grasping with application to an autonomous checkout robot. In: 2011 IEEE international conference on robotics and automation, pp 2837–2844
37. Pokorny FT, Stork JA, Kragic D (2012) Grasping objects with holes: a topological approach. In: Proceedings—IEEE international conference on robotics and automation, pp 1100–1107
38. Montesano L, Lopes M, Bernardino A, Santos-Victor J, Melo FS, Martinez-Cantin R (2009) Learning grasping affordances from local visual descriptors. In: IEEE 8th international conference on development and learning, ICDL 2009. pp 1–6
39. Detry R, Başeski E, Popović M, Touati Y (2010) Learning continuous grasp affordances by sensorimotor exploration. From motor learning to, Jan 2010
40. Bohg J, Kragic D (2010) Learning grasping points with shape context. Robot Auton Syst 58:362–377
41. Saxena A, Driemeyer J, Ng AY (2008) Robotic grasping of novel objects using vision. Int J Robot Res 27:157–173
42. Bierbaum A, Rambow M (2009) Grasp affordances from multi-fingered tactile exploration using dynamic potential fields. Humanoid Robots, Jan 2009
43. Herzog A, Pastor P, Kalakrishnan M, Righetti L, Bohg J, Asfour T, Schaal S (2013) Learning of grasp selection based on shape-templates. Auton Robots 36(1–2):51–65

Chapter 9
A Learn by Demonstration Approach for Closed-Loop, Robust, Anthropomorphic Grasp Planning

Minas V. Liarokapis, Charalampos P. Bechlioulis, George I. Boutselis and Kostas J. Kyriakopoulos

Abstract This chapter presents a learn by demonstration approach, for closed-loop, robust, anthropomorphic grasp planning. In this respect, human demonstrations are used to perform skill transfer between the human and the robot artifacts, mapping human to robot motion with functional anthropomorphism [1]. In this work we extend the synergistic description adopted in Chaps. 2–6 for human grasping, in Chap. 8 for robotic hand design and, finally, in Chap. 15 for hand pose reconstruction systems, to define a low-dimensional manifold where the extracted anthropomorphic robot arm hand system kinematics are projected and appropriate Navigation Function (NF) models are trained. The training of the NF models is performed in a task-specific manner, for various: (1) subspaces, (2) objects and (3) tasks to be executed with the corresponding object. A vision system based on RGB-D cameras (Kinect, Microsoft) provides online feedback, performing object detection, object pose estimation and triggering the appropriate NF models. The NF models formulate a closed-loop velocity control scheme, that ensures humanlikeness of robot motion and guarantees convergence to the desired goals. The aforementioned scheme is also supplemented with a grasping control methodology, that derives task-specific, force closure grasps, utilizing tactile sensing. This methodology takes into consideration the mechanical and geometric limitations imposed by the robot hand design and enables stable grasps of a plethora of everyday life objects, under a wide range of uncertainties. The efficiency of the proposed methods is verified through extensive experimental paradigms, with the Mitsubishi PA10 – DLR/HIT II 22 DoF robot arm hand system.

M.V. Liarokapis (✉)
Mason Laboratory, Yale University, 9 Hillhouse Avenue, New Haven, CT 06520, USA
e-mail: minas.liarokapis@yale.edu

C.P. Bechlioulis · K.J. Kyriakopoulos
National Technical University of Athens, 9 Heroon Polytechniou, Zografou 15780, Greece
e-mail: chmpechl@mail.ntua.gr

K.J. Kyriakopoulos
e-mail: kkyria@mail.ntua.gr

G.I. Boutselis
Georgia Institute of Technology, Atlanta, USA
e-mail: gbouts@gatech.edu

M. Bianchi and A. Moscatelli (eds.), *Human and Robot Hands*,
Springer Series on Touch and Haptic Systems, DOI 10.1007/978-3-319-26706-7_9

9.1 Introduction

As stated by its title, the main challenge of this book is to bridge the gap between neuroscience and robotics with the twofold goal of (i) advancing the design of artificial systems and (ii) increase comprehension of biological systems. Considering point (i), one of the most challenging topics is the analysis and exploitation of anthropomorphism of robot motion, to improve the effectiveness of Human-Robot Interaction (HRI) applications. In Chap. 8, the synergistic description of the kinematics of human hand grasp was exploited to devise design guidelines for the anthropomorphic robotic Pisa/IIT SoftHand. In this chapter, such an analysis is extended to the arm-hand system and considerations on the concept of anthropomorphism are reported.

Over the last decades, we experienced an increasing demand for Human Robot Interaction (HRI) applications that require anthropomorphism of robot motion, as also discussed in Chap. 8. Anthropomorphism is derived from the greek word *anthropos* that means human and the greek word *morphe* that means form. More than 140 years ago Charles Darwin suggested anthropomorphism as a necessary tool for understanding efficiently nonhuman agents [2]. The essence of anthropomorphism as described in [3], *is to imbue the imagined or real behavior of nonhuman agents with humanlike characteristics, motivations, intentions and emotions*.

Anthropomorphism is usefull for two main reasons: (1) it guarantees safety in HRI applications (see also Chap. 10) and (2) it increases robot likeability, helping robots establish social "connections" with humans. Regarding safety in HRI, when humans and robots cooperate advantageously for the execution of certain tasks, anthropomorphic robot motion can easily be predicted by humans, to comply their activity/motion and avoid injuries. Regarding robot likeability, the more human-like a robot is in terms of appearance, motion, expressions and perceived intelligence, the more likely it is to establish a social connection with humans. More details regarding the social implications of anthropomorphism, can be found in [4–6].

In [7], the authors discriminated functional and structural anthropomorphism, for the development of technical devices that assist disabled people. A functional way of developing such a device is to provide a human function independently of the structural form, while the structural way is to accurately imitate some part of the human body. Recently, we proposed a distinction between the different notions of anthropomorphism [1] and introduced functional anthropomorphism, for mapping human to robot motion. *Functional Anthropomorphism* has as priority to guarantee the execution of a specific functionality in task-space and then—having accomplished such a prerequisite—to optimize anthropomorphism of robot motion. Functional anthropomorphism can be used to transform human trajectories to humanlike robot trajectories, that have similar profiles in task space but different trajectories in joint space, executing with accuracy the same tasks.

The field of Learn by Demonstration (LbD) or Robot Programming by Demonstration (PbD) has also received increased attention over the last 30 years. Learn by demonstration *moves from purely preprogrammed robots to very flexible user-based interfaces* according to Billard et al. [8] and it is a well known approach, that has been used for various HRI applications. The concept of LbD is based on a very simple idea: that an appropriate robot controller can be derived from human observations. Thus, the ultimate scope of LbD is to formulate robot control methodologies that can be easily adapted to new environments, generalize for new tasks and perform efficiently without extra programming by the users. Some characteristic studies are those proposed by Dillman et al. in [9–12], as well as those proposed by Schaal et al. in [13–15]. All of these studies focus on learning and generalization of motor skills by robotic artifacts, utilizing human demonstrations.

Nowadays, the majority of HRI applications involve some kind of interaction with the environment. Grasping and manipulation of everyday life objects are the most common interactions and some of the most challenging areas in robotics. Thus, roboticists seek always inspiration for facilitating these interactions, at the nature's most versatile and dexterous end-effector, the human hand. When humans grasp objects through the concept of synergies (see e.g., Chaps. 2–4 and 8), they tend to adapt their hand posture according to: (1) the object to be grasped and (2) the task to be executed with the grasped object. In [16], it was shown that humans adopt postures that maximize the force and velocity transmission ratios, along the directions imposed by the desired task. In [17], authors searched for optimal grasps using the branch-and-bound method, on a required external set. In [18] Teichmann et al. minimized the number of contact points required, to balance any external force and moment. Chiu [19] proposed an index, that measures the compatibility of a manipulator to perform a given task, Li and Sastry [20] introduced the concept of the task ellipsoid, while Mavrogiannis et al. [21] proposed a task specific grasp selection scheme for underactuated robotic hands.

Although the aforementioned studies have advanced the field of robot grasping, most of the analytical approaches, make two unrealistic assumptions: (1) that the robot hand fingers are able to reach accurately the desired contact points, (2) that the object geometry and physical parameters are known. However, these assumptions are often not verified, due to control errors, low encoder resolution and backlash that lead to contact points deviation. Furthermore, when robots interact with a dynamic environment, object properties (e.g., material, roughness, shape of the object) may be roughly estimated by the state of the art of sensing systems. In [22], authors showed that the existence of such uncertainties can "violate" the force closure property of the grasp. Thus, such uncertainties should be taken into consideration during the contact points and contact forces selection procedures. An alternative approach is to use adaptive synergies in order to achieve successful grasps, for a wide variety of objects (see Chaps. 8, 12 and 13 for technical and theoretical details).

In [23–25] the concept of Independent Contact Regions (ICR) was introduced to compensate for such uncertainties. The ICR methodology guarantees that if each contact point is located inside the corresponding regions, then the force closure property is preserved. Recently [26, 27], we synthesized a complete human-inspired optimization framework, for deriving stable, robust grasps under different task specifications and under a wide range of uncertainties. In these works, we utilized the concept of Q distance—originally proposed by Zhu et al. [28]—in a novel fashion, to incorporate the task specificity in the grasp selection algorithm.

In this chapter we present a learn by demonstration approach for closed-loop, robust, anthropomoprhic grasp planning. For doing so, we use human data in a "Learn by Demonstration" manner to perform skill transfer, between the human and the robot arm hand system. A human to robot motion mapping scheme transforms human trajectories to anthropomorphic robot trajectories, using a criterion of functional anthropomorphism [1]. The generated anthropomorphic robot trajectories are projected in low dimensional manifolds exploiting a synergistic reduction of human arm-hand kinematics, to train appropriate Navigation Function models [29], leveraging on a synergistic organization of the human upper limb as observed in [30]. These models formulate a closed-loop control scheme that embeds anthropomorphism and guarantees convergence to the desired goals. Regarding generalization, the NF models are trained in a task-specific fashion, using three different task features as described in [31, 32]: (1) the subspace to move towards, (2) the object to be grasped and (3) the task to be executed with the grasped object. The final scheme is able to produce adaptive behavior similar to humans by switching to different grasping primitives based on online feedback from a vision system. The vision system used, employs a RGB-D (Kinect, Microsoft) camera in order to perform object recognition and object pose estimation. Finally, a set of human-inspired optimization principles are incorporated in the proposed scheme, in order to facilitate the execution of robust, task-specific grasps under a wide range of uncertainties, utilizing tactile sensing.

The effectiveness of the proposed methods is experimentally verified using the 15 DoF DLR/HIT II five fingered robot hand attached at the 7 DoF Mitsubishi PA10 robot manipulator. The 4256e Grip System (Tekscan) tactile sensor setup, was used in order to: (1) measure the forces exerted by the robot fingertips, (2) minimize the level of uncertainty on the contact points, (3) facilitate the computation of sufficient contact forces. The proposed approach can be used by a robot arm hand system like the 22 DoF Mitsubishi PA 10 DLR/HIT II, to reach and grasp anthropomorphically a wide range of everyday life objects.

The rest of the chapter is organized as follows: Sect. 9.1 describes the apparatus and the kinematic models, Sect. 9.2 presents a "Learn by Demonstration" approach for closed loop, anthropomorphic grasp planning based on Navigation Function models, Sect. 9.3 presents the optimization schemes formulated to achieve task-specific, robust grasps under a wide range of uncertainties, Sect. 9.4 validates the efficiency of the proposed methods through extensive simulated and experimental paradigms, while Sect. 9.5 concludes the chapter and discusses the results.

9.2 Apparatus and Kinematic Models

9.2.1 Mitsubishi PA 10 DLR/HIT II Robot Arm Hand System

The robot arm hand system used in this work, consists of a Mitsubishi PA10 7 DoF robot manipulator and a DLR/HIT II five fingered 15 DoF robot hand (Fig. 9.1).

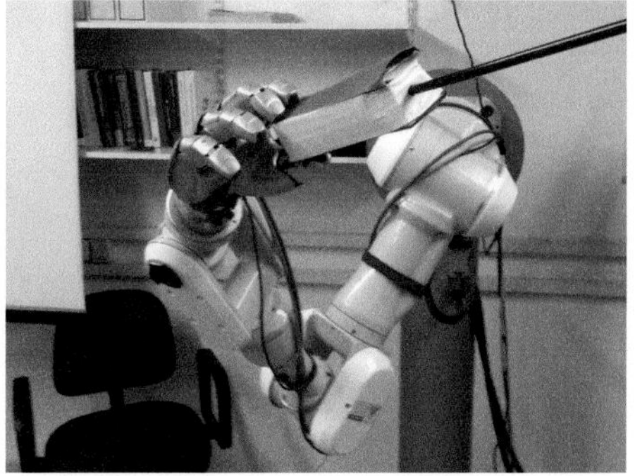

Fig. 9.1 The Mitsubishi PA10 DLR/HIT II robot arm hand system

The Mitsubishi PA 10 is a redundant robotic manipulator, which has seven rotational DoF arranged in an anthropomorphic manner: two DoF at the shoulder, two DoF at the elbow, and three DoF at the wrist. The robot servo controller communicates via the ARCNET protocol with a dedicated PC running soft real-time linux (Gentoo). The Planner PC establishes a TCP-based communication with the robot controller PC, allowing for position, velocity and torque control modes. More details regarding the Mitsubishi PA10, can be found in [33].

The DLR/HIT II is a five fingered dexterous robotic hand, with 15 DoF which was jointly developed by DLR (German Aerospace Center) and HIT (Harbin Institute of Technology). DLR/HIT II has five kinematically identical fingers with three DoF per finger, two for flexion/extension and one for abduction/adduction. The last joint of each finger (Distal Inter-Phalangeal – DIP joint analogous), is coupled with the middle one (Proximal Inter-Phalangeal – PIP joint analogous), using a mechanical coupling based on a steel wire with transmition ratio 1:1. The dimensions of the robot hand are considered to be human-like and the total weight is 1.6 kg. More details regarding the kinematics and the control of the DLR/HIT II can be found in [34].

9.2.2 *Tactile Sensors*

In order to capture the forces exerted by the robot fingertips we use the 4256e Grip System® (Tekscan), which is depicted in Fig. 9.2. The Grip System is an ultra thin (0.15 mm) tactile sensor that consists of 320 sensing elements (sensels) and which is able to measure the pressure magnitude of each sensel, using piezo-resistive technology. The Grip System® tactile arrays are mounted on the robot fingertips using appropriate rubber tape. The Planner PC establishes a TCP communication with a PC (Windows OS) that collects the forces from the tekscan system, at the frequency of 100 Hz.

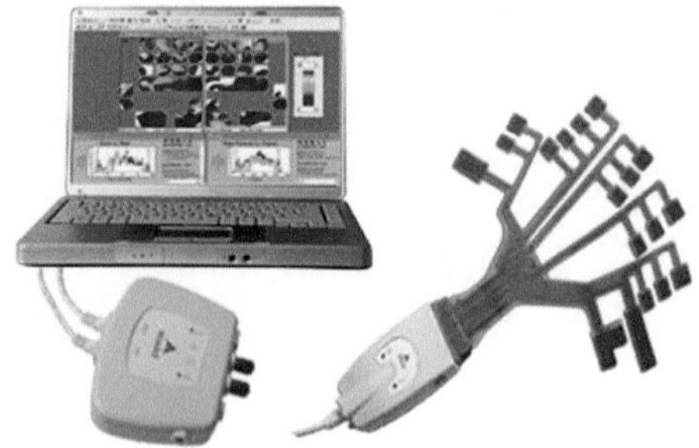

Fig. 9.2 The Grip System® tactile sensors (Tekscan)

9.2.3 Motion Capture Systems

In order to record the motion of the human arm hand system, we use a magnetic position tracking system and a dataglove. The magnetic position tracking system is the Liberty® (Polhemus Inc.) which is equipped with four position tracking sensors and a reference system. In order to capture human arm kinematics, three sensors are placed on: (1) the shoulder, (2) the elbow, (3) the wrist. More details regarding the computation of the kinematics, are presented in [35]. In order to measure the rest 22 DoF of the human hand and the wrist, we use the Cyberglove II® (Cyberglove Systems). The Cyberglove II has 22 flex sensors, capturing all twenty DoF of the human hand and the two DoF of the human wrist. More specifically, the flexion/extension of all three joints of each finger, the abduction between the fingers, as well as the abduction/adduction and flexion/extension of the wrist, can be measured. The motion capture systems are depicted in Fig. 9.3.

The position measurements are provided by the Liberty system at the frequency of 240 Hz. The Liberty system provides high accuracy in both position and orientation, with 0.03 in. and 0.15° respectively. The acquisition frequency of the Cyberglove II dataglove is 90 Hz and the nominal accuracy is less than 1°.

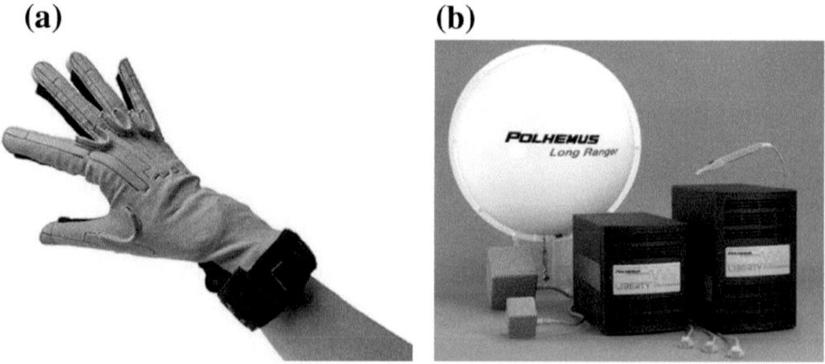

Fig. 9.3 Motion capture systems used to capture human kinematics. **a** Cyberglove (Cyberglove Systems). **b** Liberty (Polhemus)

9.2.4 *Kinematic Model of the Human Arm Hand System*

In order to describe the motion of the human upper limb in 3D space, we use three rotational DoF to model the shoulder joint, one rotational DoF for the elbow joint, one rotational DoF for pronation/supination, two rotational DoF for the wrist and finally twenty rotational DoF for the human fingers. For index, middle, ring and pinky fingers, we use three DoF for flexion/extension and one DoF for abduction/adduction, while for the thumb we use two DoF for flexion/extension, one DoF for abduction/adduction and one DoF to model the opposition to other fingers. The proposed methodology can be used with a more sophisticated human hand model, like the one proposed [36], in case there is a motion capture system available, that can measure all DoF variations of such a complex model.

9.3 Learn by Demonstration for Closed Loop, Anthropomorphic Grasp Planning

In this section we present a learn by demonstration approach for closed-loop, anthropomorphic grasp planning.

9.3.1 *Learn by Demonstration Experiments*

Experiments were performed by five (4 male, 1 female) healthy subjects 22, 25, 28, 29 and 41 years old. Subjects gave informed consent of the experimental procedure and the experiments were approved by the Institutional Review Board of the National Technical University of Athens. All subjects, were instructed to perform multiple reach to grasp movements towards different positions and objects in 3D space. Each subject performed all trials, with the dominant upper limb (the right arm hand system for all subjects). The experiments were performed for 22 positions in 3D space, marked on 5 different shelves. Four different objects were used for the experiments: a marker, a rectangular box, a small ball and a bottle. Different grasps were executed per object (e.g., front, side and top grasps) as described in [31]. For each object and object position combination 10 reach and grasp movements were executed. Adequate resting periods were used between the trials in order for the subjects to avoid fatigue. An image presenting the bookcase used, as well as the positions marked on different shelves of a bookcase, appears in Fig. 9.4.

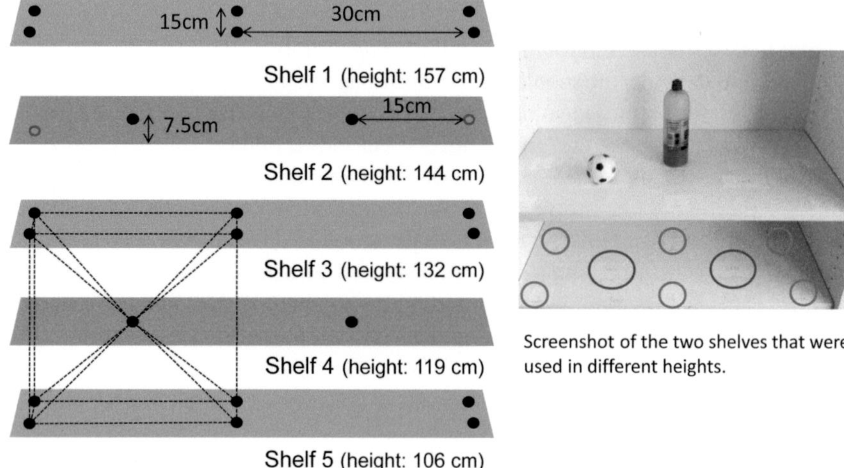

Fig. 9.4 Image depicting the object positions marked on different shelves of a bookcase

9.3.2 Mapping Human to Robot Motion with Functional Anthropomorphism

A human to robot motion mapping procedure, is used to map human kinematics to anthropomorphic robot kinematics. Various human to robot motion mapping procedures have been proposed in the past, to guarantee anthropomorphism using specific metrics of functional anthropomorphism [37], as it is also discussed in Chap. 12. In this chapter we formulate the mapping as a non-linear constrained optimization problem for the whole arm hand system, considering as end-effectors the robot fingertips.

More specifically, let $\mathbf{x}_{RAH} = f_{RAH}(\mathbf{q}_{RAH})$ denote the forward kinematics mapping from joint to task space for each robot arm hand system's finger, let m be the number of the fingers, $\mathbf{x}_{RAH}, \mathbf{x}_{RAHgoal} \in R^3$ the current and desired fingertip positions and $\mathbf{h}_c = (a_c, b_c, c_c, d_c), \mathbf{h}_g = (a_d, b_d, c_d, d_d) \in R^4$ the current and desired fingertip orientations (expressed using quaternions, to avoid singularities). Then the distance in $\mathbb{S}^3$, between human and robot fingertip orientations is defined as:

$$\bar{d}_{RAHo}(\mathbf{h}_c, \mathbf{h}_d) = cos^{-1}(a_c a_d + b_c b_d + c_c c_d + d_c d_d) \tag{9.1}$$

Taking into account the antipodal points [38], we formulate the following distance metric:

$$d_{RAHo}(\mathbf{h}_c, \mathbf{h}_d) = min\{\bar{d}_{RAHo}(\mathbf{h}_c, \mathbf{h}_d), \bar{d}_{RAHo}(\mathbf{h}_c, -\mathbf{h}_d)\}. \tag{9.2}$$

Thus we can define the following objective function under both position and orientation goals:

$$F^{xo}_{RAH}(\mathbf{q}_{RAH}) = w_{RAHx} \sum_{i=1}^{m} \left\| x_{RAH_i} - x_{RAHgoal_i} \right\|^2 + w_{RAHo} \sum_{i=1}^{m} d_{RAHo_i}(\mathbf{h}_{c_i}, \mathbf{h}_{d_i}) \quad (9.3)$$

where w_{RAHx} and w_{RAHo} are the weights that adjust the relative importance of the translation and rotation goal for each finger. These weights can be set according to the specifications of each study.

Moreover, in order to generate anthropomorphic robot motion, we incorporate in the objective function a criterion of functional anthropomorphism. Let $\mathbf{s}_{elbow} \in R^3$ denote the position of human elbow and $\mathbf{s}_j$ the vector of the robot joint positions in 3D space. For n points $s_1, s_2, \ldots, s_n$, the sum of distances between the human elbow and the robot joints positions (excluding "shoulder" and the end-effector), is given by:

$$D = \sum_{j=1}^{n} \left\| \mathbf{s}_{elbow} - \mathbf{s}_j \right\|^2 \quad (9.4)$$

The objective function F_{RAH} for the whole arm hand system, can be defined under position, orientation and anthropomorphism goals, as follows:

$$\begin{aligned} F_{RAH}(\mathbf{q}_{RAH}) = {} & w_{RAHx} \sum_{i=1}^{m} \left\| x_{RAH_i} - x_{RAHgoal_i} \right\|^2 \\ & + w_{RAHo} \sum_{i=1}^{m} d_{RAHo_i}(\mathbf{h}_{c_i}, \mathbf{h}_{g_i}) + w_D D \end{aligned} \quad (9.5)$$

where w_{RAHx} and w_{RAHo} are weights that adjust the relative importance of the translation and rotation goals (for each finger) and w_D denotes the weight that adjusts the importance of the anthropomorphism criterion. The aforementioned weights can be selected according to the specifications of each study.

Thus, the problem of mapping human to robot motion with functional anthropomorphism for the case of arm hand systems, can be formulated as:

$$\min F_{RAH}(\mathbf{q}_{RAH}) \quad (9.6)$$

s.t.

$$\mathbf{q}^-_{RAH} < \mathbf{q}_{RAH} < \mathbf{q}^+_{RAH} \quad (9.7)$$

where $\mathbf{q}_{RAH} \in R^n$ is the vector of the joint angles and $\mathbf{q}^-_{RAH}$, $\mathbf{q}^+_{RAH}$ are the lower and upper limits of the joints, respectively. More details, regarding the proposed mapping scheme, can be found in [37].

9.3.3 Learning Navigation Function Models in the Anthropomorphic Robot Low-D Space

In this work, we choose to control a robot arm hand system in a closed-loop fashion, in order to reach and grasp anthropomorphically a series of everyday life objects. In this respect, we propose Navigation Function (NF) based controllers that use "fictitious" obstacle functions (Fig. 9.5). The "fictitious" obstacles are learned in the low dimensional space of the anthropomorphic robot kinematics and apply repulsive effects on the robot arm hand system, so as to reach anthropomorphic configurations. A scheme based on NF models, is able to produce humanlike robot motion guaranteeing at the same time convergence to the desired goal [29]. Navigation Functions (NF) were first proposed by Rimon and Koditschek [39, 40]. Some characteristics of the NF based models are the following: (1) they provide closed-loop motion planning, (2) guarantee convergence to the desired goals, (3) have highly nonlinear learning capability, (4) provide continuous and smooth trajectories, (5) embed anthropomorphism (synthesizing appropriate, "fictitious" obstacle functions), (6) can generalize to similar, neighboring configurations (goal positions).

The initial formulation of the NF is for a priori known sphere worlds, however, application to geometrically more complicated worlds is achieved using diffeomorphisms which map the actual obstacles to spheres. In this work, B-splines are used to learn the structure of the NF obstacle functions. More precisely, given a desired configuration q_d for the robot arm or hand, the control law is constructed as follows:

$$u(t) = -K_p \left(\nabla_q \phi\right)(x_t) \tag{9.8}$$

where ϕ is the navigation function responsible for: (1) driving the arm or hand to its final configuration and (2) generating new anthropomorphic robot trajectories, similar to those used for training. $K_p > 0$ is a constant gain matrix and x is the system's state. The navigation function is given from the following relationship:

$$\phi = \frac{\gamma_d(x)}{\left(\gamma_d^k(x) + \beta\right)^{\frac{1}{k}}} \tag{9.9}$$

where x is the configuration, $\gamma_d(x) = \|x - x_d\|^2$ is the paraboloid attractive effect, $\beta = \prod_{i \in I_0} \beta_i$ is the aggregated obstacle repulsive effects and $k \in N \setminus \{0, 1\}$ is a tuning parameter. More precisely, for the training of the NF models we use the anthropomorphic robot motion that we derived from the human to robot motion mapping scheme. These data are represented in a lower dimensional manifold using the Principal Components Analysis (PCA),[1] as a standard dimensionality reduction technique. Such a technique is commonly used to cope with the redundancy of human hand architecture and to describe its synergistic organization, as discussed in

[1]The first 3 principal components extracted using the PCA method, describe for both the arm and the hand case, more than 88 % of the total variance.

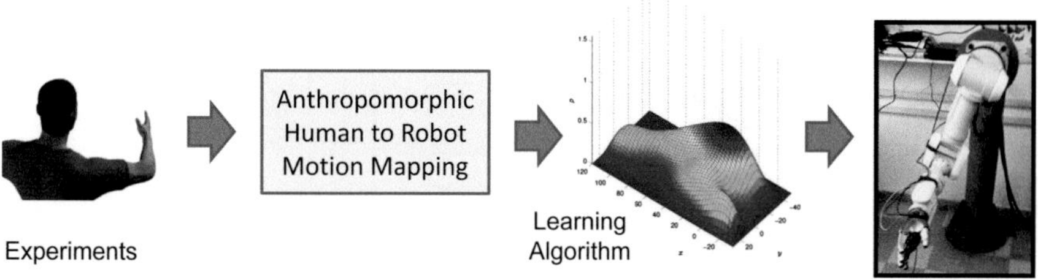

Fig. 9.5 The training procedure for the NF based models

Chaps. 2, 3, 5 for the analysis of human hand control, and in Chaps. 8 and 15 to design under-actuated robotic systems and under-sensed human hand pose reconstruction devices, respectively. However, synergistic coordination is not only limited to the hand: coordinated synergistic movements can be observed for the whole arm-hand system [30] and dimensionality reduction techniques can also be profitably employed in this case.

The output of the NF models, is then back-projected in the high dimensional space in order to control the robot arm hand system. In this case, no online human to robot motion mapping is required and computational effort diminishes. Moreover, we manage to guarantee anthropomorphism as well as to transfer skills from humans to the robot arm hand system, using a learn by demonstration approach.

In this work, NF models are trained in a task-specific way. For doing so, we use the approach described in [31, 32], discriminating the following task features: (1) subspace to move towards, (2) object to be grasped and (3) task to be executed with the grasped object. Thus, different NF models are trained offline for different tasks and then "stored". Furthermore, different NF models are trained for the robot arm and the robot hand. All models require as input the "goal" position in the low-d space of the anthropomorphic robot kinematics. The goal position can be provided by a vision system. The final scheme is able to produce adaptive robot behavior—similar to humans—by switching to different grasping primitives, based on online feedback (from the vision system).

9.3.4 A Vision System Based on RGB-D Cameras

In order for the proposed NF based methodology to be able to update the "goal" position of the task to be executed (based on online feedback), we have developed a vision system based on RGB-Depth cameras (Kinect, Microsoft). The developed vision system performs: (1) object recognition and (2) object pose estimation. A block diagram of the NF based scheme with the vision system incorporated, is presented in Fig. 9.6. For the development of the different vision modules, the Point Cloud Library has been used [41].

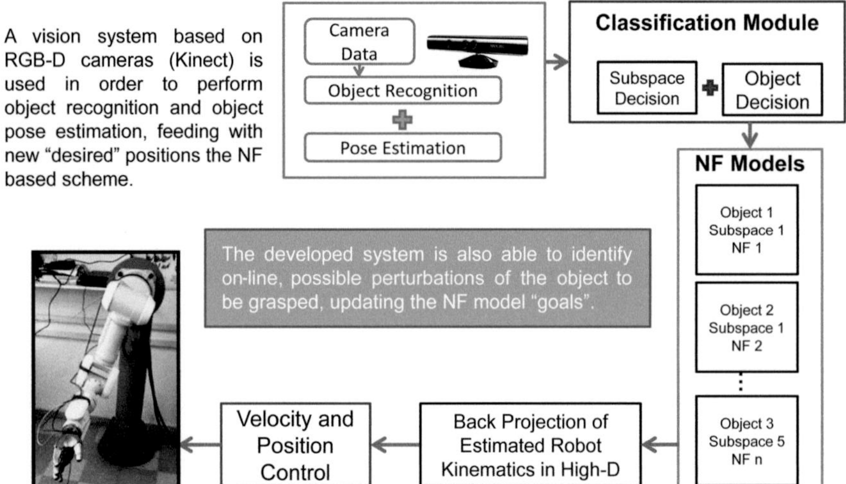

Fig. 9.6 Block diagram of the NF based scheme, with the vision system included

9.4 Task Specific, Robust Grasping with Tactile Sensing

In this section we present a set of optimization schemes capable of deriving task-specific, robust grasps under a wide range of uncertainties, utilizing tactile sensing.

9.4.1 A Scheme for Deriving Task Specific Grasping Postures

In order to derive task-specific, robust grasps, we propose a grasp selection algorithm based on the concept of Q distance, originally proposed for curved objects by Zhu et al. [28]. In this work, instead of just guaranteeing the force closure property as presented in [28], we obtain configurations that compensate disturbances in particular task-specific directions, exerting low forces. More details, regarding the utilization of the Q distance metric for deriving grasps in a task-specific manner, can be found in [27].

To formulate an optimization problem that derives task-specific grasping postures, we minimize the Q distance metric, using the joint displacements ($q \in \mathbb{R}^{n_q}$) and the wrist position/orientation ($w \in \mathbb{R}^6$) as decision variables. The optimization problem becomes:

$$\min \; d_Q(\mathbf{0}, \mathbf{co}(W)) \tag{9.10}$$

s.t.

$$q_{min} \leq q \leq q_{max} \tag{9.11}$$

$$fkine(q) \in \partial O \tag{9.12}$$

$$q^{j}_{abd/add} \leq q^{j+1}_{abd/add} \tag{9.13}$$

$$p' \notin O \tag{9.14}$$

where $\mathbf{0}$ is the origin of the wrench space, $\mathbf{co}(W)$ denotes the convex hull of the primitive wrenches and $d_Q(\mathbf{0}, \mathbf{co}(W))$ the Q distance metric. Equation (9.11) sets the inequality constraints of the joint limits (q_{min}, q_{max}), Eq. (9.12) ensures that the fingertips will be in contact with the object surface (∂O) employing appropriate equality constraints, Eq. (9.13) prevents collisions between the abduction/adduction DoF of the different robot fingers $(q^{i}_{abd/add})$ and Eq. (9.14) prevents collisions between the robot hand and the object, ensuring that no point (belonging to a set p' of finite discrete points) lying on the robotic hand will penetrate the object. For the rest of this chapter, we will refer to these constraints with the abbreviation *RHC* (*Robot Hand Constraints*).

9.4.2 A Scheme that Provides Optimal Force Transmission and Robustness Against Positioning Inaccuracies

In the previous section, we discussed the task specifications and the kinematic constraints that need to be satisfied to derive a task-specific grasp, but in most cases robot hands are also subjected to joint torque constraints. In this work we employ, the force transmission ratio r_k and compatibility index c introduced in [19], in order to derive robot hand configurations that exert the required grasping forces with minimal joint torque effort. The transmission ratio and the grasp compatibility index are defined, as follows:

$$r_k = [u_k^T (J_i J_i^T) u_k]^{-1/2}$$

$$c_i = \sum_{k=1}^{l} r_k^2 = \sum_{k=1}^{l} [u_k^T (J_i J_i^T)_k^u]^{-1}$$

where $u_k,\ k = 1, \ldots, l$, denote the desired directions for the contact forces and J_i the Jacobian of the ith finger. In this work we use frictional hard contacts, restricting each force to lie inside the corresponding friction cone. Thus, for each contact point the unit vectors u_k can be chosen to be aligned with the edges of the linearized friction cone [42]. The compatibility index for the case of the robot hand, becomes:

$$c = \sum_{i=1}^{n_p} w_{f_i} c_i = \sum_{i=1}^{n_p} w_{f_i} \sum_{k=1}^{n_g} [u_k^T (J_i J_i^T) u_k]^{-1}$$

where w_{f_i} are weighting factors for each finger. The maximization of the compatibility index c yields an optimal posture with respect to the force transmission metric, however a deviation between the actual and desired joint positions may be inevitable. Thus, we have to guarantee that the robot hand will be able to perform the given task, despite the fingertip positioning inaccuracies. For doing so, we utilize the concept of independent contact regions (*ICR*), adopting the approach described in [25], to determine whether a point on the object boundary qualifies to be a member of the *ICR*. Finally, we formulate an optimization problem that provides optimal force transmission and robustness against positioning inaccuracies, as follows:

$$\min \frac{1}{c} \tag{9.15}$$

s.t.

$$RHC \tag{9.16}$$

$$d_Q^-(\mathbf{0}, \mathbf{co}(W)) < 0 \tag{9.17}$$

$$p_i \in ICR_i \tag{9.18}$$

The inequality $d_Q^-(\mathbf{0}, \mathbf{co}(W)) < 0$ of Eq. (9.17) ensures that the force closure property will hold. Equation (9.18) constraints the deviated contact points p_i to belong to their corresponding Independent Contact Regions. It must be noted that a task-specific grasp configuration results in larger ICRs [43], so the grasp configuration derived in previous sections is ideal to initiate this second optimization algorithm.

9.4.3 A Grasping Force Optimization Scheme Utilizing Tactile Sensing

In this section, we utilize tactile sensors that are appropriately attached on the robotic fingertips, in order to relax the magnitude of uncertainties regarding: (1) joint displacements and (2) contact points deviations. Then we use the derived/updated information regarding the contact points, in order to perform a grasping force optimization. Using this scheme we are able to generate a set of contact forces that balances external disturbances, preventing object deformations and requesting minimal joint torque effort. In this work, we use the 4256e Grip System (Tekscan) and the active region of each fingertip is covered with a 4×4 tactile array. The sensels' sensors allows us to compute the position of force/contact centroid, as follows:

$$x_{cof} = \frac{\sum_{i=0}^{3} x_i \sum_{j=0}^{3} p_{ij}}{\sum_{i=0}^{3}\sum_{j=0}^{3} p_{ij}}, \; y_{cof} = \frac{\sum_{j=0}^{3} y_j \sum_{i=0}^{3} p_{ij}}{\sum_{j=0}^{3}\sum_{i=0}^{3} p_{ij}} \tag{9.19}$$

where p_{ij} is the normal force value at each sensel, x_i is the x coordinate of ith column and y_j the y coordinate of jth row of the 4×4 array. In order to map the centroid coordinates (x_{cof}, y_{cof}) to 3-D coordinates on the robot fingertip surface, we exploit the point cloud of the robotic fingertip. All robot fingers have the same fingertip, so the following procedure applies for all fingers. Initially we match the 4 corner sensels of the tactile array with their actual position p_i^{corn}, $i = 1, \ldots, 4$, on the point cloud and we compute the distance between them and all other nodes of the point cloud. Then, assuming that the sensor firmly covers the fingertips surface and given a contact centroid (x_{cof}, y_{cof}), we determine the corresponding coordinates P(X, Y, Z) on the robot fingertip point cloud, minimizing the following function:

$$\min\{\sum_{i=1}^{4}(dist_i(X, Y, Z) - arraydist_i(x_{cof}, y_{cof}))^2\} \tag{9.20}$$

where $dist_i(X, Y, Z)$ denotes the distance from p_i^{corn} to P(X, Y, Z) on the point cloud and $arraydist_i(x_{cof}, y_{cof})$ the distance between the ith corner sensel and the contact centroid on the tactile array. Such an approach assumes that the distance between two points remains invariant both in 2D and 3D coordinates (Fig. 9.7).

Following the grasping notation, we can denote the contact forces, as follows:

$$f_c = -G^+ w_{ext} + E\lambda, \tag{9.21}$$

where w_{ext} is the external disturbance, G^+ is the pseudoinverse of grasp matrix G (see also Chaps. 8, 12 and 13), E is a matrix whose columns form a basis for the nullspace of G and λ is an arbitrary vector. The first term of (9.21) is responsible for the compensation of the external wrench w_{ext} and $E\lambda$ denotes the set of internal forces [44]. The internal forces have a null resultant wrench to the object and are very significant for grasping, as they can control the robot hand's ability to squeeze

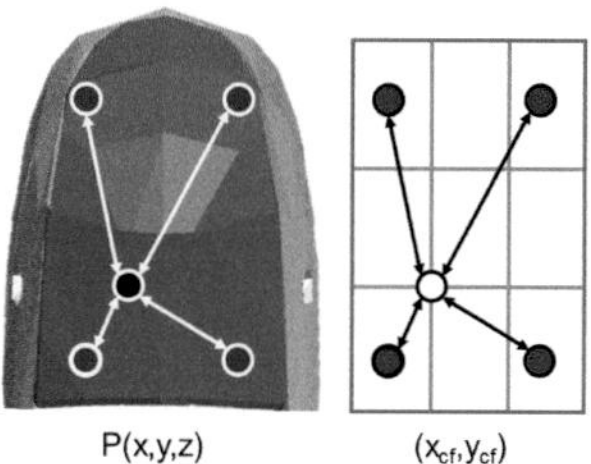

Fig. 9.7 Distances on the fingertip and the tactile array respectively

arbitrarily tight the object, ensuring stability. Thus, we formulate a linear optimization problem setting as decision variables the internal forces (vector λ), as follows:

$$\min \sum f_{n_i} \tag{9.22}$$

s.t.

$$-V_i'(f_{c_i} - n_i \|f_{c_{imax}}\|) \leq \mathbf{0} \tag{9.23}$$

$$|\tau_{i_k}| \leq |\tau_{i_{kmax}}| - |\delta \tau_{i_{kmax}}| \tag{9.24}$$

$$f_{n_i} \geq 0 \tag{9.25}$$

where f_n are the normal force components of the contact forces f_c. Equation (9.22) sets the friction cone constraints (represented by L-sided convex polyhedral cones to reduce computational complexity [45]), Eq. (9.23) sets the torque constraints and Eq. (9.24) constraints the contact forces values to be positive or zero. The presented algorithm searches for a set of internal forces that minimize the sum of the normal forces and therefore the grasp effort, satisfying simultaneously both the friction and torque constraints.

9.5 Results and Experimental Validation

In this section we present extensive experimental paradigms of the proposed methods, focusing on the two different scenarios: (1) reaching and grasping using closed loop, anthropomorphic grasp planning methodologies, (2) achieving robust, task-specific grasps under a wide range of uncertainties, utilizing bioinspired optimization principles and tactile sensing.

Using the aforementioned methods, we are now able to synthesize a complete scheme for closed-loop, robust, anthropomorphic grasp planning. The steps followed by the proposed scheme, are the following:

1. A vision system, performs object detection and object pose estimation.
2. The object shape and position information trigger a task-specific Navigation Function model (object-specific and subspace-specific).
3. The Navigation Function model produces anthropomorphic trajectories for the arm hand system to reach and grasp the identified object, ensuring convergence to the desired pose.
4. The robot fingers stop moving when the fingertips' tactile sensors detect contact with the object.
5. The actual joint positions and the contact centroids are obtained from the encoders and the tactile sensors respectively.
6. The contact centroids are mapped to the corresponding positions on the robot fingertips.

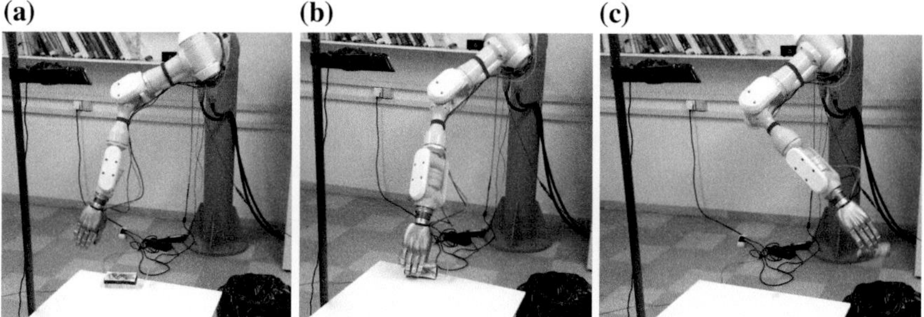

Fig. 9.8 Three instances of a reach to grasp motion, are depicted. **a** Reaching. **b** Grasping and lifting. **c** Performing the task

7. The positions of the contact points on the object are computed, solving the robot arm hand system's forward kinematics.
8. A grasping force optimization scheme is employed to compute a set of sufficient forces, for the robot hand to stably grasp the object.

9.5.1 Closed-Loop, Anthropomorphic Grasp Planning Scenario

All experiments were performed with the Mitsubishi PA10 DLR/HIT II robot arm hand system. A vision system based on RGB-D cameras (Kinect, Microsoft) was used to track everyday life objects, located in arbitrary positions and orientations in 3D space. In Fig. 9.8, the robot arm hand system is depicted while reaching and grasping anthropomorphically a rectangular object, in order to execute a specific task (e.g., to throw the object into the waste basket).

A video of the first experiment, can be found at the following url: https://www.youtube.com/watch?v=cazfjEKnsxo.

9.5.2 Task-Specific, Robust Grasping Scenario

For the task-specific robust grasping experiments we considered the stable grasp of a cylindrical object, which is filled with liquid. We showed that the proposed methodology derives robust grasps that hold the force closure property even when the object is rotated, facing disturbances caused by the center of mass changes. Four different poses of the object are depicted in Fig. 9.9. These poses are used to model the task disturbances. The rotation is implemented about the z axis and the liquid is hypothesized to be symetrically distributed about this specific axis. In the subfigures of Fig. 9.9, the black dots denote the center of mass for each pose, while the object

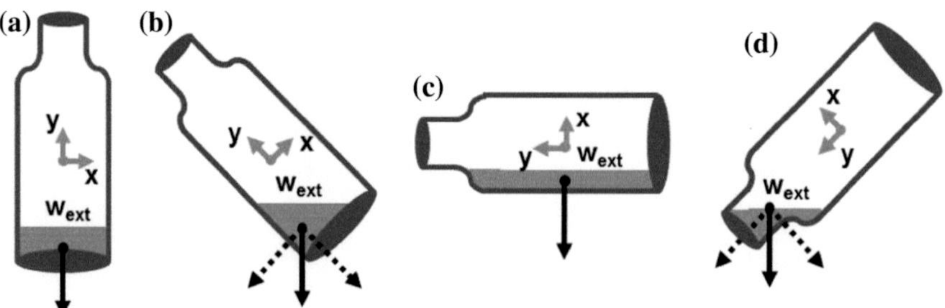

Fig. 9.9 Task description. **a** Pose I. **b** Pose II. **c** Pose III. **d** Pose IV

coordinate frame is determined by the blue axes. As it can be noticed, the object's weight causes external forces along both the x and y axis, as well as external moments about the z axis of the object coordinate frame.

The bottle used for the experiment, had a radius of 2.25 cm and a height of 13 cm. For the derived configuration we considered an 8-sided linearized friction cone. The friction coefficient selected was quite conservative with $\mu = 0.3$. The maximum contact point deviation (on the object) caused by the DLR/HIT II joint errors, was found to be 4 mm. For the *ICRs* computation, four deviated contact points were considered, at a distance of 4 mm from the nominal contact point. The friction coefficient and object model uncertainties, were considered prior to the *ICRs* computation, as discussed in [46]. Finally in order for the robot hand to perform the specified task, a set of internal forces were computed that satisfy both the friction and joint torque constraints, as presented in Sect. 9.4.3.

In Table. 9.1 we present the derived angles q and torques τ. For the computation of internal forces the uncertainty of the contact points δp_{max} is 1 cm and the center of mass uncertainty is 3 cm. The robot hand dynamic model was considered to be the flexible joint model presented in [47] and was utilized in order to exert the derived forces. Thus, we have:

$$\tau = g(q) - \tau_{ext} = K(\theta - q) \tag{9.26}$$

where q denotes the link position vector, θ denotes the motor position vector in link coordinates and $g(q)$ the gravity term. Moreover, K is the stiffness matrix and τ_{ext} the external torque vector. After contact detection for a given q we may calculate the motor displacements required to exert the desired internal forces f_{ext_d}, as:

$$\theta = K^{-1}(g(q) - J_i^T f_{ext_d}) + q \tag{9.27}$$

The term $g(q)$ can be computed using the DH parameters and the nominal masses of the DLR/HIT II robot hand [47]. A video of the second experiment (see Fig. 9.10), can be found at the following URL:

https://www.youtube.com/watch?v=lkpSgamV0b8.

Table 9.1 Experimental data (q: degrees, τ: Nm, "a/a": abduction/adduction DoF, "f/e": flexion/extension DoF)

Thumb	q	τ	**Index**	q	τ	**Middle**	q	τ	**Ring**	q	τ	**Pinky**	q	τ
a/a	−12.4	0.12	a/a	−9.1	−0.07	a/a	−4.3	-0.01	a/a	0.3	0	a/a	5.2	0
f/e 1	8.9	0.34	f/e 1	22.8	0.13	f/e 1	12	0.07	f/e 1	10.8	0.09	f/e 1	12.5	0.03
f/e 2	10.9	0.17	f/e 2	13	0.07	f/e 2	35.2	0.04	f/ext 2	33.2	0.07	f/e 2	21	0.01

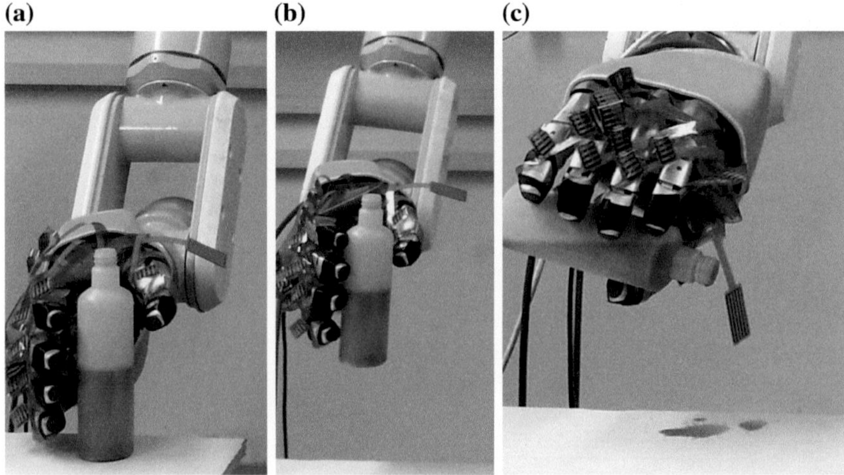

Fig. 9.10 Three snapshots of the reaching, grasping–lifting and task execution phases. **a** Reaching. **b** Grasping–Lifting. **c** Performing the task

9.6 Conclusions and Discussion

In this chapter we presented a complete scheme for closed-loop, robust, anthropomorphic grasp planning, following a learn by demonstration approach. For doing so, we captured multiple reach to grasp movements of the human arm hand system and we used them in order to generate anthropomorphic robot trajectories. Then, task-specific Navigation Function models were trained in the low-d space of the anthropomorphic robot kinematics, mimicing the synergistic organization of the human hand-arm system. The NF models were used to formulate a closed loop control scheme, that ensures humanlikeness of robot motion and guarantees convergence to the desired goals. Human-inspired optimization principles were proposed in order to derive task-specific, robust grasp configurations. Tactile sensors mounted on the robot fingertips were used to confront uncertainties regarding the joint displacements and the contact points positioning, relaxing also the computation of sufficient contact forces which result to stable grasps. A vision system based on RGB-D cameras was used to provide online feedback, perform object detection and object pose estimation and trigger appropriate task-specific NF models.

The proposed approach can be used by various robot arm hand systems, in order to reach and grasp anthropomorphically a plethora of everyday life objects, even under a wide range of uncertainties.

Acknowledgments This work has been partially supported by the European Commission with the Integrated Project no. 248587, THE Hand Embodied, within the FP7-ICT-2009-4-2-1 program Cognitive Systems and Robotics.

References

1. Liarokapis MV, Artemiadis PK, Kyriakopoulos KJ (2012) Functional anthropomorphism for human to robot motion mapping. In: 21st IEEE international symposium on robot and human interactive communication (RO-MAN), pp 31–36, Sept 2012
2. Darwin C The Expression of the Emotions in Man and Animals, Anniversary ed., P. Ekman, Ed. Harper Perennial, 1872/2009
3. Epley N, Waytz A, Cacioppo JT (2007) On seeing human: a three-factor theory of anthropomorphism. Psychol Rev 114:864–886
4. Bartneck C, Kulic D, Croft EA, Zoghbi S (2009) Measurement instruments for the anthropomorphism, animacy, likeability, perceived intelligence, and perceived safety of robots. Int J Soc Robot 1(1):71–81
5. Duffy BR (2002) Anthropomorphism and robotics. In: The society for the study of artificial intelligence and the simulation of behaviour–AISB
6. Riek LD, Rabinowitch TC, Chakrabarti B, Robinson P (2009) How anthropomorphism affects empathy toward robots. Proceedings of the 4th ACM/IEEE international conference on human robot interaction. ACM, New York, NY, USA, pp 245–246
7. Tondu B, Bardou N (2009) Anthropomorphism in robotics engineering for disabled people. World Academy of Science, Engineering and Technology
8. Billard A, Calinon S, Dillmann R, Schaal S (2008) Robot programming by demonstration. In: Siciliano B, Khatib O (eds) Handbook of robotics. Springer, Berlin Heidelberg, pp 1371–1394
9. Friedrich H, Münch S, Dillmann R, Bocionek S, Sassin M (1996) Robot programming by demonstration (RPD): supporting the induction by human interaction. Mach Learn 23(2–3):163–189. http://dx.doi.org/10.1007/BF00117443
10. Pardowitz M, Knoop S, Dillmann R, Zollner R (2007) Incremental learning of tasks from user demonstrations, past experiences, and vocal comments. IEEE Trans Syst Man Cybern Part B: Cybern 37(2):322–332
11. Jakel R, Schmidt-Rohr S, Xue Z, Losch M, Dillmann R (2010) Learning of probabilistic grasping strategies using programming by demonstration. In: IEEE international conference on robotics and automation (ICRA), pp 873–880
12. Jakel R, Schmidt-Rohr S, Losch M, Kasper A, Dillmann R (2010) Learning of generalized manipulation strategies in the context of programming by demonstration. In: IEEE-RAS international conference on humanoid robots (Humanoids), pp 542–547
13. Billard A, Schaal S (2001) Robust learning of arm trajectories through human demonstration. In: IEEE/RSJ international conference on intelligent robots and systems (IROS)
14. Ijspeort AJ, Nakanishi J, Schaal S (2002) Learning rhythmic movements by demonstration using nonlinear oscillators. In: IEEE/RSJ international conference on intelligent robots and systems (IROS), vol 1, pp 958–963
15. Pastor P, Hoffmann H, Asfour T, Schaal S (2009) Learning and generalization of motor skills by learning from demonstration. In: IEEE international conference on robotics and automation (ICRA)
16. Friedman J, Flash T (2007) Task-dependent selection of grasp kinematics and stiffness in human object manipulation. Cortex 43(3): 444–460. http://www.sciencedirect.com/science/article/pii/S0010945208704696
17. Watanabe T, Yoshikawa T (2007) Grasping optimization using a required external force set. IEEE Trans. Autom Sci Eng 4(1):52–66
18. Teichmann M, Mishra B (2000) Probabilistic algorithms for efficient grasping and fixturing. Algorithmica 26(3–4):345–363. http://dx.doi.org/10.1007/s004539910017
19. Chiu SL (1988) Task compatibility of manipulator postures. Int J Robot Res 7(5), 13–21. http://ijr.sagepub.com/content/7/5/13.abstract
20. Li Z, Sastry S (1988) Task-oriented optimal grasping by multifingered robot hands. IEEE J Robot Autom 4(1):32–44

21. Mavrogiannis C, Bechlioulis C, Liarokapis M, Kyriakopoulos K (2014) Task-specific grasp selection for underactuated hands. In: IEEE international conference on robotics and automation (ICRA), pp 3676–3681
22. Zheng Y, Qian W-H (2005) Coping with the grasping uncertainties in force-closure analysis. Int J Robot Res 24(4):311–327. http://ijr.sagepub.com/content/24/4/311.abstract
23. Ponce J, Sullivan S, Sudsang A, Boissonnat J-D, Merlet J-P. On computing four-finger equilibrium and force-closure grasps of polyhedral objects. Int J Robot Res 16(1):11–35. http://ijr.sagepub.com/content/16/1/11.abstract
24. Roa M, Suarez R (2009) Computation of independent contact regions for grasping 3-D objects. IEEE Trans Robot 25(4):839–850
25. Krug R, Dimitrov D, Charusta K, Iliev B (2010) On the efficient computation of independent contact regions for force closure grasps. In: IEEE/RSJ international conference on intelligent robots and systems (IROS)
26. Boutselis G, Bechlioulis C, Liarokapis M, Kyriakopoulos K (2014) An integrated approach towards robust grasping with tactile sensing. In: IEEE international conference on robotics and automation (ICRA), pp 3682–3687
27. Boutselis G, Bechlioulis C, Liarokapis M, Kyriakopoulos K (2014) Task specific robust grasping for multifingered robot hands. In: IEEE/RSJ international conference on intelligent robots and systems (IROS), pp 858–863
28. Zhu X, Wang J (2003) Synthesis of force-closure grasps on 3-D objects based on the Q distance. IEEE Trans Robot Autom 19(4):669–679
29. Filippidis I, Kyriakopoulos K, Artemiadis P (2012) Navigation functions learning from experiments: application to anthropomorphic grasping. In: IEEE international conference on robotics and automation (ICRA), pp 570–575
30. Mason CR, Gomez JE, Ebner TJ (2001) Hand synergies during reach-to-grasp. J Neurophysiol 86(6):2896–2910
31. Liarokapis MV, Artemiadis PK, Kyriakopoulos KJ (2013) Task discrimination from myoelectric activity: a learning scheme for EMG-based interfaces. In: IEEE international conference on rehabilitation robotics (ICORR)
32. Liarokapis M, Artemiadis P, Kyriakopoulos K, Manolakos E (2013) A learning scheme for reach to grasp movements: on EMG-based interfaces using task specific motion decoding models. IEEE J Biomed Health Inform 17(5):915–921
33. Bompos N, Artemiadis P, Oikonomopoulos A, Kyriakopoulos K (2007) Modeling, full identification and control of the mitsubishi PA-10 robot arm. In: IEEE/ASME international conference on advanced intelligent mechatronics (AIM)
34. Liu H, Wu K, Meusel P, Seitz N, Hirzinger G, Jin M, Liu Y, Fan S, Lan T, Chen Z (2008) Multisensory five-finger dexterous hand: the DLR/HIT Hand II. In: IEEE/RSJ international conference on intelligent robots and systems (IROS), pp 3692–3697
35. Artemiadis PK, Katsiaris PT, Kyriakopoulos KJ (2010) A biomimetic approach to inverse kinematics for a redundant robot arm. Autonom Robots 29(3–4):293–308
36. Liarokapis MV, Artemiadis PK, Kyriakopoulos KJ (2013) Quantifying anthropomorphism of robot hands. In: IEEE international conference on robotics and automation (ICRA)
37. Liarokapis MV, Artemiadis PK, Bechlioulis CP (2013) Kyriakopoulos KJ (2013) Directions, methods and metrics for mapping human to robot motion with functional anthropomorphism: a review". National Technical University of Athens, Technical Report, School of Mechanical Engineering Sept
38. LaValle SM (2006) Planning algorithms. Cambridge University Press, Cambridge. http://planning.cs.uiuc.edu/
39. Koditschek D, Rimon E (1990) Robot navigation functions on manifolds with boundary. Adv Appl Math 11(4):412–442
40. Rimon E, Koditschek D (1992) Exact robot navigation using artificial potential fields. IEEE Trans Robot Autom 8(5):501–518
41. Rusu RB, Cousins S (2011) 3D is here: Point Cloud Library (PCL). In: IEEE international conference on robotics and automation (ICRA)

42. Hester RD, Cetin M, Kapoor C, Tesar D (1999) A criteria-based approach to grasp synthesis. In: IEEE international conference on robotics and automation (ICRA), vol 2, pp 1255–1260
43. Kirkpatrick D, Mishra B, Yap C-K (1992) Quantitative steinitz's theorems with applications to multifingered grasping, vol 7, no 1. Springer, pp 295–318. http://dx.doi.org/10.1007/BF02187843
44. Prattichizzo D, Trinkle J (2008) Grasping. In: Siciliano B, Khatib O (eds) Handbook of robotics. Springer, Berlin Heidelberg, pp 671–700
45. Kerr J, Roth B (1986) Analysis of multifingered hands. Int J Robot Res 4(4):3–17. http://ijr.sagepub.com/content/4/4/3.abstract
46. Roa M, Suarez R (2011) Influence of contact types and uncertainties in the computation of independent contact regions. In: IEEE international conference on robotics and automation (ICRA), pp 3317–3323
47. Chen Z, Lii N, Wimboeck T, Fan S, Jin M, Borst C, Liu H (2010) Experimental study on impedance control for the five-finger dexterous robot hand DLR-HIT II. In: IEEE/RSJ international conference on intelligent robots and systems (IROS), pp 5867–5874

Chapter 10
Teleimpedance Control: Overview and Application

Arash Ajoudani, Sasha B. Godfrey, Nikos Tsagarakis and Antonio Bicchi

Abstract In previous chapters, human hand and arm kinematics have been analyzed through a synergstic approach and the underlying concepts were used to design robotic systems and devise simplified control algorithms. On the other hand, it is well-known that synergies can be studied also at a muscular level as a coordinated activation of multiple muscles acting as a single unit to generate different movements. As a result, muscular activations, quantified through Electromyography (EMG) signals can be then processed and used as direct inputs to external devices with a large number of DoFs. In this chapter, we present a minimalistic approach based on teleimpedance control, where EMGs from only one pair of antagonistic muscle pair are used to map the users postural and stiffness references to the synergy-driven anthropomorphic robotic hand, described in Chap. 7. In this direction, we first provide an overview of the teleimpedance control concept which forms the basis for the development of the hand controller. Eventually, experimental results evaluate the effectiveness of the teleimpedance control concept in execution of the tasks which require significant dynamics variation or are executed in remote environments with dynamic uncertainties.

A. Ajoudani (✉) · S.B. Godfrey · N. Tsagarakis · A. Bicchi
Department of Advanced Robotics, Istituto Italiano di Tecnologia, Via Morego 30, 16163 Genoa, Italy
e-mail: Arash.Ajoudani@iit.it

S.B. Godfrey
e-mail: Sasha.Godfrey@iit.it

N. Tsagarakis
e-mail: Nikos.Tsagarakis@iit.it

A. Bicchi
Research Center "E. Piaggio", Università di Pisa, Pisa, Italy
e-mail: bicchi@centropiaggio.unipi.it

M. Bianchi and A. Moscatelli (eds.), *Human and Robot Hands*,
Springer Series on Touch and Haptic Systems, DOI 10.1007/978-3-319-26706-7_10

10.1 Teleimpedance Control

The need to execute tasks in unstructured or hostile environments has lead to the development of several Master-Slave teleoperation interfaces, commonly recognized by two classes: unilateral, position-based and bilateral, force-reflecting teleoperation systems. The most basic teleoperation interface receives position commands from the master and replicates them on the slave side in an open-loop fashion. Despite the stability and simplicity of such systems, generation of high interaction forces between the rigid manipulator's end-effector and the uncertain environment has severely limited their application in real-world scenarios.

Later generations of teleoperation interfaces thus investigated new techniques to provide the master with information about the interaction forces between the manipulator and the remote environment. Although it has been demonstrated that force-reflecting teleoperation interfaces outperform the unilateral ones [26, 27], several drawbacks such as imposed additional costs (due to the need for force measurement devices), transparency, or even stability issues (due to the latencies in the communication channel between the master and slave robot) reduce the efficiency of such systems [18, 31, 41]. In fact, despite the continuous advancement in hardware design and software architecture of the bilateral teleoperation interfaces, still a large class of tasks (e.g. drilling and chipping) which are intuitively executed by humans cannot be effectively performed.

Indeed, humans are able to establish a reliable contact between the limb endpoint and the object/environment by generating task-efficient restoring forces in response to the environmental displacements [23, 24]. This behavior arises from effective modulation of the task impedance which appears to be carried out using different techniques. One way to achieve this is by the co-contraction of muscle groups acting on the limb. Alternatively, it is performed through adaptations in the sensitivity of reflex feedback [5] or selective control of the limb configuration [43].

To realize a similar interaction performance in a teleoperation setup, firstly, rigid slave robots must be replaced by compliant ones to enable task impedance modulations. In this direction, drawing inspiration from the compliant structure of the human limb, the soft robotics design (either by torque control techniques [7, 9] or using passively adaptive joints [10, 30, 42] has provided the possibility of teleoperating compliant slave robots to accomplish a task in an uncertain environment. However, the second, and probably bigger, issue relates to the planning of the variable impedance slave robot to accomplish the task.

To further address this issue, the concept of teleimpedance control [2], as an alternative approach to the force-reflecting teleoperation has been proposed. In teleimpedance, a compound reference command which includes the desired motion and impedance profiles of the operator, is obtained using a suitable Human-Machine Interface (HMI) and realized by a compliant slave robot in real-time. Therefore, from one side, the need for an appropriate and real-time modeling/measurement of the operator's stiffness and position trajectories is highlighted. On the other hand, robust and effective impedance control techniques must be implemented to replicate the operator's reference commands on the slave robot in real-time.

Concerning the modeling of the master's impedance trajectories, in our studies, electromyography (EMG) signals, which are formed by superimposed patterns of activations of involved motor units, are considered as the process input. This is due to the high correlations between the muscle activations, generated muscular forces and the consequent joint torques. Furthermore, easy accessibility, fast adaptivity and stability of EMG signals are other advantages which motivated our choice of adopting EMGs in the real-time control of our teleimpedance system. An alternative approach to teleimpedance for processing multiple EMG signals through machine learning as a direct input to external devices with a large numebr of DoFs is presented in the next chapter (Chap. 11).

EMGs can also be used to provide information on the limb posture, which has been used e.g. for classification of hand gestures [13, 20] or arm movements [44]. However, since EMGs directly relate to muscle forces and not limb configurations, their application to extract position references has to be indirect. A classical way to achieve this is to relate muscle forces to the limb postures using inverse dynamics methods [17]. The problem becomes even more intricate once the external forces (e.g. object mass) act on the master's limb which would affect the position estimation accuracy. Therefore, we tend to maximize the use of the external tracking systems to extract the position profile of the master, and use EMGs to estimate the task-appropriate stiffness profiles (the static component of an impedance profile) in real-time.

While on the slave side, relying on the task requirements and the slave robot's compliant structure, robust Cartesian or joint impedance control techniques can be implemented to realize the operator's reference commands. For instance, in [6], some techniques for the Cartesian impedance control of torque controlled robots are provided. Additionally, in [3, 8], the role of robot configuration in Cartesian stiffness control is discussed, particularly for robots with passive elastic joints or the ones in which not enough degrees of freedom are available to realize a full desired Cartesian stiffness matrix.

In this chapter, we review some of the work done within The Hand Embodied (THE) project regarding teleimpedance control of a robotic arm and an anthropomorphic robotic hand. In particular, in Sect. 10.2.1, a 3D model of the human arm endpoint force/stiffness will be introduced. Consequently, a Cartesian stiffness controller is developed to replicate the estimated stiffness profile and tracked wrist trajectories of the master by a 7-DoF torque controlled robot in real-time. Experimental results are provided to evaluate the efficiency of the proposed algorithm in rendering a desired interaction performance while performing dynamic tasks or the ones executed by the slave robot in a remote uncertain environment.

Meanwhile, in Sect. 10.2.2, teleimpedance control of the anthropomorphic and synergy-driven robotic hand, described in Sect. 6, is studied. In this setup, the hand postural and synergy reference commands (as defined in Chaps. 2, 3 and 5) are estimated using an antagonistic pair of muscles on the forearm. Two tactile interfaces, namely mechano- and vibrotactile, are developed to provide the user with some information about the grasping forces and the environment/object's texture. Grasp robustness and improved interaction performance using teleimpedance control are evaluated through grasping experiments.[1]

10.2 Application

10.2.1 Teleimpedance Control of a Robotic Arm

This section reviews some of the work done in [1, 2] aimed at remote impedance control of a 7-DoF torque controlled robot arm in real-time. Here, the compound reference command consists of master's wrist position and stiffness profiles. Corresponding to the high priority given in teleoperation interfaces to position accuracy, our teleimpedance interface uses accurate measurement of arm position references through an optical tracking system. Meanwhile, we acquire and process eight EMG channels to estimate the 3D arm endpoint stiffness[2] profile in real-time, as we will elaborate in the following section.

10.2.1.1 Arm Endpoint Impedance Modeling in 3D

It has been demonstrated that variations in viscoelastic components of the human arm endpoint strongly correlate with the patterns of activations of the involved muscles in task execution [19, 36, 40]. While this dependency appears to be highly nonlinear in general (due to the nonlinear nature of the EMG-to-Force mapping [32] and the joint-angle dependency of the moment arms), it can be safely and almost accurately implemented by a linear mapping in a constant configuration of the arm [19, 40]. To that end, the overall mapping between EMG measurements and consequent arm endpoint force and stiffness variations in Cartesian coordination in a constant joint configuration of the arm can be described by

$$\begin{bmatrix} F \\ \sigma \end{bmatrix} = \begin{bmatrix} T_F \\ T_\sigma \end{bmatrix} P + \begin{bmatrix} 0 \\ \sigma_0 \end{bmatrix}, \tag{10.1}$$

where $F, \sigma \in \mathbb{R}^3$ are the endpoint force and stiffness vectors, respectively, σ_0 is the intrinsic stiffness in relaxed conditions. $P \in \mathbb{R}^n$ is the vector of muscular activities of the n considered muscles, as obtained from preprocessing EMG signals which

[1]Teleimpedance control concept has also been used for assistive control of a knee exoskeleton device [28, 29]. The proposed controller captures the user's intent to generate task-related assistive torques by means of the exoskeleton while performing daily tasks.

[2]It is important to note here that this model only takes into account the effect of muscular co-contractions in endpoint stiffness modulations. As regards the role of arm configuration in endpoint stiffness geometry in teleimpedance control, readers may refer to [3].

includes high-pass filtering, full-rectification, low-pass filtering and normalization stages.

In an ideal condition, force (T_F) and stiffness (T_σ) mappings can be experimentally identified through a rich and varied set of data samples both for force and stiffness measurements. While the first measurement can be easily and accurately carried out using 6 axis force/torque sensors, accurate identification of the EMG-to-stiffness map T_σ is more difficult [36]. The reason for that lies in the difficulty of the endpoint stiffness measurements which is commonly and traditionally carried out by perturbing the wrist and probing the restoring forces [35].

To address this problem, we identify a basis for T_σ using straightforward and accurate force measurements, and acquire a smaller set of the endpoint stiffness data to calibrate this mapping. This is achieved by taking into account that, in general, end-point impedance has three components, depending on posture, force, and co-contraction, respectively. While the first two components may be large and even dominating [33] in a large enough range of variations, an ample literature reports the existence and independence of co-contraction contribution to stiffness: e.g. [5, 16, 34]. In addition, in our experiments, the master will perform the tasks in a fixed arm configuration with no significant generation of the endpoint force. Therefore, we consider a decomposition of the space of muscular activations $\mathscr{P} \ni P$ as the direct sum of a force-generating subspace $\mathscr{P}_F$ and the force-map null space $\mathscr{P}_k = \ker\{T_F\}$, i.e.

$$\mathscr{P} = \mathscr{P}_F \oplus \mathscr{P}_k.$$

By choosing a right-inverse T_F^R of T_F, i.e. any $n \times 3$ matrix[3] such that $T_F T_F^R = I$, we also affix a system of coordinates to these subspaces. In these coordinates, we can decompose the vector of muscular activations P in a force-generating component P_F and a null-space component P_k as

$$P = T_F^R T_F \, P + \left(I - T_F^R T_F\right) \, P \overset{def}{=} P_F + P_k.$$

The null space component P_k contains information on the co-contraction component of stiffness generation. It is convenient to give an alternative description of P_k as follows. Let N_F denote a basis matrix for the kernel of T_F, and let $\lambda = N_F^+ P_k = QP$ be the coordinates in that basis of P_k, where $Q \overset{def}{=} N_F^+ \left(I - T_F^R T_F\right)$. Hence the model of cartesian stiffness regulation through co-contraction is written as

$$\sigma - \sigma_0 = M_\sigma \, Q \, P \qquad (10.2)$$

where $M_\sigma \in \mathbb{R}^{3\times 5}$ is a mapping from the kernel of T_F (the set of muscle activations that do not change endpoint force, in the selected coordinate frame) to stiffness

[3]The existence of a right inverse is guaranteed by the fact that in nonsingular configurations T_F is full row-rank. Because $n > 3$, there exist infinite right-inverses: a particular choice is for instance $T_F^+ = T_F^T (T_F T_F^T)^{-1}$, i.e. the pseudoinverse of T_F.

variations. The map M_σ can then be identified and calibrated once, based on direct measurements of human arm end point stiffness, at different coactivation levels as described in the following section.

10.2.1.2 Stiffness Model Calibration-Identification

Identification of M_σ is carried out through two sets of experiments. The first set concerns the identification of the EMG-to-Force mapping and is performed by the measurement of endpoint forces and eight channel EMG electrodes (see muscle names in Table 10.1). In this set, a KUKA LWR was serving only as a support structure for a 6 axis F/T sensor (ATI Inc.) mounted at endpoint of the arm (see Fig. 10.1). The subject was asked to apply constant forces of ±5N, ±10N, ±15N and ±20N, respectively, along 6 directions ($\pm x$, $\pm y$ and $\pm z$) while holding the handle (isometric conditions). A graph with three colored bars on the screen was used to provide the user with the information about the measured force components. Each trial was 60 s long. Data from the first 10 s were discarded to eliminate transient force fluctuations. For each direction and force level, 4 trials were executed and recorded (for an overall number of $4 \times 3 \times 2 \times 3$ trials) in EMG-to-force map identification experiments. Consequently, the mapping (T_F) was identified by means of a least-squared-error algorithm, and a basis of its nullspace and the projector matrix Q were computed.

In the second set, with the purpose of off-line calibration of the EMG-to-Stiffness mapping, the subject's arm endpoint impedance profile was measured in different levels of muscular co-contraction. Following Perreault et al. [38], continuous stochastic perturbations with the maximum peak-to-peak value of 20 mm were applied to the subject's wrist through the handle in x, y and z directions (see [2] for details). A KUKA robot with fast research interface [39] was programmed and controlled in position to apply the desired perturbation profile. Subject's wrist position and restoring force profiles were synchronously measured using an optical tracking system (NaturalPoint, Inc.) and a FT sensor, respectively. A rough co-contraction indicator was graphically shown consisting of a bar of length proportional to the norm $|P|$ of the vector of muscle activations. Four levels of stiffness reference were provided in

Table 10.1 Muscles used for EMG measurements

Flexors		Extensors	
Monoarticular	Biarticular	Monoarticular	Biarticular
Deltoid clavicular part (DELC)	Biceps long head (BILH)	Deltoid scapular part (DELS)	Triceps long head (TRIO)
Pectoralis major clavicular part (PMJC)		Triceps lateral head (TRIA)	
Brachioradialis (BRAD)		Triceps medial head (TRIM)	

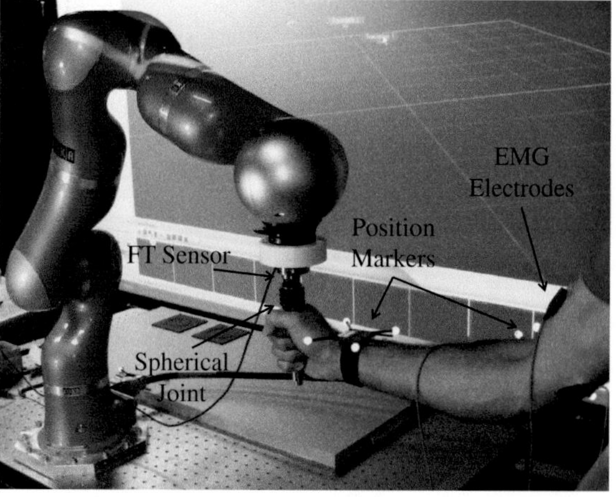

Fig. 10.1 Experimental setup used for the first calibration experiments. Subject applies constant forces in 6 directions while holding the handle attached to an idle spherical joint

different trials, where the first level (minimum muscle activity), was aimed at the identification of the intrinsic stiffness profile σ_0.

Identification of the endpoint impedance profiles in different levels $|P|$ consisted of two non-parametric and parametric identification procedures. Firstly, multiple-output (MIMO) dynamics of the endpoint impedance was decomposed into the linear subsystems associating each input to each output. Based on this assumption, and indicating with $F_x(f)$, $F_y(f)$ and $F_z(f)$ the Fourier transforms of the endpoint force along the axes of the Cartesian reference frame, with $x(f)$, $y(f)$ and $z(f)$ the transforms of the human endpoint displacements, the dynamic relation between the displacements and force variations can be described by

$$\begin{bmatrix} F_x(f) \\ F_y(f) \\ F_z(f) \end{bmatrix} = \begin{bmatrix} G_{xx}(f) & G_{xy}(f) & G_{xz}(f) \\ G_{yx}(f) & G_{yy}(f) & G_{yz}(f) \\ G_{zx}(f) & G_{zy}(f) & G_{zz}(f) \end{bmatrix} \begin{bmatrix} x(f) \\ y(f) \\ z(f) \end{bmatrix} \tag{10.3}$$

A non-parametric algorithm was adopted to identify the empirical transfer function of each of the SISO subsystems described above in frequency domain (MATLAB, The MathWorks Inc.). Consequently, we adopted a parametric, second order, linear model of each impedance transfer function of the type

$$G_{ij}(s) = I_{ij}\mathrm{s}^2 + B_{ij}\mathrm{s} + K_{ij},\ \mathrm{s} = 2\pi f\sqrt{-1} \tag{10.4}$$

where I, B and K denote the endpoint inertia, viscosity and stiffness matrices, respectively. The parameters of the second order linear model were identified based on least squares algorithm in frequency range from 0 to 10 Hz.

Eventually, experimental EMG vectors P were mapped in the EMG-to-force map nullspace through the previously computed projector matrix Q. The elements of the

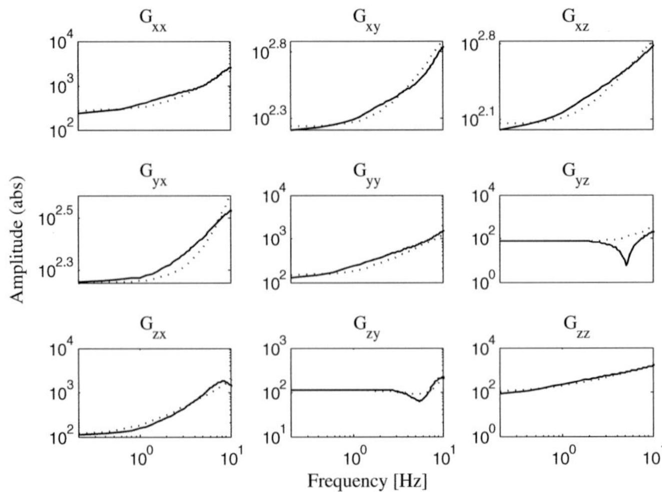

Fig. 10.2 Non-parametric (*solid lines*) and parametric second order (*dotted lines*) transfer functions of SISO impedance subsystems obtained from stochastic perturbations

stiffness matrix K were used as estimates for the components of σ, and the map M_σ was estimated by applying a least-squared-error method to (10.2).

The strength of linear dependency between measured force signals and estimates via the least-squared-error identification of the components of T_F was evaluated by Pearson's product-moment correlation coefficient. The coefficient is defined as

$$R_k = \frac{\sum \hat{f}_k f_k - \frac{\sum \hat{f}_k \sum f_k}{N}}{\sqrt{(\sum {f_k}^2 - \frac{(\sum f_k)^2}{N})(\sum \hat{f}_k^2 - \frac{(\sum \hat{f}_k)^2}{N})}}, \quad k = x, y, z \tag{10.5}$$

where f_k and $\hat{f}_k$ are measured and estimated values of force in the Cartesian directions, and N is the number of pairs of data. The fit was consistently good in the three directions, resulting in average $R^2 = 81\,\%$.

Figure 10.2 demonstrates the results of non-parametric and second order model identification of the hand impedance transfer functions in the frequency range from 0 to 10 Hz, according to methods described above. The second order parametric impedance models presented 69.7 % of the data variance across all directions in minimum muscular activity trials in the frequency range of 0–10 Hz.

Once the EMG-to-stiffness mapping was calibrated, it was used for the real-time estimation of the 3D endpoint stiffness matrix of the operator using EMG measurements of the muscles as illustrated in Table 10.1.

10.2.1.3 Experiments

The efficiency of the teleimpedance approach in rendering a desired interaction performance while executing tasks with dynamic requirements was evaluated in a ball reception task. In this experiment, two identical rigid balls ($m = 0.92$ kg, radius 52.5 mm) were suspended at the same distance above the human and robotic arm endpoints. The subject was prepared to receive the ball and instructed to hold his arm in a posture very close to that used during calibration experiments. The slave arm position, under gravity compensation, was corresponding.

The subject was instructed to receive the ball and stabilize its position in a natural way. The position of the slave endpoint was controlled along the master's wrist trajectory while executing the task, whereas the Cartesian stiffness values were commanded and controlled in three different approaches: In the first approach, the Cartesian stiffness of the slave endpoint was set to a relatively high, constant level ($K = [1200, 1200, 1200]$ N/m) throughout the task. The second one was analogous, with low constant stiffness values ($K = [120, 120, 120]$ N/m). In the third approach, variable impedance was implemented in three directions, as estimated from the stiffness model. Damping coefficient in all experiments was set to a constant value of 0.7 N.s/m.

The experimental setup and information flow are shown in Fig. 10.3. A body marker was attached to the wrist aiming at reference trajectory calculation for robot motion. The robot base frame was considered as the overall reference frame for other frames (tracking system and FT sensor). The position path of the human wrist was measured, low-pass filtered (cutoff 15 Hz) and used for trajectory planning. At the same time, EMG signals were acquired from the master arm and used to evaluate its endpoint stiffness based on the model and calibration described in the previous section. All processing and control algorithms were performed in real-time in C++ environment. KUKA interface was similar to the ones explained in identification trials.

10.2.1.4 Experimental Results

The measured forces at the endpoint of the slave robotic arm while executing the task in the three stiffness control modes (constantly high, constantly low, and teleimpedance) are reported in Fig. 10.4, while the corresponding deviation errors from the reference equilibrium position are in Fig. 10.5. The regulation of the human arm muscle activations and resulting endpoint stiffness modifications during the catching experiment are shown in Fig. 10.6. Increased stiffness at the time of impact and its progressive decrease afterward are the results of explicit muscular activity regulation by the subject.

As expected, the stiffer the arm, the smaller the deviation, as seen in the experimental results under constant high stiffness (Fig. 10.5, left). The tradeoff for the accuracy and reduced deviation from equilibrium position with high values of endpoint stiff-

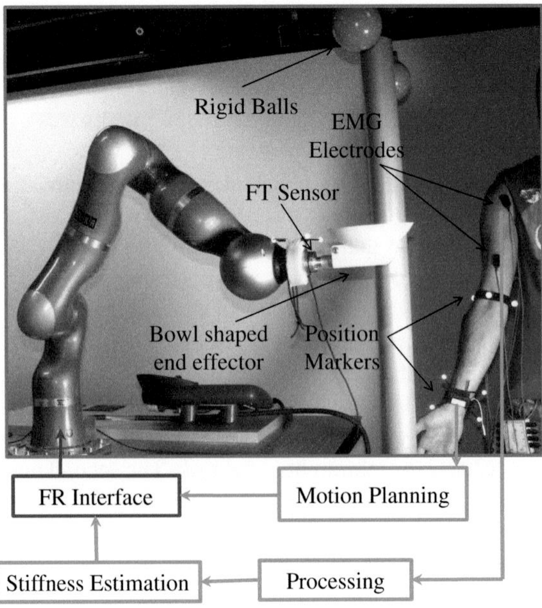

Fig. 10.3 Experimental setup of the ball-catching experiments. The slave KUKA LWR arm, EMG electrodes, position tracking markers and F/T sensor are shown

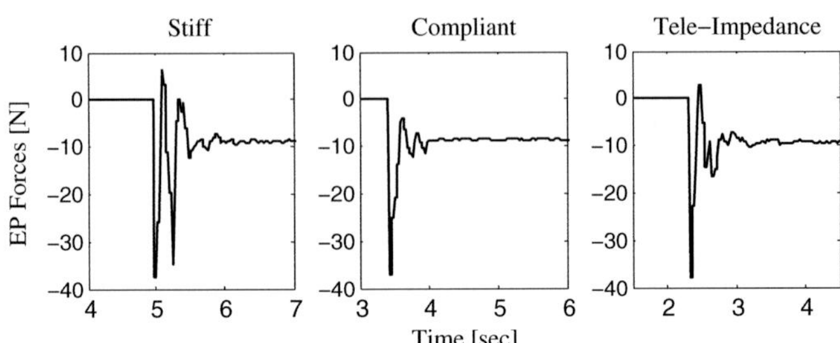

Fig. 10.4 Measured force values in z direction during the task with the slave robotic arm under constantly high, constantly low, and teleimpedance stiffness control

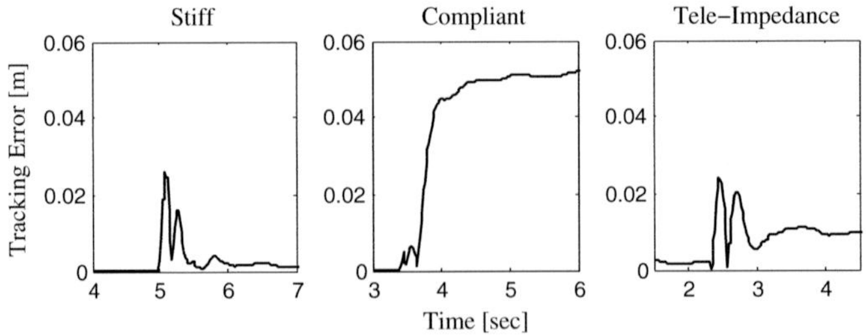

Fig. 10.5 Absolute tracking position error in z direction during the task with the slave robotic arm under constantly high, constantly low, and teleimpedance stiffness control

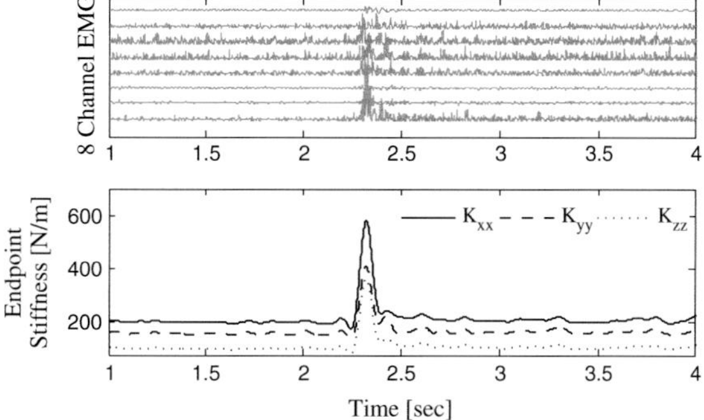

Fig. 10.6 Fully rectified eight channel raw EMGs (*upper plot*) and estimated and mapped endpoint stiffness (*lower plot*) in real-time for ball-reception task

Table 10.2 Performance indexes

Lift off index (LOI)	$\int_{\Delta t} \lvert f_z - f_w \rvert dt$
Position error index (PEI)	$max_{t \in \Delta t} \lvert e_z \rvert (t)$
Damping ratio index (DRI)	$\frac{1}{\sqrt{1+\left(\frac{2\pi}{\delta}\right)^2}} \quad \delta = \log\left(\frac{f_{z_{p,1}}}{f_{z_{p,2}}}\right)$
Bouncing time index (BTI)	$\Sigma \Delta t_B$ if $f_z \geq 0$

ness is the occurrence of bouncing: indeed, the second force peak (at $t \approx 5.26$ s) in the stiff case (Fig. 10.4, left) shows a second impact of the ball (see also Extension 1).

To obtain a more stable contact is to reduce the endpoint stiffness values; however, using constantly low stiffness directly affects the position deviation, which may grow to very large, possibly unacceptable values (Fig. 10.5, middle). Another drawback of such compliant control is the insufficiency of generated torques for repositioning the ball to its equilibrium even after transient end.

The transient behavior of the system under teleimpedance appears to benefit from the active control of stiffness, increasing at the very first moment of impact (from $t \approx 2.3$ s to $t \approx 2.4$ s), leading to a reduced deviation from reference equilibrium position. Also, the bouncing phenomenon appears to be avoided due to the subsequent stiffness reduction phase (between $t \approx 2.4$ s to $t \approx 2.7$ s, see Fig. 10.6). This behavior is in accordance with previous studies which have shown the capabilities of the human body to minimize soft-tissue vibrations and impact transitions by means of increased damping or decreased stiffness (modified resonance frequency) within involved tissues (see e.g. [45]). In addition, other behavioral studies demonstrated an increase of cocontraction levels in human arm while performing tasks which need quick torque generations and/or to cancel components of torques orthogonal to the desired direction [25].

A comparative performance analysis of the three control methods was done by defining different indexes, which are summarized in Table 10.2. LOI is computed as the integral of the difference between the vertical component of wrist force F_z and its steady-state value (i.e., the hand plus ball weight F_w), where Δt is the time interval

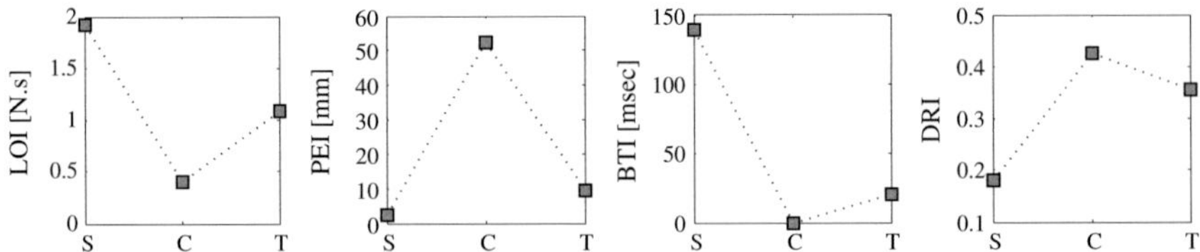

Fig. 10.7 Performance index plots over different elastic endpoint profiles (*S* Stiff, *C* Compliant and *T* Teleimpedance)

duration between the first impact and steady stabilization. A high value of the "lift off index" LOI indicates a reception with multiple bouncing and/or long underdamped ball trajectories.

The second index is the maximum deviation from the equilibrium position in z direction at steady state. As a third index, we consider an estimate of the damping ratio of the bouncing phenomenon, experimentally estimated (see e.g. [21]) using the logarithmic decrement between the first and second force peaks. Fourth and last, the "bouncing time index" BTI was introduced as the duration of the interval during which contact between the ball and robot's end effector is completely lost. The value is calculated by summing the intervals Δt_B along which f_z is zero (complete disconnection) or positive (as result of acceleration of bowl) after the first impact.

Figure 10.7 shows the values obtained in experiments for the four indices in the three different stiffness regulation modes. Teleimpedance control appears to strike a good compromise among the two extremes, consistently scoring close to the best performance obtained by either of the two constant settings, thus enabling the human ability to to be effectively transferred to the slave arm.

10.2.2 Teleimpedance Control of a Robotic Hand

Following the implementation of teleimpedance control for teleoperation of a robotic arm described above, we began to explore translating this approach to the control of a prosthetic hand. Although using a prosthesis is not typically thought of as a teleoperation scenario, the user is driving a terminal device in real-time often with only visual feedback as guidance. The control of these devices is suboptimal and research strategies including incorporating feedback, machine learning, and peripheral technology are being investigated. This section reviews the initial steps towards implementing a teleimpedance prosthetic controller [22] and further refinement [4] of this technique.

10.2.2.1 Pisa/IIT SoftHand

The Pisa/IIT SoftHand was developed through a collaboration between the University of Pisa and the Istituto Italiano di Tecnologia. The SoftHand was used in the experiments presented below and will be described here in brief. For a more detailed description, please see [14] and Chap. 8.

The goal of the SoftHand was to design and build a robotic hand that is highly functional yet simple and robust. This was achieved by combining the soft synergies approach [11] with underactuation [12]. The former uses human hand grasping synergies as a reference position for a virtual hand. The virtual hand position or stiffness profile connecting the virtual and real hands can thus be varied to control the interaction forces between the hand and the environment. The latter employs fewer actuators than available degrees of freedom, thus lowering cost, weight, and complexity of the device. Underactuation also imparts a degree of adaptability to the hand, thus the combination of these techniques was termed "adaptive synergies." Additionally, to make the hand more robust and safer in human-robot interaction scenarios, the hand was designed with soft robotics principles in mind: the fingers can be bent, twisted, struck, etc., and will deform out of the way and then return to their original conformation, protecting both the hand and the environment from damage in the event of a collision. The SoftHand is anthropomorphic and contains a single motor. This motor pulls a tendon that winds through the fingers and thumb to simultaneously flex and abduct the fingers. To enable testing of the hand with non-amputee subjects, a forearm adapter was employed, see Fig. 10.8.

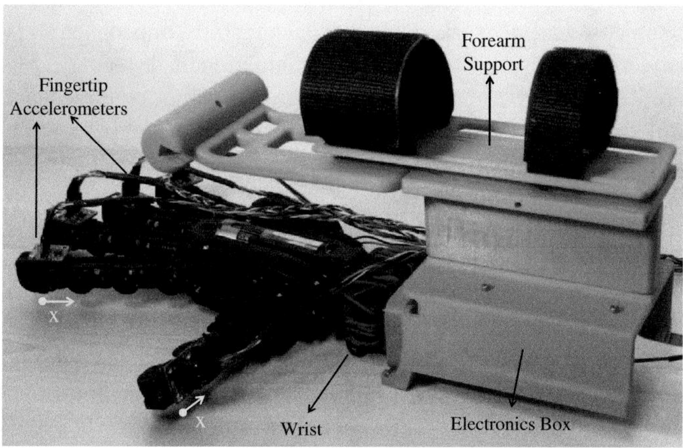

Fig. 10.8 SoftHand equipped with the able-bodied adapter (forearm support) and fingertip accelerometers

10.2.2.2 Initial Evaluation

A pilot experiment is presented in [22]. In this study, we implemented a standard proportional and a novel impedance controller with and without vibrotactile feedback using MATLAB Simulink and the Real-Time Windows Target (Mathworks, Inc.). The main finger flexor and extensor muscles (the flexor digitorum superficialis (FDS) and extensor digitorum communis (EDC), respectively) were sampled using surface electromyography (EMG) electrodes. With both controllers, the reference position of the hand was proportional to EMG amplitude. With impedance, the stiffness with which the proportional control was applied was based on the average of the flexor and extensor EMG signals. Both control modes were also tested with vibrotactile feedback applied using a small (7 × 2 mm) eccentric mass motor (Precision Microdrives Ltd.). The amplitude and frequency of the feedback was proportional to the grasping force. When an object is grasped, interaction forces occur as the reference position moves inside the object. In this way, the error between the reference and measured position can be used to estimate the grasping force.

In testing, each subject attempted to grasp four everyday objects of varied size and weight (water bottle, screwdriver, spray bottle, and ball; see Table 10.3) with each type of control mode. In total, four controls modes were tested: standard (proportional), impedance, vibrotactile (standard with feedback), and vibrotactile-impedance (VI). Each grasp was attempted three times for each of the objects and control modes, resulting in 48 grasps per subject. Mode and grasp order was fixed, but subjects were allowed a brief familiarization period in each condition to minimize learning effects. After each condition, subjects were also asked to evaluate the amount of physical and mental exertion required using a 5-point Likert scale. After all conditions were completed, subjects were also asked whether each feature (impedance and feedback) made the hand easier to use and whether the combination made the hand easier to use.

While only a pilot experiment, results suggested using teleimpedance in prosthetic control could provide an improvement in control of the prosthetic hand and subsequently the user's experience. Grasp success rate was above 90 % in all conditions, implying that the SoftHand was generally easy to control with minimal training and its conformal grasp was effective. Figure 10.9 shows the quantitative EMG results including duration of EMG activity (left) and cumulative and average EMG amplitudes (right). Subjects spent the longest time above minimum EMG thresholds in standard mode, less time with impedance and vibrotactile modes, and finally the shortest time in VI mode. Cumulative EMG was used as a proxy for physical exertion. This was highest again in standard mode, lower in vibrotactile, and lowest in impedance and VI modes. Average EMG amplitudes were similar across

Table 10.3 Dimensions and weights of test objects

Object	Water bottle	Screwdriver	Spray bottle	Ball
Dims (mm)	307 × 55 × 55	294 × 25 × 25	275 × 84 × 47	94 × 94 × 94
Weight (g)	250	50	500	500

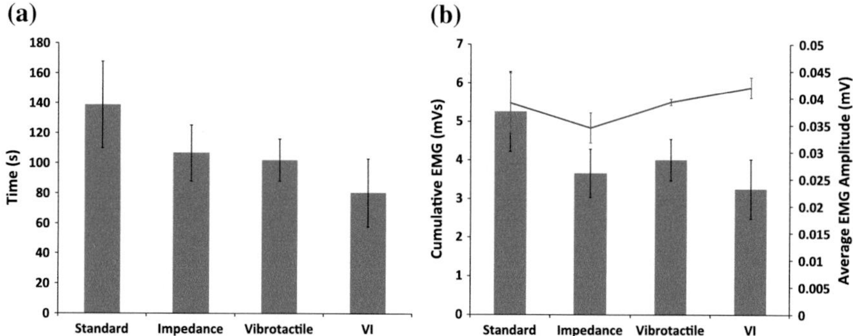

Fig. 10.9 Time spent above threshold averaged across subjects (*top*). Average FDS EMG amplitude (*bottom, bars*) and cumulative EMG (*bottom, line*). **a** Duration of EMG activity, **b** EMG activity

the conditions; the variations showed that subjects had a tendency to contract more with feedback in either control modality, and less in impedance mode. Because the motor used to provide feedback had low resolution, it is possible subjects produced larger contractions to increase their benefit from this feedback. Finally, the qualitative results from the Likert surveys mirrored the quantitative results: subjects reported lower mental and physical effort with impedance and vibrotactile modes compared to standard and lowest with the VI mode. These results suggest that both impedance control and vibrotactile feedback provide improved prosthetic control and user satisfaction. It is worth noting, however, that grasp success rates were still high without these features and that order effects had a potential influence on the results.

10.2.2.3 Extension of the Hand Controller

Following the results of the pilot experiment, more advanced versions of the above proportional and teleimpedance controllers were developed. The goal was to map the FDS and EDC EMG signals to position and stiffness models to a achieve more accurate control for each subject. Subjects then attempted to grasp everyday objects with three types of controllers: stiff, using the position model and a fixed, high impedance value; compliant, using the position model and a fixed, low impedance value; and teleimpedance, using a varying control gain based on the users' stiffness profile. A block diagram of the control scheme employed is presented in Fig. 10.10. Further, two haptic interfaces were included. In the first, a mechanical version of the force feedback described above was developed. A mechanical cuff tightened around the upper arm as grasp force increased so as to provide modality-matched feedback to the user. In the second, surface texture was measured by placing accelerometers on the SoftHand fingertips and then replicating the measured vibrations with a bracelet of eccentric mass motors on the forearm. This setup was used in combination with the teleimpedance controller for a blind surface discrimination and grasping experi-

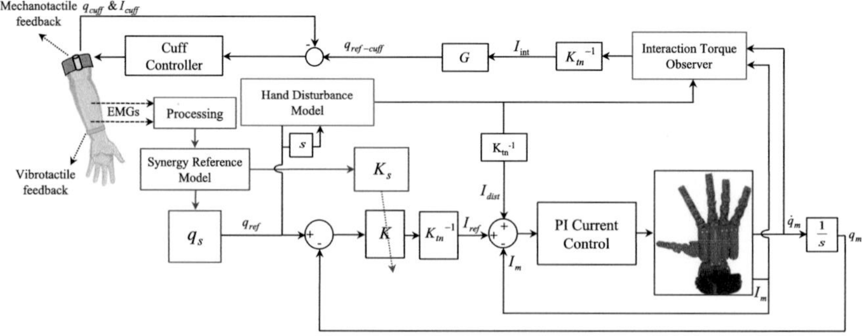

Fig. 10.10 Block diagram of the synergy-driven hand teleimpedance

ment. For further information on these haptic experiments, please see [4]. Below we describe the modeling, parameter identification, and evaluation.

Muscle force increases with muscular activity, and as individual muscle forces increase, they affect the torque at the joints they cross. Cocontraction of antagonist pairs, however, affects the impedance of the joint [37]. To begin, we consider the forward dynamics of the first grasp synergy and write

$$\begin{aligned} \tau &= a_\tau \delta, \\ a_\tau \delta &= I\ddot{q}_s + c\dot{q}_s + K_s(q_s - q_0) + \tau_E, \end{aligned} \tag{10.6}$$

where τ, a_τ and τ_E denote the torque synergy, its gain, and external torques, respectively; q_s and q_0 are the position of the hand and the object along the first synergy; δ is a function of the difference in activation of the antagonistic muscles $(FDS - EDC)$, and I, c and K_s are the inertia, damping and stiffness of the hand along the first synergy, respectively. The effects of inertia and external torques can be neglected, leaving us with

$$\dot{q}_s = \frac{-K_s}{c}(q_s - q_0) + \frac{a_\tau}{c}\delta. \tag{10.7}$$

Next, we can use T and k to represent the time step and iteration number and estimate the dynamics in discrete time as follows

$$q_{s_{k+1}} = (1 - \frac{K_s T}{c})q_{s_k} + \frac{T a_\tau}{c}\delta + \frac{K_s T}{c}q_0. \tag{10.8}$$

Finally, we use two modified hyperbolic tangents [15] to map the position and stiffness synergy references:

$$\delta = \frac{a_q[1 - \mathrm{e}^{-b_q(\mathrm{FDS}-\mathrm{EDC})}]}{[1 + \mathrm{e}^{-b_q(\mathrm{FDS}-\mathrm{EDC})}]}, \tag{10.9}$$

$$K_s = \frac{a_k[1 - e^{-b_k(\text{FDS}+\text{EDC})}]}{[1 + e^{-b_k(\text{FDS}+\text{EDC})}]}, \tag{10.10}$$

where FDS and EDC are the processed EMG signals of the corresponding muscles, and the gains a_q, b_q, a_k, and b_k are identified experimentally. To do so, the subject was asked to open and close his or her hand while FDS and EDC activity were recorded; meanwhile, the SoftHand opened and closed as a visual reference. Twenty natural, self-paced open/close movements were recorded to determine the parameters of the position synergy model. A further 20 movements were recorded while asking the subject to maintain various levels of cocontraction. Subjects were given visual feedback of their cocontraction levels and asked to perform 4 movement cycles at 5 different levels.

Half of each set of trials was used to identify the parameters of the models and the other half to evaluate the modeling accuracy. Averaged across subjects, we found normalized root-mean-squared error (NRMSE) values of 17.6 and 13.4 % for the postural and stiffness test trials, respectively. Ultimately, mental imagery, bilateral action using a mirror box, or similar techniques could be used to identify these parameters in persons with amputations.

After parameter identification, subjects attempted to grasp everyday objects with each type of controller: stiff, compliant, and teleimpedance. A sample grasp of a rigid object (a mug) with each of the controllers is presented in Fig. 10.11. Subjects were highly successful with all controllers. However, grasp quality and stability was highest with teleimpedance. With the stiff controller, subjects would occasionally apply excessive force and damage deformable objects. In contrast, with the compliant controller, subjects would occasionally lose the grip on and drop heavier objects. The teleimpedance controller seemed to mitigate both of these problems. While these results are preliminary, they suggest teleimpedance control of a prosthetic hand is both functional and intuitive. Future work will apply this novel controller in a clinical setting.

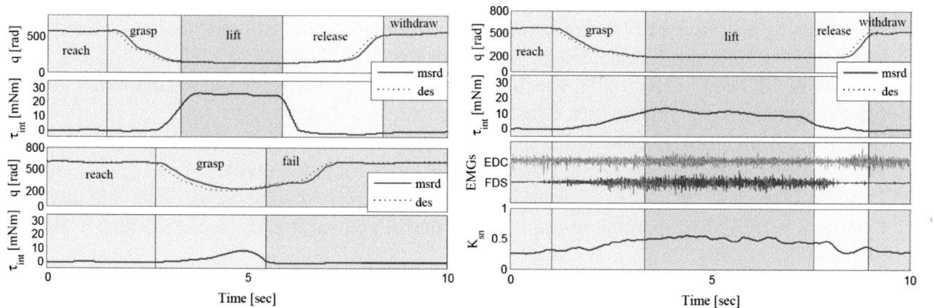

Fig. 10.11 Sample results of the SoftHand grasping a hard object (mug), with the controller under high, fixed stiffness gain (*left top pair*, $K = 40\,\text{Nm/rad}$), low, fixed stiffness gain (*left mid pair*, $K = 10\text{Nm/rad}$), and teleimpedance (right four, $a_{q_{norm}} = 1, b_q = 5.03, a_k = 1.87$, and $b_k = 0.579$)

References

1. Ajoudani A, Tsagarakis N, Bicchi A (2011) Tele-Impedance: preliminary results on measuring and replicating human arm impedance in tele operated robots. In: IEEE international conference on robotics and biomimetics-ROBIO, pp 216–223
2. Ajoudani A, Tsagarakis NG, Bicchi A (2012) Tele-Impedance: Teleoperation with impedance regulation using a body-machine interface. Int J Robot Res 31(13):1642–1655
3. Ajoudani A, Gabiccini M, Tsagarakis NG, Albu-Schäffer A, Bicchi A (2012) TeleImpedance: exploring the role of common-mode and configuration-dependant stiffness. In: IEEE-RAS international conference on humanoid robots
4. Ajoudani A, Godfrey S, Bianchi M, Catalano M, Grioli G, Tsagarakis NG, Bicchi A (2014) Exploring teleimpedance and tactile feedback for intuitive control of the Pisa/IIT softhand. IEEE Trans Haptics 7:203–2015
5. Akazawa K, Milner T, Stein R (1983) Modulation of reflex EMG and stiffness in response to stretch of human finger muscle. J Neurophisiol 49:16–27
6. Albu-Schaffer A, Hirzinger G (2002) Cartesian impedance control techniques for torque controlled light-weight robots. In: IEEE international conference on robotics and automation, 2002. Proceedings. ICRA'02, vol 1. IEEE, pp 657–663
7. Albu-Schäffer A, Ott C, Hirzinger G (2007) A unified passivity-based control framework for position, torque and impedance control of flexible joint robots. Int J Robot Res 26(1):23–39
8. Albu-Schaffer A, Fischer M, Schreiber G, Schoeppe F, Hirzinger G (2004) Soft robotics: what Cartesian stiffness can obtain with passively compliant, uncoupled joints? In: IEEE/RSJ international conference on intelligent robots and systems, vol 4. IEEE, pp 3295–3301
9. Albu-Schäffer A, Haddadin S, Ott C, Stemmer A, Wimbock T, Hirzinger G (2007) The DLR lightweight robot lightweight design and soft robotics control concepts for robots in human environments. Ind Robot J 34:376–385
10. Bicchi A, Tonietti G (2004) Fast and soft arm tactics: dealing with the safety-performance trade off in robot arms design and control. IEEE Robot Autom Mag 11(2):22–33
11. Bicchi A, Gabiccini M, Santello M (2011) Modelling natural and artificial hands with synergies. Philos Trans Royal Soc B: Biol Sci 366(1581):3153–3161
12. Birglen L, Gosselin C, Laliberté T (2008) Underactuated robotic hands. Springer
13. Castellini C, Van Der Smagt P (2009) Surface EMG in advanced hand prothetics. Biol Cybern 100(1):35–47
14. Catalano MG, Grioli G, Farnioli E, Serio A, Piazza C, Bicchi A (2014) Adaptive synergies for the design and control of the pisa/iit softhand. Int J Robot Res 33(5):768–782
15. Chen C-T, Chang W-D (1996) A feedforward neural network with function shape autotuning. Neural Netw 9(4):627–641
16. De Serres S, Milner T (1991) Wrist muscle activation patterns and stiffness associated with stable and unstable mechanical loads. Exp Brain Res 86:451–458
17. Erdemir A, McLean S, Herzog W, van den Bogert AJ (2007) Model-based estimation of muscle forces exerted during movements. Clin Biomech 22(2):131–154
18. Eusebi A, Melchiorri C (1998) Force reflecting telemanipulators with timedelay: stability analysis and control design. IEEE Trans Robot Autom 14(4):635–640
19. Franklin D, Osu R, Burdet E, Kawato M, Milner T (2003) Adaptation to stable and unstable dynamics achieved by combined impedance control and inverse dynamics model. J Neurophysiol 90(5):3270–3282
20. Fukuda O, Tsuji T, Kaneko M, Otsuka A (2003) A human-assisting manipulator teleoperated by EMG signals and arm motions. IEEE Trans Robot Autom 19(2):210–222
21. Genta G (1995) Vibration of structures and machines. Springer, New York
22. Godfrey S, Ajoudani A, Catalano M, Grioli G, Bicchi A (2013) A synergy-driven approach to a myoelectric hand. In: IEEE international conference on rehabilitation robotics-ICORR, pp 1–6
23. Gomi H, Osu R (1998) Task-dependent viscoelasticity of human multi joint arm and its spatial characteristics for interaction with environments. J Neurosci 18:65–78

24. Gribble P, Ostry D (2000) Compensation for loads during arm movements using equilibrium-point control. Exp Brain Res 4(135):474–482
25. Gribble P, Mullin L, Cothros N, Mattar A (2003) Role of cocontraction in arm movement accuracy. J Neurophysiol 89:2396–2405
26. Hannaford B, Anderson R (1988) Experimental and simulation studies of hard contact in force reflecting teleoperation. In: International conference on robotics and automation, pp 584–589
27. Imaida T (2004) Yokokohji Y, Doi MOT, Yoshikawa T Ground space bilateral teleoperation of ETS-VII robot arm by direct bilateral coupling under 7-s time delay condition. IEEE Trans Robot Autom 20(3):499-511
28. Karavas N, Ajoudani A, Tsagarakis N, Caldwell D (2013) Human-inspired balancing assistance: application to a knee exoskeleton. In: IEEE international conference on robotics and biomimetics-ROBIO
29. Karavas N, Ajoudani A, Tsagarakis N, Saglia J, Bicchi A, Caldwell D (2015) Tele-impedance based assistive control for a compliant knee exoskeleton. Robot Auton Syst 73:78–90
30. Laffranchi M, Tsagarakis N, Cannella F, Caldwell D (2009) Antagonistic and series elastic actuators: a comparative analysis on the energy consumption. In: IEEE/RSJ international conference on intelligent robots and systems, IEEE. IEEE, pp 5678–5684
31. Leung G, Francis B, Apkarian J (1995) Bilateral controller for teleoperators with time delay via mu-synthesis. IEEE Trans Robot Autom 11(1):
32. Lloyd D, Besier T (2003) An EMG-driven musculoskeletal model to estimate muscle forces and knee joint moments in vivo. J Biomech 36(6):765–776
33. Milner T (2002) Contribution of geometry and joint stiffness to mechanical stability of the human arm. Exp Brain Res 143:515–519
34. Milner T, Cloutier C, Leger A, Franklin D (1995) Inability to activate muscle maximally during cocontraction and the effect on joint stiffness. Exp Brain Res 107:293–305
35. Mussa-Ivaldi F, Hogan N, Bizzi E (1985) Neural, mechanical, and geometric factors subserving arm posture in humans. J Neurosci 5(10):2732–2743
36. Osu R, Gomi H (1999a) Multijoint muscle regulation mechanism examined by measured human arm stiffness and EMG signals. J Neurophysiol 81:1458–1468
37. Osu R, Gomi H (1999b) Multijoint muscle regulation mechanisms examined by measured human arm stiffness and emg signals. J Neurophysiol 81(4):1458–1468
38. Perreault E, Kirsch R, Crago P (2004) Multijoint dynamics and postural stability of the human arm. Exp Brain Res 157:507–517
39. Schreiber G, Stemmer A, Bischoff R (2010) The fast research interface for the KUKA lightweight robot. In: IEEE conference on robotics and automation (ICRA)
40. Selen L, Beek P, Dieen JV (2005) Can co-activation reduce kinematic variability? A simulation study. Biol Cybern 93:373–381
41. Sheridan T (1993) Space teleoperation through time delay: review and prognosis. IEEE Trans Robot Autom 9(5):592–606
42. Tonietti G, Schiavi R, Bicchi A (2005) Design and control of a variable stiffness actuator for safe and fast physical human/robot interaction. In: IEEE international conference on robotics and automation, pp 526–531
43. Trumbower R, Krutky M, Yang B, Perreault E (2009) Use of self-selected postures to regulate multijoint stiffness during unconstrained tasks. PLoS One 4(5):
44. Vogel J, Castellini C, Der Smagt P (2011) EMG-based teleoperation and manipulation with the DLR LWR-III): design and modelling. In: IEEE/RSJ international conference on intelligent robots and systems (IROS), pp 672–678
45. Wakeling J, Liphardt A, Nigg B (2003) Muscle activity reduces soft-tissue resonance at heel-strike during walking. J Biomech 36:1761–1769

Chapter 11
Incremental Learning of Muscle Synergies: From Calibration to Interaction

Claudio Castellini

Abstract In the previous chapter it has been shown how sEMG gathered from only two loci of muscular activity with opposite mechanical actions can be used to control the synergy-inspired robotic hand described in Chap. 8. Here, the problem of simplifying the control of a multi-DoF, multi-DoA mechatronic system—more specifically a prosthetic hand—is tackled from the opposite perspective, i.e. by leveraging the information contained in the sEMG gathered from multiple sources of activity. Natural, reliable and precise control of a dexterous hand prosthesis is a key ingredient to the restoration of a missing hand's functions, to the best extent allowed for by the current technology. However, this kind of control, based upon machine learning applied to synergistic muscle activation patterns, is still not reliable enough to be used in the clinics. In this chapter we propose to use *incremental* machine learning to improve the stability and reliability of natural prosthetic control. Incremental learning enforces a true, endless adaptation of the prosthesis to the subject, the environment, the objects to be manipulated; and it allows for the adaptation of the subject to the prosthesis in the course of time, leading to the exploitation of reciprocal learning. If proven successful in the large, this idea will prepare the shift from prostheses, which need to be calibrated, to prostheses that interact with human beings.

11.1 Introduction

One of the simplest ways to characterize the animal kingdom is to consider the typically animal ability of voluntarily *moving* [31, 38]. Animals move in the world to survive, feed, mate, adapt to the environment and adapt the environment to their needs—basically, for everything they do. Mammals, in particular, move and act by activating their muscles, which are an extremely smart product of evolution. Actually, from the point of view of the modern engineer, a muscle is an incredibly energy-efficient, light and versatile actuator.

C. Castellini (✉)
Institute of Robotics and Mechatronics, DLR - German Aerospace Center,
Muenchener Strasse 20, 82234 Oberpfaffenhofen, Germany
e-mail: claudio.castellini@dlr.de

M. Bianchi and A. Moscatelli (eds.), *Human and Robot Hands*,
Springer Series on Touch and Haptic Systems, DOI 10.1007/978-3-319-26706-7_11

Such a marvelous set of actuators requires an equally marvelous control system, which however does not resemble much any standard *control system* as found in Control Theory or on modern robots. Each muscle is in fact composed of up to thousands smaller actuators called Motor Units (MUs), each one producing contractile force on a joint, each one in principle independently controlled [34, 39]; precisely this redundancy, coupled with a smart recruitment mechanism, enables mammals achieve their spectacular performances while running, climbing, swimming, nurturing their offspring, mating, etc. In fact, essentially every action involves most, if not all, the MUs in a certain musculoskeletal region. It therefore seems that MUs are always controlled in large batches altogether at the same time, in a coordinated fashion. Since the total number of MUs in the musculoskeletal system is too large to be directly consciously controlled MU by MU [3], a simplifying paradigm is needed.

In parallel to the concept of kinematic synergies widely discussed in the first part of this book, starting from 1998, the idea of *muscle synergies* as a solution to this problem was introduced [4, 14–16, 53, 54]. Muscle synergies, as traditionally defined, are basic coordinated muscle activations that can be extracted using, e.g., Principal Component Analysis (PCA) from kinematic or sEMG data. The strong compression factors uniformly obtained by PCA on data gathered from human subjects while performing large sets of everyday-living tasks seem to indicate that only a few synergies (three or four) are required to perform most such actions (see the discussion carried out elsewhere in this book, e.g., in Chaps. 2–4, 6, 8 and 15). The situation becomes less clear when training is involved, for instance when a subject learns to play the piano (see, e.g., [58]). When additional motion finesse is required, it is likely that more and more synergies must consciously be controlled. It is quite possible, anyway, that this paradigm works as a general control schema for the mammalian motion: maybe grabbing a pen, caressing one's partner, playing the piano, breaking an egg, carrying a 50 kg. weight, all these actions are performed via muscle synergies control [25]. But dealing with this problem is not in the scope of this chapter, and as well, different definitions of this concept exist [16, 32, 52]; in fact, we hereby adopt the simplest possible definition of a muscle synergy: a coordinated, task-directed activation of a set of MUs. In this sense, any voluntary action, for instance the act of flexing one's index finger to a determined amount of the maximum voluntary contraction, corresponds to a specific synergy.

Now, any specific synergy corresponds to a signal pattern that can be detected by employing an adequate array of sensors and a signal processing system—what we call the Human-Machine Interface or HMI. Such an HMI is the ideal basis of modern, dexterous prosthetic control. A prosthesis is needed whenever a person has lost a limb, be it due to a traumatic event, planned surgery or congenital deficiency; the loss of a limb leads to a severe degradation of the quality of life [37, 44], therefore it is very desirable to restore the lost body functions to the best extent allowed for by the technology. The main idea is that of employing the HMI to let the amputee directly control a robotic prosthesis in the most natural way, that is, "by desiring so" [19, 28]. Our main object of study is, in particular, hand prostheses, given that the human hand is one of the most wonderful tools ever evolved by Nature, and the loss of a hand is a very disabling condition in the modern world.

In Chap. 10, an approach based on minimalistic sEMG mapping has been introduced, showing how such a strategy can be successfully exploited to control the robotic hand described in Chap. 8, under-actuated according to the concept of *adaptive synergies*. On the other hand, to deal with artificial hands with many DoFs and DoAs, machine learning techniques have been developed and academically tested in hundreds; still, at the time of writing they are essentially not used in the clinics,[1] the main problem being the unreliability of the HMI [5, 8, 18, 19, 44]. In this case "unreliable" means that the control signals generated by the HMI are not stable with respect to the user's intent or, equivalently, that the patterns to be recognized are too diverse or change in time. All in all, the prosthetic control system needs to be improved.

We deem necessary a paradigm shift here. In particular, with the advent of multi-fingered hand prostheses, complete arm/hand prosthetic systems and advanced surgery methods such as Targeted Muscle Reinnervation [2, 30, 55], the standard prosthetic control does not suffice any more. Among the several advancements called for by the community [26], we push the *incrementality* of the control system [20, 57]. Incrementality of a machine learning system is the possibility of updating the model obtained so far whenever required, without recalibrating, without loosing previous information and without waiting for the calibration time; in the context of using a prosthesis, this concept directly leads to *interaction* between the human subject and the system. We believe that the chance that the subject continually teaches the system new patterns as they arise in real life is paramount to improve the *reliability* of the prosthetic artifact.

The rest of this chapter is a series of recommendations and ideas on how to pursue this goal. In particular, Sect. 11.2 sets the background, describing the current flaws and limitations of natural prosthetic control and stating a list of requirements for the new kind of control system we are advocating; in Sect. 11.3 we describe our own solution to the problem and show a couple successful applications of a system based upon the ideas described; and lastly Sects. 11.4 and 11.5 contain final remarks.

11.2 Background

Current prosthetic control systems are in the vast majority based upon machine learning applied to patterns of synergistic muscle activations voluntarily generated by the user. We hereby argue that a specific characteristic of the control system that has been so far neglected might represent a solution to the notorious problem of the unreliability of such kinds of human-machine interfaces, namely incrementality.

---

[1] As of today, the only commercially available machine-learning-based myocontrol system is manufactured by Coapt LLC (www.coaptengineering.com) and no statistics on its effectiveness are available.

11.2.1 Muscle Activations in Prosthetic Control

Let us concentrate on voluntary muscle activations, that is, movements enacted by a precise, conscious, coordinated muscle contraction—as two typical examples, let us consider playing a C-major chord on a piano, and carrying an egg from A to B in one's kitchen, for example to prepare an omelet. Each action requires an extremely fine control of the activation of thousands of motor units in the hand, forearm, upper arm and even, in the case of carrying the egg around, of the whole body. Given the right level of granularity of an action (more on this point will be said later on), one can mathematically say that the intent of performing an action generates a dynamic pattern of muscle activation, $\mathbf{a}(t)$, where the vector $\mathbf{a}$ denotes the activation level of each motor unit involved in the action. Without loss of generality one can think of $\mathbf{a}$ being expressed in normalized coordinates, for instance as a fraction of the maximum possible activation of each motor unit; in this case $\mathbf{a}(t) \in \mathbb{R}^M$ where M is the number of motor units involved. In the above example of the C-major chord, and considering the hand and wrist only, simultaneous activation of the wrist, thumb, middle finger and little finger is required to hit the C, E and G keys at the same time, using the right amount of force to produce the desired volume[2]; this would correspond to, say, $\mathbf{a}_1(t)$. In the second example, at least the thumb and index finger (and usually much more) must be activated, again, simultaneously and to the right amount in order to pick up the egg, and carry it without letting it slip and without crushing it. We could denote this action as $\mathbf{a}_2(t)$. And so on, for each required action.

Notice that an exceptionally fine control over $\mathbf{a}$ is required *along time*. The value of $\mathbf{a}$ must remain as stable as possible in time, notwithstanding any disturbance, external and/or internal to the body. Such disturbances are quintessentially unavoidable, as they include, e.g., other movements required at the same time; for instance, playing a bass line with the other hand on the piano, or walking while carrying the egg. Clearly—and this is the problem of granularity of an action, mentioned above—there is a particular range of values within which $\mathbf{a}$ must remain in order to achieve the desired goal; for instance, the egg-carrying action can be stably performed across an interval of time Δt only as long as $\mathbf{a}_2^{min} < \mathbf{a}_2(t) < \mathbf{a}_2^{max}$ for all $t \in \Delta t$. This directly leads to the definition of a *muscle activation pattern*, which enforces the desired action. From the point of view of the engineer, such a pattern can be represented, in the simplest instance, by the average of the values obtained while the subject repeats the action over and over again: $\overline{\mathbf{a}_2}$. (More complex representations can include, for instance, a probabilistic description of the distribution of the signal obtained across these repetitions.) Such a pattern is a time-abstracted simultaneous muscle activation, which matches our previous definition of a muscle synergy, precisely the synergy that enables the subject carry the egg in a stable way. Such synergies are also the patterns that a machine-learning based prosthetic control system will try to recognize: as long as the subject keeps her/his activation levels close to that pattern (given a certain distance metric—see also the concept of *good variance* vs. *bad variance* in

[2]At least according to the standard piano-playing technique as told in most modern musical methods.

the synergy definition given in [32]), the system will issue the right commands to the prosthesis and enforce stable grasping.

On the other hand, failure by the subject to maintain the synergistic MU activations "close enough" to the reference pattern $\overline{\mathbf{a}_2}$, or more likely, failure by the system to effectively recognize that pattern, will result in the egg being dropped or crushed; in general, the inability of reliably recognizing a pattern $\overline{\mathbf{a}_i}$ will lead to unstable grasping of type i, which can have dramatic consequences. Intact human subjects employ a wide array of sensors to close the loop over the MU activations, but this feedback stream of information is precisely what an amputated subject lacks. It is then no wonder that the biggest cause of abandonment of upper-limb prostheses is their *unreliability*, that is, the inability of the control system to correctly, stably detect the intent of the patient [5, 8, 18].

11.2.2 Unreliability

Since the 1950s sEMG (surface Electro-Myography), originally a muscle disorder diagnosis technique, has been used to enforce muscle-activation-based control of one-DoF hand prostheses [44]: traditionally, two sites of large residual activity (see also Chap. 10) would be identified on the patient's stump, usually corresponding to flexion and extension of the wrist; these two sites would be used to determine the speed of opening and closing the prosthesis. With the advancement of prosthetic technology, more sophisticated arrangements of sEMG electrodes have been used (with higher sensitivity, better noise-rejection properties and/or higher spatial resolution) and novel kinds of signals have been explored as potential replacements/augmentations of sEMG. Among these, tactile [47] and pressure sensors [50, 61] detecting the stump surface deformation corresponding to muscle activations; ultrasound imaging [22, 57] detecting the displacement of the remnant anatomical structures in the stump; strain sensors to detect the same kind of deformations; and computer vision [33] to aid the prosthetic control by putting prior information on the decision regarding the action to be performed. Moreover, sophisticated statistical methods belonging to the class of *machine learning* (ML, also denoted as "pattern matching" or "pattern recognition" in the rehabilitation community) algorithms have been applied to these signals.

In general, once an educated guess has been made about a certain type of bodily signals (sEMG, ultrasound, etc.) to be meaningful of the underlying muscle activity, a ML method works as follows: given a set of pairs $S = \{(\mathbf{x}_i, y_i)\}$ in which $\mathbf{x}_i$ is a sample of the signal, and y_i is an integer (a "label") abstractly denoting a required action, or a real position/force value directly denoting the required control signals for the DoFs of a prosthesis, a map between signals and actions will be created via some kind of statistical approximation: $y = f(\mathbf{x})$. The approximant f is usually found by minimization of a cost functional, which makes the operation computationally costly (as is the case, e.g., of Support Vector Machines [6, 56, 59]) and/or unsafe due to the presence of local minima (as with artificial neural networks). Anyway, in the machine-learning lingo, S is called *training set*, since it is the data set used to "train" the machine to recognize a certain set of patterns in the input signal; similarly, the creation of the map f using S, which an engineer would call "calibration", is called in this case *training phase*, as opposed to the *prediction phase* in which f is actually used to guess y from the signal $\mathbf{x}$.[3]

Now, the quality of the obtained control function f strictly depends on the "quality" of S (in machine learning, in general, apart from the choice of the basic functions used to compose f, e.g., linear or not, there is little more than S to determine what f looks like); in turn, how to define the quality of a training set is a matter of debate. If the map f, which represents the prosthetic control system, is supposed to stably and reliably recognize a set of patterns (muscle synergies) $P_i, i = 1, \ldots, N$ corresponding to the required actions $\overline{\mathbf{a}_i}$, then those patterns

1. must appear in S correctly associated to the required output values;
2. must be repeatable; and
3. they must be stable.

Item (1) is not problematic; as opposed to this, items (2) and (3) can be, and usually are. The gathering of the training set can be long and psychologically challenging for the subject, mainly since (s)he has no control on what (s)he is doing, due to the above-mentioned lack of sensory feedback (this issue is also tackled in Chap. 10). A pattern $\overline{\mathbf{a}_i}$ can be very different from a pattern $\overline{\mathbf{a}_i'}$ gathered at some later point in time *but representing the same desired action* i, due to a number of competing factors such as, e.g., electrical external disturbances; slightly different muscle activations leading to very similar actions; muscular fatigue and sweat, which are well known to significantly alter the sEMG signal [34, 35]; and so on. On top of that, one must notice that in order to guarantee stability of prosthetic control, a pattern $\overline{\mathbf{a}}$ contained in the training set must represent the corresponding action *in all possible conditions subsequently encountered by the subject*. This includes all musculoskeletal configurations requiring a different activation for the same action, such as, e.g., all possible weights one might want to carry, all possible pronation/supination configurations of the wrist, all possible activation artifacts due to walking, etc. Often, a prosthetic control system, which was properly trained in the beginning will miserably fail later on, because the subject is standing instead of sitting, or because she is carrying a one-kilogram bottle of water, which was the very purpose of grasping it! (See, e.g., [9] for a study in slightly less lab-controlled conditions.)

To make the situation worse, most ML methods enforce what we call "monolithic learning": S is gathered at the beginning of the experiment, then f is created (training/calibration), then the prediction starts; there is no chance of updating f once the prediction has begun, unless one stops the prediction, updates S to some new (larger) training set and trains anew. This is unacceptable since S is potentially unlimited in size; as well, particularly whenever it is required that f be non-linear (as is mostly

[3]Notice that from the point of view of the clinician, this term represents a bizarre semantic twist, since normally it is the human subject which must be "trained" to use a prosthesis and not vice-versa!

the case in hand prosthetics) training takes a long time and depends on the size of S. This entails that the pattern $\overline{\mathbf{a}}$ must be gathered correctly once and for all during the training phase.

In other words, the quality of the control strictly depends on S, and S must be gathered optimally since the start. As one might guess, this is an essentially impossible task. We claim that it is mainly for this reason that ML-based prosthetic control is still unreliable nowadays, after 80 years of research, and it is essentially not yet used in the clinics.

11.2.3 Building a More Detailed Model or Learning More?

Let us call the space of all signals that the prosthetic hardware can gather, the *input space* I. In practice, only a "useful" subset $I_U \subseteq I$ is what we are interested in; actually, the task outlined above boils down to building a sensible f *restricted to* I_U. I_U can be either defined by the tasks to be correctly carried out, for instance power grasp, pinch grip and stretched hand, or by the prosthetic hand at our disposal, say that we want to control each single motor of the prosthesis. As previously mentioned, this latter idea offers a different perspective to the synergy-inspired simplification strategy discussed in Chaps. 8, 10 and 13. It represents the starting point of the research of the author of the current chapter, see [20] for instance, and stems from the idea of *simultaneous and proportional control* [27]. Anyway, f should *always* work correctly on I_U, where "correctly" is defined by the three items in the previous subsection, and may ignore what is outside it. Since S is all we have at our disposal to properly build f, it follows that S must somehow contain I_U, or at least a relevant fraction of it, given the generalization power of f.

Now, if I_U is too badly structured, or simply too vast to be captured by S, no proper f can be built, and there are two possibilities at hand to improve the situation: either we try and map I_U onto the space "captured" by S, or we expand S itself.

The first option means that one must have a *model* of the physical process being approximated via f. A remarkable example of such an attempt is the psycho-physical modeling of muscular fatigue and its effect on the sEMG signal (see, e.g., [36]) which has lead to several systems in which fatigue is detected (e.g., [1]) and somehow "corrected". We see this as an instance of the first possibility above: given a desired action $\overline{\mathbf{a}}$, I_U contains necessarily all of its instances under fatigue, say $\overline{\mathbf{a}'}$, $\overline{\mathbf{a}''}$, etc. Since it is impossible to gather an example of each of these instances in S, some kind of preprocessing $\mathcal{P}$ is applied to I, with the hope that it will project all fatigue-ridden instances $\overline{\mathbf{a}'}$, $\overline{\mathbf{a}''}$ and so on, onto $\overline{\mathbf{a}}$ itself. In set-theoretic terms, the operator $\mathcal{P}$ projects I_U back onto I'_U, where I'_U is the reduced portion of I_U, which has originally been captured by S.

Our opinion is that the results achieved by such methods are in practice never guaranteed to make the control system really reliable. The size and extent of the useful input space I_U is essentially unpredictable, and one is never guaranteed that $\mathcal{P}$, which must be evaluated a priori, will correctly project the entire I_U onto I'_U.

A model always necessarily represents a limited view of the world, and especially in the case of a prosthesis, in principle actively worn twelve hours a day, the number of situations in which $\mathcal{P}$ will fail to reduce I_U to what f already knows is fundamentally unlimited. Think of the action required to carry the egg, namely a precision grip, but performed while the subject is running, doing something else with the other arm or lifting the arm to place the object of interest in a cupboard. Compressing all this information in I'_U would entail having at our disposal a complete dynamic model of the musculoskeletal system. This is very likely unfeasible.

The second possibility, and in our opinion the only one left, is that of "learning more", that is, that of expanding S until it induces a useful subset I'_U, which virtually coincides with I_U, that is, it contains all possible instances of each action of interest. This method seems at first as unfeasible as the previous one, for at least two reasons:

1. the size of S is now extremely large—in principle unlimited;
2. again, the initial gathering of S must take into account all possible future situations.

Item 1 can only be solved by using an approach that is bounded in space and time, that is, whose time and space complexity *do not depend on the number of samples in* S. To solve Item 2 one possibility is that of gathering S piecewise, "on-demand", only whenever a new situation arises. We propose that incremental learning represents a solution to both problems, having the potential to radically advance the state of the art in prosthetic control.

11.2.4 Incremental/Interactive Learning

Before we move on to describe our own solution to this problem, that is a working incremental/interactive learning system for hand prosthetics (Sect. 11.3), let us try and enumerate a few characteristics such a system must enjoy. By *incremental learning* it is hereby meant an adaptive system able to update its own model whenever required. That the system must be adaptive stems from the observations of the previous subsection. In particular we speculate that

- the range of possible situations in which the control system must be able to reliably work is too large for a monolithic system;
- a full model of the human arm/hand musculoskeletal system would be too complex to be of any practical usefulness, and anyway unfeasible for miniaturization on a prosthetic device.

Requirement #1 The system must be *adaptive*. It must be possible to calibrate it specifically for each subject. In other words, it must be possible to build a specific *model* for each subject.

We believe that machine learning is the way ahead. Potentially, each subject needs a different f to be tailored ("calibrated", "trained") for her/him; in particular this is the case, in the literature found so far, for amputees, who present an extremely wide range of stumps and remnant muscle structures to the outside world [8]. As the calibration in this case is represented by the gathering of the training set S, we also require that

Requirement #2 The system must be *quickly calibrated.* In this case "quickly" means, fast enough not to distract the subject from the task (s)he is performing, without imposing too high a cognitive burden, and without forcing her/him into a potentially distressing or dangerous activity.

Moreover, to take into account the potentially endless range of different situations the subject might want to have the system work correctly, and since no machine can feasibly stand an endless flow of data, we require that

Requirement #3 The system must be *bounded in space and time*. The model generated by the system must be independent from the size of the training set S and, in general, it must not depend on the time it has been active.

Lastly, whenever a new situation "worth learning" appears, we need the system to be able to update its own model, maintaining the three requirements above. This is our own definition of incrementality:

Requirement #4 The system must be *incremental.* The model generated by the system must be updatable on-the-fly, whenever required, whenever new information is available and whenever the subject deems that the prediction is no longer reliable (for instance, due to muscular fatigue).

Notice that this last requirement entails the ability to both "correct" previously learned patterns, which change their appearance in time (e.g., because of muscle fatigue), and to learn new patterns the subject deems interesting and that the system has never seen before. Actually, the two cases are completely equivalent from the machine learning point of view, given that the right target values are assigned to each new pattern—old ones in the case of pre-existing patterns found in a new situation, and brand new ones in the case of totally new patterns.

A system which enforced all four requirements above would constitute a new way of coupling a human subject and a complex robotic artifact. Adequate speed and easiness of calibration, united with accuracy of the prediction (a requirement that we assume as already present and do not even list above, of course) and incrementality, leads to the possibility for the subject to stop the prediction whenever required; correct the system's mistakes or show it a new pattern to be learned; and then go back to prediction.

11.3 A Practical Method of Incremental Learning

In this section a natural prosthetic control system is described, fulfilling the four requirements set out in Sect. 11.2. The system we describe enforces *regression* rather than classification, yielding in general approximated values for the *activations* of each DoF of a prosthesis instead of a label denoting a predefined action. Notice the difference between the two approaches: whereas classification is essentially a *decision* system, imposing artificial hard boundaries on regions of the input space, regression outputs values in real-valued range, enabling control over an infinite manifold of configurations (of positions, forces and so on).

In each of the following subsections the system is introduced in successive steps. Firstly a simple, monolithic linear method, then its non-linear extension and then its incremental variant. Lastly, a few optimizations are introduced, which improve its practical usability.

11.3.1 Monolithic Learning in the Linear Case

Machine learning is essentially about building a function approximation starting from a training set S (supervised, non-parametric learning). From this point of view, one of the simplest ML approaches is represented by Least-Squares Regression, which we employ in the regularized form called Ridge Regression (RR from now on, [24]). Given a training set of N (sample, target) pairs, $S = \{(\mathbf{x}_i, y_i)\}_{i=1}^N$, RR builds a linear approximation $\hat{y}_i = \mathbf{w}^T \mathbf{x}_i$ in a numerically stable way, such as to minimize the Mean-Squared Error between $\hat{y}_i$ and y_i, for all pairs in S. We hereby assume that the input space be represented by d-dimensional feature vectors somehow extracted from the (possibly preprocessed) signals, $\mathbf{x} \in \mathbb{R}^d$ (this implies that $\mathbf{w} \in \mathbb{R}^d$, too). We also assume that $y \in \mathbb{R}$. Notice that this does not restrict the possibility of having many RR machines in parallel, each one yielding a value for a DoF of the prosthesis.

Let X be a matrix representing S, that is, $X \in \mathbb{R}^{N \times d}$ is the ordered juxtaposition of all signal samples collected so far; similarly, the vector $\mathbf{y} \in \mathbb{R}^N$ orderly collects all target values. Then the RR model $\mathbf{w}$ is given by

$$\mathbf{w} = (X^T X + \lambda I)^{-1} X^T \mathbf{y} \tag{11.1}$$

where I is the identity matrix of order d and $\lambda > 0$.

RR is a good candidate as a monolithic learning approach, whenever it can be safely assumed that there exists a *linear* relationship between the samples and the target values. Notice that both the time and space complexity of RR, in turn $O(d^3 + Nd^2)$ and $O(d^2 + Nd)$, depend on the size of the training set N—this is clearly the case since the matrix X must be stored somewhere and used, e.g., to evaluate $X^T X$. However, this dependency is only linear; the dominating terms, d^3 and d^2, only depend on the *dimension of the input space*. For instance the d^3 time complexity is due to the matrix inversion in the expression of $\mathbf{w}$—but the matrix to be inverted, $(X^T X + \lambda I)$, is only $d \times d$.

Simple as it is, and limited to the linear case, RR already fulfills Requirement #1 (adaptivity) and partially fulfills Requirement #2 (fast calibration) in the case N is not exceedingly large, since it only depends linearly on it. As opposed to that, it does not *not* fulfill Requirement #3 (boundedness). Lastly, notice that the model $\mathbf{w}$ is *calculated* directly from S (that is from X and $\mathbf{y}$) without the need of minimizing a cost functional—actually, the minimum of the regularized Mean-Squared Error cost functional

$$\arg\min_{\mathbf{w}} \sum_{i=1}^{N} (y_i - \mathbf{w}^T \mathbf{x}_i)^2$$

is found exactly for the above-mentioned value of $\mathbf{w}$. Being able to directly evaluate $\mathbf{w}$ has the non-negligible advantage of getting rid of local minima, guaranteeing that $\mathbf{w}$ is consistently *the optimal model* (in the sense of the MSE) given the assumption of linearity and the training set S.

11.3.2 Extension to the Non-linear Case

In case the assumption of linearity must be lifted, the simplest way of extending RR is that of employing a linear combination of non-linear basis functions to build the approximant f, in other words $\hat{y}_i = \mathbf{w}^T \phi(\mathbf{x}_i)$. This is essentially a variant of the kernel trick. One very convenient method to build such a theoretically solid extension is given by Random Fourier Features (RFFs, [48, 49]). As opposed to other, more popular and established kernel methods such as, e.g., Support Vector Machines [6, 56], using RFFs one is able to *directly* compute the mapping ϕ, whereas in most kernel-based approaches only the product of two applications of ϕ, that is $k(x, y) = \phi(x)\phi(y)$ can be evaluated. This is a direct consequence of the fact that RFFs represent a finite-dimensional approximation to the Gaussian kernel. The number of RFFs, $D > 0$, which must be decided a priori, controls the accuracy of this approximation and, not incidentally, dominates the computational complexity of RFFs when applied to RR. In the standard case, as D grows the prediction becomes more accurate but the computational requirement grows, too—one must find a trade-off.

Another way to describe RFFs is that they represent a non-linear extension to RR, which can be "plugged into" it. We now give an informal description of the approach, suggesting that the reader interested in the mathematical details should consult the seminal papers [48, 49] as well as [20, 21] for some applications. Here, suffice it to say that according to Bochner's Theorem (plus some inessential assumptions), any shift-invariant kernel is the expected value of the inner product of two applications of ϕ_ω,

$$k(\mathbf{x}, \mathbf{y}) = \mathbb{E}[\phi_\omega(\mathbf{x})\phi_\omega(\mathbf{y})] \approx \phi_\omega(\mathbf{x})\phi_\omega(\mathbf{y})$$

where $\boldsymbol{\omega}$, a d-dimensional vector of real numbers, is drawn randomly from a probability distribution corresponding to the kernel being approximated. Intuitively, this means that any kernel can be approximated by a sort-of finite Fourier expansion of its own probability distribution; in the case of the Gaussian kernel, $k(\mathbf{x}, \mathbf{y}) = e^{-\gamma||\mathbf{x}-\mathbf{y}||^2}$ where $\gamma > 0$, $\boldsymbol{\omega}$ can be simply drawn from a normal distribution with zero mean value and covariance $2\gamma I$, getting to a closed-form expression for ϕ_ω,

$$\phi_\omega(\mathbf{x}) = \sqrt{2}\cos(\boldsymbol{\omega}^T\mathbf{x} + \beta)$$

(additionally, β is drawn from a uniform distribution in $[0, 2\pi]$.) This particular ϕ_ω maps an input vector $\mathbf{x}$ to a real number, associated to a particular $\boldsymbol{\omega}$; it is however standard to create D vectors $\boldsymbol{\omega}_i$ rather than just one, in order to reduce the variance associated with a random distribution. In the end (dropping the $\boldsymbol{\omega}$ subscript to simplify the notation), the RFF approach works by non-linearly mapping each and every input sample $\mathbf{x} \in \mathbb{R}^d$ into a D-dimensional vector:

$$\phi(\mathbf{x}) = \frac{1}{\sqrt{D}}[\cos(\boldsymbol{\omega}_1^T\mathbf{x} + \beta_1) \ \dots \ \cos(\boldsymbol{\omega}_D^T\mathbf{x} + \beta_D)]^T$$

The operator ϕ induces a D-dimensional space called *feature space* by projecting $\mathbf{x}$ onto a manifold of $\mathbb{R}^D$, namely the surface of the $\frac{1}{D}$-radius D-dimensional hypersphere. This particular mapping is guaranteed by Bochner's theorem to converge to the Gaussian kernel approach as D grows.

Given then ϕ, as is standard in kernel-based methods, we hope to be able to linearly solve the originally non-linear problem by pushing all the linear machinery (RR in our case) in the feature space. In order to compute the model $\mathbf{w}$, which is now D-dimensional, one simply plugs ϕ back into Eq. 11.1, obtaining

$$\mathbf{w} = (\phi(X)^T\phi(X) + \lambda I)^{-1}\phi(X)^T\mathbf{y}$$

where, with a slight abuse of notation, we denote by $\phi(X)$ the application of ϕ to each row of X; therefore, $\phi(X) \in \mathbb{R}^{N\times D}$. This method has several useful properties:

1. it only involves drawing the $\boldsymbol{\omega}$s and βs from two random distributions, once and for all at the beginning. Given a reasonably large value of D, all "runs" of the approach will yield comparable results;
2. its time and space complexities are $O(D^3 + ND^2)$ and $O(D^2 + ND)$ analogously to the linear case; that means that the additional computational burden with respect to RR only depends on the choice of D;

3. as a consequence, the grid search necessary to tune the two additional hyperparameters D and γ is in practice very fast; usually D is set at a "reasonable" value around 500, or anyway to the maximum value that can be afforded, given the computational constraints.

RFFs, coupled with RR, represent a cheap and surprisingly simple non-linear approximant; the computational machinery required is limited to algebraic matrix manipulation plus matrix inversion, provided that in the beginning the ω_i, β_i are generated.

11.3.3 Incrementality

The naive way of making such a method incremental is, of course, to store S and add to it every new (sample, target) pair that is gathered. This is clearly unacceptable since in the long run S will make any finite memory bank overflow, let alone the computational burden required to evaluate $\mathbf{w}$ every time, a task which depends on N. An alternative approach is that of limiting the size of S, keeping it fixed at some predetermined value N_{max} entailing a computationally bearable evaluation of $\mathbf{w}$; this idea has been explored, e.g., in [17, 29, 40]. In our case, a very convenient solution is that of considering the arrival of a new (sample, target) pair as a *perturbation* to the inverse matrix $(X^T X + \lambda I)^{-1}$. Using a rank-1 update method directly on it, the explicit inversion can be avoided. In practical terms, it is convenient to redefine Eq. 11.1 as the product of a matrix A and a vector $\mathbf{b}$:

$$\mathbf{w} = (X^T X + \lambda I)^{-1} X^T \mathbf{y} := A\mathbf{b} \tag{11.2}$$

where A is $(X^T X + \lambda I)^{-1}$ and $\mathbf{b}$ is $X^T \mathbf{y}$—notice that A *already is the inverse* of a matrix. Given a new (sample, target) pair $(\mathbf{x}', y')$, the updated model $\mathbf{w}' = A'\mathbf{b}'$ is given by applying the Sherman-Morrison formula [23]:

$$A' = A - \frac{A\mathbf{x}'\mathbf{x}'^T A^T}{1 + \mathbf{x}'^T A\mathbf{x}'} \quad \text{and} \quad \mathbf{b}' = \mathbf{b} + \mathbf{x}' y'$$

In practice, one starts by setting $A = \frac{1}{\lambda} I$ and $\mathbf{b} = 0$, so that $\mathbf{w} = 0$; as new $(\mathbf{x}', y')$ pairs arrive, the updated model $\mathbf{w}'$ is built. It is easy to prove that the model $\mathbf{w}$ obtained after, say, N such steps is exactly the same that would have been calculated one-shot, having at our disposal the whole training set S containing N (sample, target) pairs. Notice that, as no explicit matrix inversion is required by the above formula, the computational complexity of the update step is only $O(d^2)$ *both in time and space*. As a matter of fact, in this case X and $\mathbf{y}$ need not be explicitly stored anywhere: as soon as $\mathbf{w}'$ has been evaluated, there is no further need of keeping $(\mathbf{x}', y')$. The Sherman-Morrison formula gives us an effective tool to perform RR incrementally (iRR), without any danger of exhausting the computational resources of the control system.

As a last step, consider the application of RFFs to iRR. Again, the non-linear mapping operator ϕ can be simply applied to $\mathbf{x}$ wherever it appears in the Sherman-Morrison formula, finally yielding

$$A' = A - \frac{A\phi'\phi'^T A^T}{1 + \phi'^T A\phi'} \quad \text{and} \quad \mathbf{b}' = \mathbf{b} + \phi' y'$$

where we denote by ϕ' the application $\phi(\mathbf{x}')$ in order to keep the notation light. Again, one can start by setting $A = \frac{1}{\lambda}I$ (this time I is the identity matrix of order D rather than d) and $\mathbf{b} = 0$. As one can easily guess, the computational complexity of the model update step is, in this case, $O(D^2)$ both in time and space.

11.3.4 Obtaining Ground Truth

Joining RFFs to iRR (call the new approach iRR-RFF) as described in the previous subsection constitutes a practical tool for natural prosthetic control, in the sense outlined by the four Requirements of Sect. 11.2. A detailed summary of this match is given at the end of this section. Before that, as a last remark, let us notice two further factors that potentially limit its applicability, in particular to amputated subjects:

1. amputated subjects cannot operate any position/force sensor, therefore the experimenter has the problem of gathering sensible ground truth, i.e., the target values $\mathbf{y}$ in S. One partial solution is that of having them use the remaining limb in a bilateral fashion [10, 41], but one is never sure how much the two limbs match each other—bottom line, not even the amputee is!
2. In general, an amputation deprives the subject of sensory feedback (including visual feedback); as a consequence of this, amputated subjects are usually unable to perform finely graded tasks. The experimenter cannot sensibly expect, e.g., that an amputee imagines flexing the middle finger with 50 % of the maximum voluntary contraction.

Additionally, the initial data gathering phase can be tiresome and stressful for the subject—it must be kept as short as possible. To counter these problems, a couple simple strategies can easily be put into place.

Firstly, the usage of *goal-directed stimuli* in order to have the subject generate sensible ground truth for the system. In practice, rather than relying on data sampled from sensors, the experimenter puts the subject in a maximally comfortable situation and then asks for a specific voluntary muscle contraction. It can either be the activation of a single DoF of the prosthesis, such as, e.g., flexing the index finger or the wrist, as well as enacting a specific type of grasp (power, cylindrical, precision grip, etc.). In order to foster the production of a sensible input signal, a visual stimulus can be presented to the subject, such as a 3D-generated model of the missing limb assuming the required posture; or, the experimenter can vocally instruct the subject while showing the required posture with her/his own limb; or even, the stimulus can be delegated to the prosthesis itself, which can be commanded a specific movement to be imitated by the subject. In some cases, even looking at some graphical representation of the input signal itself (for instance, a radial graph showing the voltages recorded be the sEMG electrodes) can help. As already remarked, there is no assurance that the subject will be doing what (s)he is required to do; not even the subject her/himself can be sure of that. The hope is that the input signal, possibly when stable, faithfully represents the intent of the subject.

Secondly, coherently with the reduction-oriented approach suggested by the notion of synergies as widely discussed throughout the book, it is convenient to only gather minimal and maximal activations and then let the regression machines interpolate the rest. This makes the data gathering phase shorter and more suitable for an amputated subject. These two strategies have been successfully employed together for the first time in [57] in the linear case, where they were collectively termed "realistic approach". In a further analysis and practical demonstration [20], the approach has been proved successful in the non-linear case, too.

To sum up, here is how iRR-RFF matches the four Requirements outlined at the end of the previous section. As all machine learning approaches, it is adaptive (Requirement #1), meaning that it builds its own model based upon data gathered by a human subject engaged in a goal-oriented task. The significant differences found in the human anatomy of different subjects, as well as the fact that each amputation produces a very different final layout of muscle remnants, suggest that it will be a very hard, if not impossible task, to build such a universal system. The hope is therefore that of making the (machine) adaptation, already called calibration or training phase, as short as possible; possibly, resilient to the daily donning and doffing of the prosthesis—this seems a much more doable task, as the electrode layout in a prosthetic socket never changes along time.

Requirements #2 and #3 are matched by the time complexity of iRR-RFF, as well as by the easiness of the data gathering if one enforces the two last strategies outlined above; and by the fact that iRR-RFF is also bounded in space, the only space requirement being the storage of a $D \times D$ matrix. Experimental results (see the next subsection for more details) reveal that iRR-RFF can be implemented in practice in a mid-level imperative programming language such as, e.g., C, on standard hardware, achieving a constant update time in the order of magnitude of the tens of milliseconds.

Finally, Requirement #4 is exactly realized by the usage of a rank-1 matrix update technique—in the case outlined above, the Sherman-Morrison formula. It is worth remarking once again that incrementality in this case still yields the theoretically optimal model that would have been achieved using the same data in a batch fashion. iRR-RFF can therefore serve as the basis for a theoretically well-founded, fast, incremental intent gathering system. The next subsection describes two of its practical applications.

11.3.5 Applications

11.3.5.1 The Ultrapiano/Ultraharmonium

Using an instrumented glove and a commercial ultrasound machine, in [12] it was first proved that first-order spatial averages of the gray levels in ultrasound images of the human forearm are linearly related to the metacarpo-phalangeal angles, i.e., the angles formed by the first phalanx of the fingers with the palm. (A deeper analysis appears in [13].) This unexpected phenomenon was first exploited in [57] to outline and practically show that the above-mentioned averages could be used as the input space to a system enforcing iRR. The update times were found to be on average 16.5 ms, while the prediction without update took only 3.7 ms; these times were ascertained to be independent of the number of samples gathered so far, and compatible with a cinema-quality visualization of a 3D hand model on a screen (30 frames per second). This paved the way to two further applications. The real-time prediction of finger angles and forces, coupled to the detection of the position of the wrist obtained via a standard magnetic tracker, was transmitted to a virtual-reality system showing in turn a piano [51] and a harmonium [11]. The system was tested on several intact subjects revealing a satisfactory level of immersion in the virtual world. The usage of ultrasound imaging as a HMI for the disabled is actually gaining momentum and its perspectives have been widely discussed in [7]. Figure 11.1 shows, and quickly describes, the setup used in [11].

11.3.5.2 Teleoperated Manipulation with a Prosthesis

In [20] an *i-LIMB Ultra* hand prosthesis by Touch Bionics[4] was used to manipulate, pick and place and carry a few everyday-life objects in a teleoperated scenario. Compared to the Pisa/IIT SoftHand described in Chap. 8 and tele-operated leveraging strategies discussed in Chap. 10, the i-LIMB Ultra has more than one DoA, and this justifies the synergy-inspired approach to cope with its control through the techniques described in the current chapter. Teleoperation in this case is used as a proxy for the real-life application of a prosthesis to an amputee: it constitutes a simpler case since all problems related to the weights to be carried can be neglected (as they are taken care of by the slave platform). As in the case outlined above, a magnetic tracker was used to track the position of the human wrist and control the position of the slave's end-effector using a high-stiffness impedance controller [43] on the humanoid platform TORO. At the same time, 10 standard sEMG electrodes by Ottobock[5] were used to gather the muscle activity of the forearm of the master. Using iRR-RFF, the sEMG signal was converted into torque (current) commands for the five motors of the prosthesis, enforcing one of four predefined grasp schemes. The offline experiment

[4]See www.touchbionics.com.

[5]Namely, *MyoBock 13E200*, see www.ottobock.com.

Fig. 11.1 The setup used in [11]. A magnetic tracker and an ultrasound transducer are fixed on the subject's forearm; an HMI based upon iRR converts local spatial features of the ultrasound images to finger forces (screen on the *right*); lastly, finger forces and wrist position are used in a virtual setup (screen on the *left*) to play a piano

performed in the paper clearly showed that non-linear, incremental regression was required to keep the prediction error at a reasonable level (see Fig. 11.2, reproduced from [20]).

In the demonstration, a success rate of 75–95 % was obtained while grasping, lifting, picking up and placing objects such as a bottle, a ball, a credit card, independent of wide ranges of hand motion and wrist rotation, and related high speeds.

11.4 Discussion

Introducing incrementality in a machine-learning-based prosthetic control system represents in our opinion a very beneficial improvement, at least in two senses:

1. it potentially solves the problem of predicting all possible situations in which an action will be performed by introducing on-demand model update;
2. it realizes a virtuous loop between man and machine, exploiting the phenomenon of reciprocal learning.

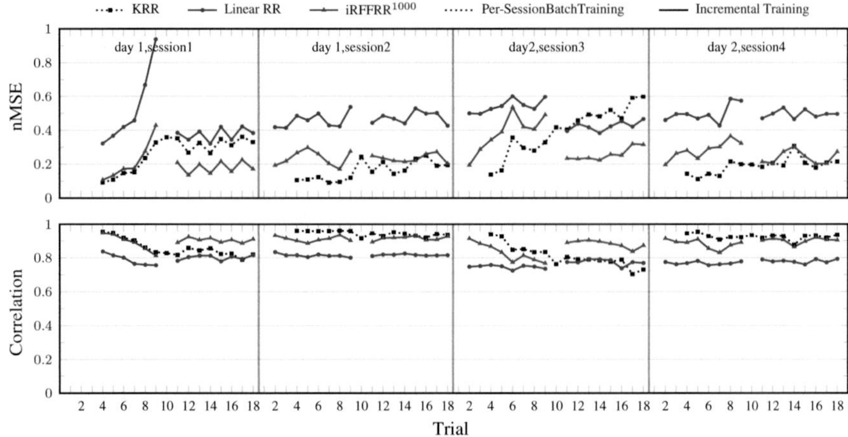

Fig. 11.2 Performance obtained by RR ("Linear RR"), Kernel Ridge Regression ("KRR") and iRR-RFF with $D = 1000$ ("iRFFRR^{1000}") while predicting five voluntary muscle contractions using sEMG. The possibility of updating the models amidst the prediction (trials 9 and 10 of each session and day) keeps the performance of iRR-RFF well above both RR, which is linear, and KRR, which is non-linear but also not incremental. Reproduced from [20]

Notice that, while the first claim is being proved in the academic world in these years, the second is a whole unexplored territory so far. That new muscle synergies (in the broad definition used in this chapter) can be learned, retained over the weeks and then re-used whenever required is the subject of a very exciting line of research (see, e.g., [25, 46]); as well, there are hints that the very usage of a prosthesis induces better signals for its own control [45], which seems to point in the direction of goal-directed stimuli claimed here in Sect. 11.3.

What the best "reciprocal training" strategy is; how to best help the subject use the control system; what kind of *games* to employ; these questions are still open and indeed fascinating. This research is also motivated by the remarkable fact that improving the embodiment of a prosthesis seems to diminish phantom-limb pain [42] and amend abnormal phantom sensations. In any case, interactive learning would represent a crucial form of help to reach this goal.

11.4.1 On the Capacity of Incremental Learning

There seems to be a paradox in the claim that a good control system (as defined in the very chapter) must be bounded in space: such a system is limited and it therefore seems that eventually, as I_U (and accordingly, S) grows, the control function f won't be able to accommodate all required patterns. This is indeed true and boils down to the question of how "large" the learning machine should be; unfortunately, to the best of our knowledge, so far no machine learning method is known that can change its own capacity (in the sense of the Vapnik-Chervonenkis dimension, see [56, 59]) and there is no substantial way of determining this a priori. To stay with our own example of iRR-RFF: how large a D is required? This is a crucial question since D cannot be sensibly altered after it has been chosen. So far the answer to this question can only be empirical: choose D as large as possible given the hardware at disposal; but a more sensible way to determine the size of a model is a very desirable achievement, and a very interesting research question.

11.4.2 Relation to Muscle Synergies as Traditionally Defined

At the time of writing we are not sure whether and how the traditional concept of task-based muscle synergy can be used in the control of dexterous prostheses. Early experiments [60] indicate that such a control is indeed possible, but will inevitably be limited to the combinations of a few synergies. If such a control can be extended to more complex control manifolds, such as, e.g., those required to play a keyboard, is unclear; it is as well unclear whether or not even an extended control based upon muscle synergies would not look quite like the one described in this chapter. All in all, in order to improve one's own dexterity, a subject must learn (think of the painful process required, e.g., to proficiently play tennis!), and that is probably tantamount to using many more synergies than those required for the classical basic set of everyday-life tasks. Here too, the question is open and fascinating.

11.5 Conclusions

The ideal (hand) prosthesis is like a pair of glasses: you wear it in the morning, it works seamlessly all day long, you take it off in the evening, and then wear it again the morning after.[6] Clearly, none of the control systems currently available in the academy, let alone in the clinics, can even hope to enforce this. One possible solution is that of simplifying the prosthesis itself: for example, the Pisa/IIT SoftHand (Chap. 8) reduces the complexity of controlling many DoFs through an innovative design with only one DoA—this motivates the minimalistic tele-impedance approach described in Chap. 10. On the other hand, most of the current prosthetic systems have several actuators to be controlled, and in this case right now *the control system is the bottleneck*. For instance, the *i-LIMB Ultra Revolution* by Touch Bionics has six independent motors, as well as Vincent System's *Vincent Hand Evolution2*; Steeper's *BeBionics* has five, while the *Michelangelo* hand/wrist system has four; and no system so far can control these DoFs independently. That means that there is more dexterity

[6]This inspirational metaphor is due to Peter J. Kyberd in a personal communication with the author of this chapter.

available than what any patient can hope to use. We also believe that machine-learning-based control is the way ahead, but its reliability is still very questionable.

In this chapter we have argued that *incremental/interactive learning* would make prosthetic control radically more reliable. In one sentence: give the subjects the chance to teach their own control system what is needed. We claim that this idea could in the near future represent a leap forward.

Acknowledgments This work was partially supported by the Swiss National Science Foundation Sinergia project #132700 NinaPro (Non-Invasive Adaptive Hand Prosthetics) and by the FP7 project The Hand Embodied (FP7-248587).

References

1. Artemiadis PK, Kyriakopoulos KJ (2011) A switching regime model for the EMG-based control of a robot arm. IEEE Trans Syst Man Cybern Part B Cybern 41(1):53–63
2. Aszmann OC, Roche AD, Salminger S, Paternostro-Sluga T, Herceg M, Sturma A, Hofer C, Farina D (2015) Bionic reconstruction to restore hand function after brachial plexus injury: a case series of three patients. Lancet 9983:2783–2789
3. Bernshtein NA (1967) The coordination and regulation of movements. Pergamon Press, Oxford
4. Bicchi A, Gabiccini M, Santello M (2011) Modelling natural and artificial hands with synergies. Philos Trans R Soc Lond Ser B Biol Sci 366(1581):3153–3161
5. Biddiss E, Chau T (2007) Upper-limb prosthetics: critical factors in device abandonment. Am J Phys Med Rehabil 86(12):977–987
6. Boser BE, Guyon IM, Vapnik VN (1992) A training algorithm for optimal margin classifiers. In: Haussler D (ed) Proceedings of the 5th annual ACM workshop on computational learning theory. ACM press, pp 144–152
7. Castellini C (2014) State of the art and perspectives of ultrasound imaging as a human-machine interface. In: Artemiadis, P (ed) Neuro-robotics: from brain-machine interfaces to rehabilitation robotics. Trends in augmentation of human performance, vol 2. Springer, Netherlands, pp 37–58. doi:10.1007/978-94-017-8932-5
8. Castellini C, Artemiadis P, Wininger M, Ajoudani A, Alimusaj M, Bicchi A, Caputo B, Craelius W, Dosen S, Englehart K, Farina D, Gijsberts A, Godfrey S, Hargrove L, Ison M, Kuiken T, Markovic M, Pilarski P, Rupp R, Scheme E (2014) Proceedings of the first workshop on peripheral machine interfaces: going beyond traditional surface electromyography. Front Neurorobot 8:22. doi:10.3389/fnbot.2014.00022
9. Castellini C, Fiorilla AE, Sandini G (2009) Multi-subject/daily-life activity EMG-based control of mechanical hands. J Neuroeng Rehabil 6(41):12. doi:10.1186/1743-0003-6-41
10. Castellini C, Gruppioni E, Davalli A, Sandini G (2009) Fine detection of grasp force and posture by amputees via surface electromyography. J Physiol (Paris) 103(3–5):255–262. doi:10.1016/j.jphysparis.2009.08.008
11. Castellini C, Hertkorn K, Sagardia M, Sierra González D, Nowak M (2014) A virtual piano-playing environment for rehabilitation based upon ultrasound imaging. In: Proceedings of BioRob–IEEE international conference on biomedical robotics and biomechatronics, pp 548–554. doi:10.1109/BIOROB.2014.6913835
12. Castellini C, Passig G (2011) Ultrasound image features of the wrist are linearly related to finger positions. In: Proceedings of IROS–international conference on intelligent robots and systems, pp 2108–2114. doi:10.1109/IROS.2011.6048503
13. Castellini C, Passig G, Zarka E (2012) Using ultrasound images of the forearm to predict finger positions. IEEE Trans Neural Syst Rehabil Eng 20(6):788–797. doi:10.1109/TNSRE.2012.2207916

14. Castellini C, van der Smagt P (2013) Evidence of muscle synergies during human grasping. Biol Cybern 107(2):233–245. doi:10.1007/s00422-013-0548-4
15. d'Avella A (2009) Muscle synergies. In: Binder M, Hirokawa N, Windhorst U (eds) Encyclopedia of neuroscience. Springer, Berlin, pp 2509–2512
16. D'avella A, Lacquaniti F, (2013) Control of reaching movements by muscle synergy combinations. Front Comput Neurosci 7(42): doi:10.3389/fncom.2013.00042
17. Dekel O, Shalev-Shwartz S, Singer Y (2008) The forgetron: a kernel-based perceptron on a budget. SIAM J Comput 37(5):1342–1372. doi:10.1137/060666998
18. Farina D, Jiang N, Rehbaum H, Holobar A, Graimann B, Dietl H, Aszmann O (2014) The extraction of neural information from surface EMG for the control of upper-limb prostheses: emerging avenues and challenges. IEEE Trans Neural Syst Rehabil Eng 22(4):797–809
19. Fougner A, Stavdahl Ø, Kyberd PJ, Losier YG, Parker PA (2012) Control of upper limb prostheses: terminology and proportional myoelectric control–a review. IEEE Trans Neural Syst Rehabil Eng 20(5):663–677
20. Gijsberts A, Bohra R, Sierra González D, Werner A, Nowak M, Caputo B, Roa M, Castellini C (2014) Stable myoelectric control of a hand prosthesis using non-linear incremental learning. Front Neurorobot 8(8): doi:10.3389/fnbot.2014.00008
21. Gijsberts A, Metta G (2011) Incremental learning of robot dynamics using random features. In: IEEE international conference on robotics and automation, pp 951–956. doi:10.1109/ICRA.2011.5980191
22. Guo JY, Zheng YP, Xie HB, Koo TK (2013) Towards the application of one-dimensional sonomyography for powered upper-limb prosthetic control using machine learning models. Prosthet Orthot Int 37(1):43–49
23. Hager WW (1989) Updating the inverse of a matrix. SIAM Rev 31:221–239. doi:10.1137/1031049
24. Hoerl AE, Kennard RW (1970) Ridge regression: biased estimation for nonorthogonal problems. Technometrics 12:55–67
25. Ison M, Artemiadis P (2014) The role of muscle synergies in myoelectric control: trends and challenges for simultaneous multifunction control. J Neural Eng 11:051001
26. Jiang N, Došen S, Müller KR, Farina D (2012) Myoelectric control of artificial limbs: Is there a need to change focus? [in the spotlight]. IEEE Signal Process Mag 29(5):150–152. doi:10.1109/MSP.2012.2203480
27. Jiang N, Englehart K, Parker P (2009) Extracting simultaneous and proportional neural control information for multiple-dof prostheses from the surface electromyographic signal. IEEE Trans Biomed Eng 56(4):1070–1080. doi:10.1109/TBME.2008.2007967
28. Jiang N, Rehbaum H, Vujaklija I, Graimann B, Farina D (2013) Intuitive, online, simultaneous and proportional myoelectric control over two degrees of freedom in upper limb amputees. IEEE Trans Neural Syst Rehabil Eng 22(3):501–510. doi:10.1109/TNSRE.2013.2278411
29. Kõiva R, Hilsenbeck B, Castellini C (2013) Evaluating subsampling strategies for sEMG-based prediction of voluntary muscle contractions. In: Proceedings of ICORR–international conference on rehabilitation robotics, pp 1–7. doi:10.1109/ICORR.2013.6650492
30. Kuiken TA, Li G, Lock BA (2009) Targeted muscle reinnervation for real-time myoelectric control of multifunction artificial arms. J Am Med Assoc 301(6):619–628
31. Kumar A (2003) Movement and Locomotion in Animals. Discovery Publishing Pvt Ltd., New Delhi
32. Latash M (2008) Synergy. Oxford University Press, New York
33. Marković M, Došen S, Cipriani C, Popović D, Farina D (2014) Stereovision and augmented reality for closed-loop control of grasping in hand prostheses. J Neural Eng 11:046001
34. Merletti R, Aventaggiato M, Botter A, Holobar A, Marateb H, Vieira T (2011) Advances in surface EMG: recent progress in detection and processing techniques. Crit Rev Biomed Eng 38(4):305–345
35. Merletti R, Botter A, Cescon C, Minetto M, Vieira T (2011) Advances in surface EMG: recent progress in clinical research applications. Crit Rev Biomed Eng 38(4):347–379

36. Merletti R, Botter A, Troiano A, Merlo E, Minetto M (2009) Technology and instrumentation for detection and conditioning of the surface electromyographic signal: state of the art. Clin Biomech 24:122–134
37. Micera S, Carpaneto J, Raspopović S (2010) Control of hand prostheses using peripheral information. IEEE Rev Biomed Eng 3:48–68
38. Muybridge E, Brown LS (1957) Animals in motion (Dover anatomy for artists). Dover Publications, New York
39. Netter FH (2006) Atlas der Anatomie des Menschen, 3rd edn. Thieme, Stuttgart
40. Nguyen-Tuong D, Seeger MW, Peters J (2009) Model learning with local Gaussian process regression. Adv Robot 23(15):2015–2034
41. Nielsen JLG, Holmgård S, Jiang N, Englehart KB, Farina D, Parker PA (2011) Simultaneous and proportional force estimation for multifunction myoelectric prostheses using mirrored bilateral training. IEEE Trans Biomed Eng 58(3):681–688
42. Ortiz-Catalan M, Sander N, Kristoffersen MB, Håkansson B, Brånemark R (2014) Treatment of phantom limb pain (PLP) based on augmented reality and gaming controlled by myoelectric pattern recognition: a case study of a chronic PLP patient. Front Neurosci 8:24
43. Ott C, Eiberger O, Roa M, Albu-Schäffer A (2012) Hardware and control concept for an experimental bipedal robot with joint torque sensors. J Robot Soc Jpn 30(4):378–382
44. Peerdeman B, Boere D, Witteveen H, in 't Veld, RH, Hermens H, Stramigioli S, Rietman H, Veltink P, Misra S, (2011) Myoelectric forearm prostheses: state of the art from a user-centered perspective. J Rehabil Res Dev 48(6):719–738
45. Powell MA, Kaliki RR, Thakor NV (2014) User training for pattern recognition-based myoelectric prostheses: improving phantom limb movement consistency and distinguishability. IEEE Trans Neural Syst Rehabil Eng 22(3):522–532
46. Powell MA, Thakor NV (2013) A training strategy for learning pattern recognition control for myoelectric prostheses. J Prosthet Orthot 25(1):30–41
47. Radmand A, Scheme E, Englehart K (2014) High-resolution muscle pressure mapping for upper-limb prosthetic control. In: Proceedings of MEC–myoelectric control symposium, pp 193–197
48. Rahimi A, Recht B (2008) Random features for large-scale kernel machines. Adv Neural Inf Process Syst 20:1177–1184
49. Rahimi A, Recht B (2008) Uniform approximation of functions with random bases. In: Allerton conference on communication control and computing (Allerton08), pp 555–561
50. Ravindra V, Castellini C (2014) A comparative analysis of three non-invasive human-machine interfaces for the disabled. Front Neurorobot 8(24): doi:10.3389/fnbot.2014.00024
51. Sagardia M, Hertkorn K, Sierra González D, Castellini C (2014) Ultrapiano: a novel human-machine interface applied to virtual reality. In: Proceedings of ICRA–international conference on robotics and automation, p 2089. doi:10.1109/ICRA.2014.6907142
52. Santello M, Baud-Bovy G, Jörntell H (2013) Neural bases of hand synergies. Front Comput Neurosci 7:23
53. Santello M, Flanders M, Soechting JF (1998) Postural hand synergies for tool use. J Neurosci 18(23):10105–10115
54. Santello M, Flanders M, Soechting JF (2002) Patterns of hand motion during grasping and the influence of sensory guidance. Neuroscience 22(4):1426–1435
55. Scheme E, Englehart K (2011) Electromyogram pattern recognition for control of powered upper-limb prostheses: state of the art and challenges for clinical use. J Rehabil Res Dev 48(6):643–660
56. Shawe-Taylor J, Cristianini N (2004) Kernel methods for pattern analysis. Cambridge University Press, Cambridge
57. Sierra González D, Castellini C (2013) A realistic implementation of ultrasound imaging as a human-machine interface for upper-limb amputees. Front Neurorobot 7(17): doi:10.3389/fnbot.2013.00017
58. Tresch MC, Jarc A (2009) The case for and against muscle synergies. Curr Opin Neurobiol 19(6):601–607

59. Vapnik VN (1998) Statistical learning theory. Wiley, New York
60. Wimböck T, Jahn B, Hirzinger G (2011) Synergy level impedance control for multi-fingered hands. In: Proceedings of the IEEE/RSJ international conference on intelligent robots and systems, pp 973–979
61. Yungher D, Wininger M, Baar W, Craelius W, Threlkeld A (2011) Surface muscle pressure as a means of active and passive behavior of muscles during gait. Med Eng Phys 33:464–471

Chapter 12
How to Map Human Hand Synergies onto Robotic Hands Using the SynGrasp Matlab Toolbox

Gionata Salvietti, Guido Gioioso, Monica Malvezzi and Domenico Prattichizzo

Abstract Throughout this book, we have described how neuroscientific findings on synergistic organization of human hand can be used to devise guidelines for the design and control of robotic and prosthetic hands as well as for sensing devices (see Chaps. 8, 10, 11 and 15). However, the development of novel robotic devices open issues on how to generalize the outcomes to different architectures. In this chapter, we describe a mapping strategy to transfer human hand synergies onto robotic hands with dissimilar kinematics. The algorithm is based on the definition of two virtual objects that are used to abstract from the specific structures of the hands. The proposed mapping strategy allows to overcame the problems in defining synergies for robotic hands computing PCA analysis over a grasp dataset obtained empirically closing the robot hand upon different objects. The developed mapping framework has been implemented using the SynGrasp Matlab toolbox. This tool includes functions for the definition of hand kinematic structure and of the contact points with a grasped object, the coupling between joints induced by a synergistic control, compliance at the contact, joint and actuator levels. Its analysis functions can be used to investigate the main grasp properties: controllable forces and object displacements, manipulability analysis, grasp quality measures. Furthermore, functions for the graphical representation of the hand, the object and the main analysis results are provided.

G. Salvietti (✉) · G. Gioioso · D. Prattichizzo
Department of Advanced Robotics, Istituto Italiano di Tecnologia, Via Morego 30, 16163 Genoa, Italy
e-mail: gionata.salvietti@iit.it

G. Gioioso
e-mail: guido.gioioso@iit.it

D. Prattichizzo
e-mail: domenico.prattichizzo@iit.it

G. Gioioso · M. Malvezzi · D. Prattichizzo
Dipartimento di Ingegneria dell'Informazione e Scienze Matematiche, University of Siena, Via Roma 56, 53100 Siena, Italy
e-mail: malvezzi@dii.unisi.it

M. Bianchi and A. Moscatelli (eds.), *Human and Robot Hands*,
Springer Series on Touch and Haptic Systems, DOI 10.1007/978-3-319-26706-7_12

12.1 Introduction

Robotic hands share with the human hand some of the fundamental primitives of motion, grasping, and manipulation. As described in Chap. 8 and in the first part of the book, a deeper understanding of the human way to move their hands could suggest an approach to programming hands that allows users to more easily control the different devices that may be used in a robotic system, by encapsulating the hand hardware in functional modules, and ignoring the implementation-specific details. Human hand synergies introduced in [1] can play the role of such functional modules. Synergies capture the concept that, in the sensorimotor system of the human hand, combined actions are favored over individual component actions, with advantages in terms of simplification and efficiency of the overall system. This reduced subspace allows to design more easily and intuitively robotic hand control algorithms, due to the lower number of DoFs that has to be addressed (see Chaps. 8 and 13). Anyway, this approach can be pursued only if there exists a mapping method that allows to replicate the actions defined in the synergistic subspace.

In this chapter, we will propose a way to map human synergies onto robotic hands by using an object-based method. The target is to reproduce deformations and movements exerted by the paradigmatic human-like hand on a virtual sphere computed as the minimum sphere containing the hand fingertips. This allows to work directly on the task space avoiding a specific projection between different kinematics. Such algorithm has been implemented using the SynGrasp toolbox. Differently from other simulators like GraspIt! [2] and Opengrasp [3], SynGrasp has been developed entirely in MATLAB and offers an easy and intuitive Graphical User Interface (GUI) and script programming. One of the main feature of this programming environment is the possibility of quickly integrate other specific tools and built-in math functions enabling the exploration of multiple approaches and the integration with other analysis tools, e.g. statistical elaboration of experimental data, optimization, dynamic models and simulations etc. Moreover, Matlab is well known also outside the robotic community. This makes our toolbox for grasp analysis also useful in other fields, such as experiment design and validation in Neuroscience, physiology and haptics. The SynGrasp toolbox has been developed in the context of the EU Project "THE—The Hand Embodied". Together with the theoretical framework described in Chap. 13, it represents a useful analytical tool and it provides several functions for human and robot grasping evaluation (see Fig. 12.1) including specific functions for human hand synergies evaluation [4].

The main functions provided within the toolbox can be used for:

- Hand modeling;
- Grasp definition;
- Grasp analysis and optimization;
- Graphics.

The rest of the chapter is organized as it follows. In Sect. 12.2 the main functions of the toolbox are presented, while in Sect. 12.3 the proposed object-based mapping

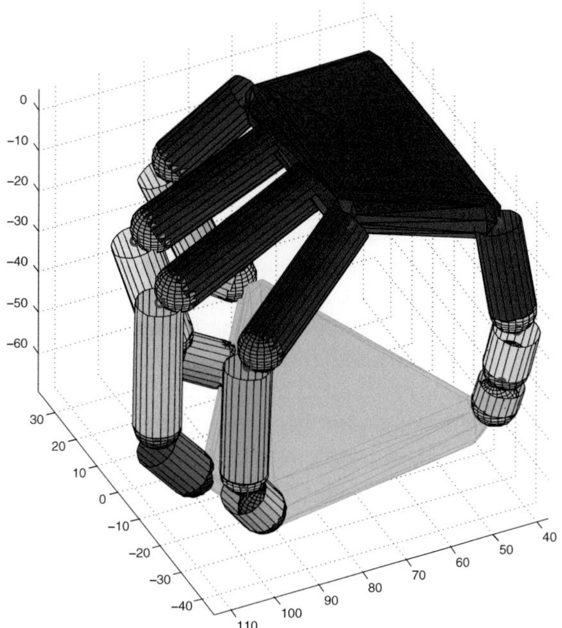

Fig. 12.1 Visualization of the SynGrasp model of an anthropomorphic hand grasping an object

together with its implementation in SynGrasp is described. Finally, Sect. 12.4 summarizes the proposed work, draws some conclusions and indicates the directions of future development of the mapping framework.

12.2 The SynGrasp Toolbox

SynGrasp is a MATLAB toolbox developed for the analysis of grasping, suitable both for robotic and human hands. It includes functions for the definition of hand kinematic structure and of the contact points with a grasped object. The coupling between joints induced by an underactuated control can be modeled. The hand modeling allows to define compliance at the contact, joint and actuator levels. The provided analysis functions can be used to investigate the main grasp properties: controllable forces and object displacement, manipulability analysis, grasp quality measures. Functions for the graphical representation of the hand, the object and the main analysis results are provided. The toolbox, all the relative documentation, and some examples can be downloaded from http://SynGrasp.dii.unisi.it and from HandCorpus (www.handcorpus.org), the open access repository created within the THE Hand Embodied project.

12.2.1 How to Use SynGrasp

The SynGrasp toolbox can be used through the provided GUI or developing Matlab scripts which embed the toolbox specific functions. The GUI allows the user to use the available hand models and interactively perform hand and grasp analysis. The interface is reported in Fig. 12.2. It consist of six separated areas. The main plot is placed in the center of the window. On the left, there are two ares that take care of loading on the main plot a hand model and an object model, respectively. Together with the available hand models, it is possible to load a hand model defined by the user. At the moment, five hands are available in the toolbox. **SG3Fingered** is a three finger hand, inspired by the Barrett hand, **SGDLRHandII** is the DLR-HIT Hand II [5], **SGmodularHand** is a three fingered modular hand presented in [6]. There are also two models of the human hand: **SGparadigmatic** is a 20 DoFs model of an anthropomorphic hand, referred to as *paradigmatic hand* in [7], while the **SGhuman24Dof** is a human hand model with 24 DoFs. Position and orientation of the hand models can be set and modified by the user. Concerning the object, at the moment only spheres, cubes and cylinders are available. Similar to the hand, also for the object it is possible to set and modify position and orientation and moreover it is possible to select the size. On the bottom area, there are sliders that can be used to modify the hand model configuration. A number of sliders equal to that of active joints in the hand model can be used to control each joint separately. The user can also select to use a set of sliders to coordinately move the joints along synergy directions. Synergies are already defined for the human hand models while for the robotic hands they can be: (i) defined directly by the user, (ii) derived from the linearization of the forward kinematic relationships, (iii) obtained from the processing of experimental data or (iv) mapped from the human hand, as reported in Sect. 12.3. If the joints are

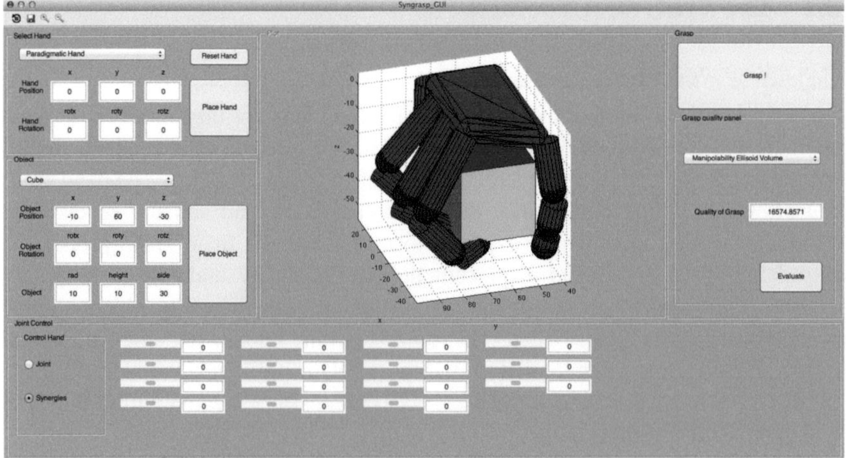

Fig. 12.2 The SynGrasp GUI

moved so be in contact with the object, a contact-detection algorithm is implemented and it is used in order to perform grasp analysis. The hand can be also automatically closed pressing the button placed in the top-right part of the GUI. The arising motion is a "close all" policy. Finally, on the bottom-right there is the zone devoted to grasp analysis. Starting from the hand in contact with the object, it is possible to analyse the quality of the obtained grasp. The desired quality index can be chosen from the drop-down menu.

Scripting mode solution is preferred if a customization is needed. The user can include his/her own functions and/or can modify those already existing. The complete description of all the toolbox functions and their usage is reported in [8]. Some of the function can be used to provide a simple graphical representation of the manipulator and the object. The function **SGplotHand()** display the hand in the joint configuration specified by the user. The function **SGhandFrames()** plots a useful scheme of the hand which highlights the kinematical structure, the joints and links and the orientation of the local frames for each link. It is also possible to draw some simple objects to be grasped using the functions **SGplotCylinder()**, **SGplotSphere()** and **SGplotCube()**.

All the functions described in the rest of the section can be used in scripting mode. Most of them are also embedded on the GUI.

12.2.2 Hand Modelling

This section groups all the functions needed to describe the kinematics of a hand. The hand structure is defined in terms of fingers, links and joints. A cell named **base**, containing as many elements as the number of fingers, collects in each cell element a 4×4 matrix representing the homogeneous transformation matrix between the wrist reference frame and a reference frame defined at the beginning of each finger kinematic chain. Denavit-Hartenberg (DH) parameters [9] have been chosen as default notation. A table containing the DH parameters of each finger has to be provided to describe a hand. A cell named **DHpars**, which has as many elements as the number of fingers, collects in each element a matrix with four columns and as many rows as the number of joints of each finger. Each row represents the DH parameters allowing to define the joint with respect to the preceding one or with respect to the base reference frame. The function **SGmakeHand()** defines a hand structure, whose arguments are defined by the function **SGmakeFinger()**.

Hand configuration is defined by the joint variables $q = [q_1 \ \dots \ q_{n_q}]^{\mathrm{T}} \in \Re^{n_q}$. The user can modify the hand configuration through the function **SGmoveHand()**.

The toolbox can be used to investigate the properties of hands in which the joint displacements are coupled, mechanically or by means of a suitable control algorithm. In the case of human hand synergies, this coupling has been described as a synergy matrix associated to hand model [4]. For the 20 DoFs model of the human hand available in the toolbox, the synergy matrix refers to the data collected by Santello et al. in [1] and it is provided through the function **SGsantelloSynergies**.

The function **SGdefineSynergies()** associates to a specific hand model the relative coupling matrix. The function **SGactivateSynergies()** activates a synergy or a combination of synergies on the hand. The function **SGplotSyn()** draws the movement corresponding to the activation of one synergy. It draws on the same plot the hand in the initial reference configuration and in the configuration obtained activating one or more synergies.

12.2.3 Grasp Definition

Grasp analysis is based on the definition of the contact between the hand and the grasped object (for a complete theoretical framework for grasp analysis of underactuated synergistic hands, the reader is invited to refer to Chap. 13). We adopted the common assumption consisting in the approximation of the contacts with discrete points. Let us denote with n_c the number of contact points. Their positions are identified w.r.t. a base reference frame, that we assume to be inertial, denoted with $\{N\}$. A local reference frame $\{B\}$ is defined on the object. The configuration (position and orientation) of $\{B\}$ frame w.r.t. $\{N\}$ is described by the vector $u \in \Re^6$. The wrench imposed by the interaction between the hand and the object at each contact point can be evaluated once a suitable contact model is defined [10]. For each contact point i, the contact model selects the contact wrench $\lambda_i \in \Re^{l_i}$ components that can be applied, l_i value depends on the type of contact (e.g., $l_i = 1$ for a single point without friction model, $l_i = 3$ for hard finger model, $l_i = 4$ for soft finger contact model [10]). The contact actions λ_i are collected in a vector $\lambda \in \Re^{n_l}$, where $n_l = \sum_{i=1}^{n_c} l_i$. For each grasp it is necessary to define a variable structure representing the grasped object and containing the object center position, the coordinates of the contact points and the unit vectors normal to the contact surface. This structure can be specified in SynGrasp using the function **SGmakeObject()**.

In SynGrasp a grasp can be defined in two ways. One way is to start from the hand, choosing on it the contact points and thus defining an object that fills them. This solution can be used, for instance, when contact point positions are acquired through an experimental setup on a robotic or human hand or if an external grasp planner is used. Alternatively, it is possible to consider a hand and an object and close the hand on it to define the contact point positions, through the provided grasp planner.

The contact point coordinates can be either defined by directly assigning the location of the contact points anywhere on the hand with the function **SGaddContact()**, or using the function **SGaddFtipContact()** to add contacts on the hand fingertip. The object center coordinates and the normal unit vectors can be either automatically computed by the software on the basis of the contact point locations or manually defined by the user. A convex object defined by the given contact points is drawn by the function **SGplotObject()**.

The toolbox contains a grasp planner. Given the hand model, an object to grasp, the number of pre-grasp positions and the metric to be used for grasp evaluation, the grasp planner function **SGgraspPlanner()** iteratively evaluates the *best* grasp. The pregrasp positions are set using the function **SGgenerateCloud()**, for each trial the hand moves **SGtransl()** toward the center of the object with a random orientation of the palm. From the pregrasp position the function **SGcloseHand()** closes the fingers until they reach the object. The procedure is open and fully customizable. The user can set specific values for several parameters, e.g. which joints are involved in the grasping action, the size of the step used to close the finger around the object and the offset between the hand and the object. The function **SGcontactDetection()** evaluates if a link of the hand is contact with the object and eventually stops the relative joint. Once all the fingers are in contact with the object or have reached their limits, the grasp quality is evaluated. This procedure is repeated for all the generated pregrasps and the obtained performance, in terms of the selected grasp quality measure, are sorted. Figure 12.3 shows an example of outcome of the grasp planner.

Grasp definition is the first step of the analysis: once the hand, the object and their contact points are defined, it is immediately possible to evaluate all the matrices relevant to grasp analysis. The selection matrix $H \in \Re^{n_l \times 3n_c}$, that extracts from the contact wrench the components effectively applied, according to the chosen contact model, is evaluated by the function **SGselectionMatrix()**. The grasp matrix $G \in \Re^{6 \times n_l}$, relates the external wrench applied to the object to the contact actions, i.e.

$$w = G\lambda \tag{12.1}$$

where $w \in \Re^6$ is the object external wrench. G matrix depends on contact point and object center positions, and on the selected contact model. It can be evaluated with the

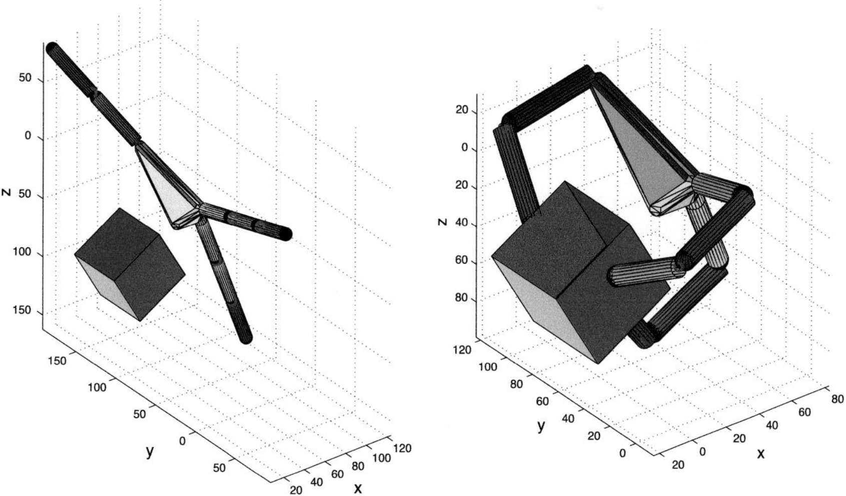

Fig. 12.3 Grasp planner output. In the *left hand side* the pregrasp position and on the *right* the obtained grasp

SynGrasp function **SGgraspMatrix()**. The hand Jacobian matrix $J \in \Re^{n_l \times n_q}$ relates the components of the contact twists constrained by the contact model $\nu_c^{hnd} \in \Re^{n_l}$ to the joint velocities $\dot{q} \in \Re^{n_q}$, i.e.

$$\nu_c^{hnd} = J\dot{q}. \tag{12.2}$$

For a given grasp, Jacobian matrix J can be evaluated by means of the function **SGjacobianMatrix()**. More details on the evaluation of grasp matrix and hand Jacobian matrix can be found in [10, 11].

The role of compliance in grasp is fundamental for the definition of contact force distribution. In SynGrasp the compliance can be modeled at different levels: at the contact, at the joints and at the synergies actuation. For each compliance type, a lumped parameter model is considered [12]. More details on stiffness modeling and evaluation in grasps with underactuated compliant hands are available in [13]. Contact stiffness model considers the local deformation of the contact surfaces and relates it to the contact force. Approximately, this relationship can be considered linear, i.e.

$$\Delta\lambda = K_c(J\Delta q - G^{\mathrm{T}}\Delta u), \tag{12.3}$$

where $K_c \in \Re^{n_l \times n_l}$ is the contact compliance matrix symmetric and positive definite. The elements of K_c matrix depend on material properties and on the contact surface geometries (curvature radii). Δq is the joint variable variation and Δu represents a variation on the object reference frame position. The SynGrasp function **SGcontactStiffness()** assigns to a grasp the corresponding contact stiffness matrix.

Often the structural stiffness of the hand and the controllable servo stiffness of the joints have the same order of magnitude of the contact stiffness, so they have to be considered in grasp modeling [14]. Considering a lumped parameter model in which the stiffness is concentrated in the joints, the difference between a reference Δq_r and the actual Δq variations of the joint displacements is proportional to the joint torque variation $\Delta\tau \in \Re^{n_q}$, i.e. to

$$\Delta\tau = K_q(\Delta q_r - \Delta q), \tag{12.4}$$

where $K_q \in \Re^{n_q \times n_q}$ is the joint stiffness matrix, symmetric and positive definite. SynGrasp function **SGjointStiffness()** assigns to a hand the corresponding joint stiffness matrix.

In [7] a *softly underactuated model* was introduced to model the joint aggregation induced by synergy definition (see also Chaps. 8 and 13). A stiff definition of synergies does not allow the hand to adapt to the actual shape of the grasped objects and then limits the applicability of this type of underactuation. A stiffness matrix is then introduced at the synergy level, relating the synergy action variation $\Delta\sigma \in \Re^{n_z}$ to the difference between a reference and the actual variation of synergy inputs

$$\Delta\sigma = K_z(\Delta z_r - \Delta z), \tag{12.5}$$

where $K_z \in \Re^{n_z \times n_z}$ is a symmetric and positive definite matrix that defines the synergy stiffness. The SynGrasp function **SGsynergyStiffness()** assigns the specified K_z matrix to a given hand.

12.2.4 Grasp Analysis

Grasp analysis tools provided by SynGrasp represent the core of the toolbox. These tools are the result of different modeling efforts carried out on both fully and under-actuated hand models. In the following, a synergy-actuated human hand model is considered for an overview of the SynGrasp functions devoted to grasp analysis.

The quasi-static model adopted to describe a grasp in SynGrasp is the result of a linear approximation of the kinematic and compliance equations in the neighborhood of an equilibrium configuration. A detailed description of this model can be found in [4, 7, 15]. Let us consider an initial equilibrium configuration of the hand/object system. A small variation of the input synergy references Δz_r can be then applied to the system and, assuming that the system reaches a new equilibrium point, the following equations can be written

$$\begin{bmatrix} -G & 0 & 0 & 0 & 0 & 0 \\ J^{\mathrm{T}} & 0 & 0 & -I & 0 & 0 \\ 0 & 0 & 0 & S^{\mathrm{T}} & -I & 0 \\ I & K_c G^{\mathrm{T}} & -K_c J & 0 & 0 & 0 \\ 0 & 0 & K_q & I & 0 & -K_q S \\ 0 & 0 & 0 & 0 & I & K_z \end{bmatrix} \begin{bmatrix} \Delta\lambda \\ \Delta u \\ \Delta q \\ \Delta\tau \\ \Delta\sigma \\ \Delta z \end{bmatrix} = \begin{bmatrix} 0 \\ 0 \\ 0 \\ 0 \\ 0 \\ K_z \Delta z_r \\ 0 \end{bmatrix}. \qquad (12.6)$$

Solving this linear system in (12.6) we can define the transfer matrices V, Q, Y and P which represent a mapping between the input Δz_r of our system and the output variables as it follows

$$\Delta u = V \Delta z_r \qquad (12.7)$$

$$\Delta q = X S Y \Delta z_r = Q \delta z_r \qquad (12.8)$$

$$\Delta z = Y \Delta z_r \qquad (12.9)$$

$$\Delta\lambda = P \Delta z_r \qquad (12.10)$$

In SynGrasp the function **SGquasistatic()** can be used to solve the linear system (12.6) for a given grasp and a given Δz_r. The corresponding variation of the output variables can then be computed according to (12.7)–(12.10).

Let us now consider the Eq. (12.10). A basis matrix E_s for the subspace of controllable internal forces [4] (i.e. the internal forces that can be produced applying a synergy reference variation Δz_r), can be computed as

$$E_s = \mathscr{R}(P). \tag{12.11}$$

The function **SGgraspAnalysis()** can be used to analyze a grasp in terms of internal forces and object motions. The relation between the input variation Δz_r and the resulting object displacement is expressed by (12.7). The object motions which do not involve visco-elastic deformations at the contact points are called *rigid-body motions*. If underactuation is considered, usually only some of the possible rigid body motions can be controlled acting on the synergy reference values. In order to compute them let us consider a synergy reference values $\Delta z_{rh} \in \mathscr{N}(P)$ that modify the hand/object configuration without changing the contact forces. The corresponding displacements of the object and the hand can be thus computed, according to (12.7) and (12.8), as $\Delta u_h = V \Delta z_{rh}$ and $\Delta q_h = Q \Delta z_{rh}$, respectively. These rigid displacements of object and hand can be computed in SynGrasp by **SGrbMotions()**.

The linear relationship between a variation of the wrench applied to the object and its resulting motion can be seen as the stiffness $K \in \Re^{6\times 6}$ of the overall grasp. The following equation holds

$$\Delta w = K \Delta u. \tag{12.12}$$

In [13] the expression of K matrix has been evaluated as a function of grasp properties as

$$K = G\left(K_c^{-1} + J K_q J^{\mathrm{T}} + J S K_z S^{\mathrm{T}} J^{\mathrm{T}}\right)^{-1} G^{\mathrm{T}}. \tag{12.13}$$

The value of K for a given grasp can be evaluated in SynGrasp using **SGgraspStiffness()**.

Finally, we report the grasp quality measures [16] that are implemented in SynGrasp:

- **SGminSVG()** evaluates the minimum singular value of the grasp matrix G: it gives a measure of the distance of the considered grasp from singular joint configurations [10, 17].
- **SGdistSingularConfiguration()** considers the distance of the finger configurations from singularities: it evaluates the smallest singular value of the matrix $Ho = G^{+} J$ [18].
- **SGmanipEllisoidVolume()** evaluates the manipulability ellipsoid [19].
- **SGunifTransf()** takes into account the transformation between the velocity domain in the joint space and the velocity domain in the task space: it checks if the contribution of each joint is the same in all the components of the object velocity, thus giving a "uniformity" measure.
- **SGgraspIsotropyIndex()** gives an additional "uniformity" measure for the grasp: it tells to the user how much the considered grasp is isotropic w.r.t. a uniform set of internal forces [20].

12.3 Object-Based Mapping Using SynGrasp

Hand models in SynGrasp can embed a synergistic joint couplings. The function **SGdefineSynergies()** allows to associate a specific S matrix to a given hand model. Moreover, the human hand synergies presented by Santello et al. in [1] can be loaded using the function **SGsantelloSynergies()** and associated to the human hand model implemented in SynGrasp (i.e. the paradigmatic hand model).

Different approaches must be used to exploit synergies onto robotic hands. New synergy matrices can be, in fact, defined by the user on the basis of experimental data or desired behaviors of the device. Another possibility is to map the human synergies onto the robotic device using the object-based approach presented in [21, 22] and implemented in the toolbox. In the following, we will describe the algorithm presenting, in parallel, the SynGrasp functions implementing it. Let us consider the paradigmatic hand model (loaded calling the function **SGparadigmatic**) and a three-fingered hand that can be loaded in SynGrasp calling the function **SG3Fingered** as the target robotic hand on which the human hand synergies have to be mapped. Let us define a set of reference points on the two hands. For the sake of simplicity we will consider, in the following, the fingertips of the two hands as reference points. Other possible choices of the reference points are possible, (e.g. the intermediate phalanges) as proposed in [6, 21]. At the beginning of the mapping procedure two virtual objects must be defined on the two hands as the minimum volume spheres containing all the reference points of the considered hand. The utility **minbound-sphere()** provided in SynGrasp can be used for this purpose. As the paradigmatic hand moves under the effect of a synergy activation z (**SGactivateSynergies()**), its reference points move accordingly, consequently changing position, orientation and radius of the virtual sphere. We can describe the motion of the hand as:

- a *rigid-body* motion, defined by linear and angular velocities of the sphere center, denoted by $\dot{o}_h$ and ω_h, respectively;
- a *non-rigid* deformation represented by the sphere radius rate of change $\dot{r}_h$.

The key idea underlying the mapping procedure is to impose the same motion $\dot{o}_h$ and ω_h and deformation $\dot{r}_h$ read on the human hand to the virtual sphere computed on the robotic hand, apart from a scaling factor introduced to deal with the possible differences on the workspace dimensions (see Fig. 12.4). Inverse kinematic techniques can be then used to compute the joint values for the robotic hand. This defines a way of controlling the robotic hand joints starting from the synergy activation vector z imposed on the human hand. In particular, the output of the mapping procedure are the velocities $\dot{q}_r$ of the robotic hand joints defined as

$$\dot{q}_r = S_r \dot{z}, \tag{12.14}$$

where S_r is the robotic synergy matrix obtained by mapping the human synergy matrix (denoted in the following by S_h).

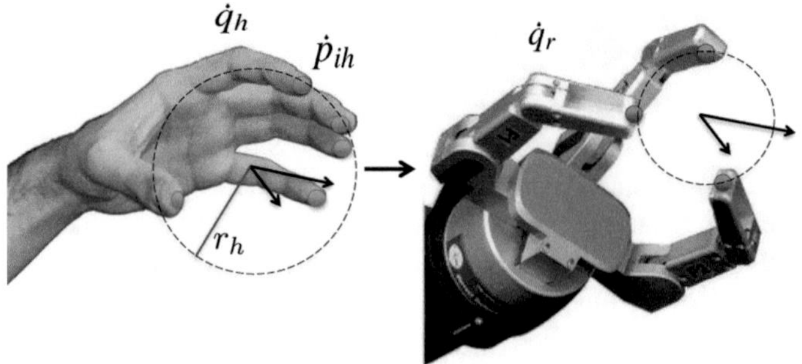

Fig. 12.4 Mapping synergies from the human hand model to the robotic hand. The reference points on the paradigmatic hand p_h (*blue dots*) allows to define the *virtual sphere*. Activating the human hand synergies, the *sphere* is moved and strained; its motion and strain can be evaluated from the velocities of the reference points $\dot{p}_h$. This motion and strain, scaled by a factor depending on the *virtual sphere* radii ratio, is then imposed to the *virtual sphere* relative to the robotic hand, defined on the basis of the reference points p_r (*red dots*)

In the following how S_r can be computed analytically is reported. We assume that the two hands are in given starting configurations denoted by q_{0h} and q_{0r}, respectively. The velocity of the generic reference point of the human hand can be written as

$$\dot{p}_{ih} = \dot{o}_h + \omega_h \times (p_{ih} - o_h) + \dot{r}_h (p_{ih} - o_h). \tag{12.15}$$

Grouping all the reference point in a single vector, we obtain

$$\dot{p}_h = A_h \begin{bmatrix} \dot{o}_h \\ \omega_h \\ \dot{r}_h \end{bmatrix}, \tag{12.16}$$

where matrix $A_h \in \Re^{n_{ch} \times 7}$ is defined as

$$A_h = \begin{bmatrix} I & -s(p_{1h} - o_h) & (p_{1h} - o_h) \\ \cdots & \cdots & \cdots \\ I & -s(p_{ih} - o_h) & (p_{ih} - o_h) \\ \cdots & \cdots & \cdots \end{bmatrix} \tag{12.17}$$

and $s()$ is the skew operator. The matrix A_h can be easily computed in SynGrasp by **mapping_A**.

Now, the motion of the virtual sphere can be expressed as function of the derivative of the human synergy activation vector $\dot{z}$ as

$$\begin{bmatrix} \dot{o}_h \\ \omega_h \\ \dot{r}_h \end{bmatrix} = A_h^{\#} \dot{p}_h = A_h^{\#} J_h S_h \dot{z}, \tag{12.18}$$

where $A_h^{\#}$ denotes the pseudo-inverse of the matrix A_h. Let us denote by k_{sc} the ratio between the radii of the two spheres (i.e. $k_{sc} = \frac{r_r}{r_h}$). Note that k_{sc} not only depends on the dimensions of the two hands but also on their configurations. Motions and deformations of the robotic virtual sphere can be then computed scaling motions and deformations of the sphere defined on the paradigmatic hand as

$$\begin{bmatrix} \dot{o}_r \\ \omega_r \\ \dot{r}_r \end{bmatrix} = K_c \begin{bmatrix} \dot{o}_h \\ \omega_h \\ \dot{r}_h \end{bmatrix}, \tag{12.19}$$

where the matrix $K_c \in \Re^{7\times 7}$ is defined as

$$K_c = \begin{bmatrix} k_{sc} I_{3,3} & 0_{3,3} & 0_{3,1} \\ 0_{3,3} & I_{3,3} & 0_{3,1} \\ 0_{1,3} & 0_{1,3} & 1 \end{bmatrix}. \tag{12.20}$$

According to Eq. (12.16), the velocities of the robot reference points $\dot{p}_r$ are given by

$$\dot{p}_r = A_r \begin{bmatrix} \dot{o}_r \\ \omega_r \\ \dot{r}_r \end{bmatrix}, \tag{12.21}$$

where matrix $A_r \in \Re^{n_{cr}\times 7}$ is defined as in Eq. (12.17). At this point, we are able to express the velocities $\dot{p}_r$ of the robotic reference points as a function of the synergy velocities $\dot{z}$

$$\dot{p}_r = A_r K_c A_h^{\#} J_h S_h \dot{z}. \tag{12.22}$$

Considering the robot hand differential kinematics $\dot{p}_r = J_r \dot{q}_r$ (where $J_r \in \Re^{n_{cr}\times n_{qr}}$ is its Jacobian matrix), we can compute the robotic hand joint velocities as

$$\dot{q}_r = J_r^{\#} A_r K_c A_h^{\#} J_h S_h \dot{z}. \tag{12.23}$$

Finally the synergy mapping S_r in (12.14) for the robotic hand can be written as

$$S_r = J_r^{\#} A_r K_c A_h^{\#} J_h S_h, \tag{12.24}$$

where $J_r^{\#}$ is the pseudoinverse of the Jacobian of the robotic hand and J_h is the Jacobian of the human hand model. The obtained synergy matrix can be assigned to the robotic hand using the function **SGdefineSynergies()**.

The described procedure is implemented in the **SGmappingExample** script included in the toolbox.

12.4 Conclusion

This chapter presented how the SynGrasp toolbox can be used to implement the object-based mapping algorithm introduced in [21]. The toolbox is entirely open-source and can be easily customized by the users. It is organized with a modular structure allowing the evaluation of hand kinematics and dynamics, grasping properties like force distribution at the contacts and quality measures. Specific functions based on synergies enable the control of the hand through a reduced number of inputs according to a synergistic approach. The possible applications of the SynGrasp range from Neuroscience, where tools for human hand analysis are needed, to the design and optimization of robotic hands. The mapping algorithm can be used to determine a synergistic underactuation of robotic hands with very dissimilar kinematics (see also Chaps. 8 and 13). This allows to envisage a control framework where the motion of the human hand is reproduced with different robotic hand models using the same mapping framework [23]. We are currently working in exploiting the mapping algorithm in a real-time scenario enabling teleoperation between different kinematics. Preliminary results are presented in [24]. However the abstraction brought by the mapping procedure has to be compensated by the robotic device to guarantee the stability of the grasp. In [25] a passive controller is used as a first attempt in this direction.

Acknowledgments The research leading to these results has received funding from the European Union Seventh Framework Programme FP7/2007-2013 under grant agreement No. 248587, "THE Hand Embodied", within the FP7-ICT- 2009-4-2-1 program "Cognitive Systems and Robotics" and the Collaborative EU—Project "Hands.dvi" in the context of ECHORD (European Clearing House for Open Robotics Development).

References

1. Santello M, Flanders M, Soechting J (1998) Postural hand synergies for tool use. J Neurosci 18(23):10105–10115
2. Miller A, Allen P (2004) Graspit! a versatile simulator for robotic grasping. IEEE Robot Autom Mag 11(4):110–122
3. León B, Ulbrich S, Diankov R, Puche G, Przybylski M, Morales A, Asfour T, Moisio S, Bohg J, Kuffner J, et al (2010) OpenGRASP: a toolkit for robot grasping simulation. In: Simulation, modeling, and programming for autonomous robots, pp 109–120
4. Prattichizzo D, Malvezzi M, Bicchi A (2010) On motion and force controllability of grasping hands with postural synergies. In: Proceedings of robotics: science and systems, Zaragoza, Spain. June 2010
5. Butterfass J, Grebenstein M, Liu H, Hirzinger G (2001) DLR-Hand II: next generation of a dextrous robot hand. In: Proceedings of the IEEE international conference on robotics and automation, pp 109–114
6. Gioioso G, Salvietti G, Malvezzi M, Prattichizzo D (2013) An object-based approach to map human hand synergies onto robotic hands with dissimilar kinematics in Robotics: science and systems VIII, MIT Press, 96–104
7. Gabiccini M, Bicchi A, Prattichizzo D, Malvezzi M (2011) On the role of hand synergies in the optimal choice of grasping forces. Auton Robots 31(2):235–252

8. Malvezzi M, Gioioso G, Salvietti G, Prattichizzo D (2014) Syngrasp: a matlab toolbox for underactuated and compliant hands, user's guide. http://sirslab.dii.unisi.it/syngrasp
9. Siciliano B, Sciavicco L, Villani L, Oriolo G (2009) Robotics: modelling, planning and control. Springer Science and Business Media
10. Prattichizzo D, Trinkle J (2008) Grasping. In: Siciliano B, Kathib O (eds) Handbook on robotics. Springer, Heidelberg, pp 671–700
11. Murray R, Li Z, Sastry S (1994) A mathematical introduction to robotic manipulation. CRC Press, Boca Raton
12. Chen S, Kao I (2000) Conservative congruence transformation for joint and Cartesian stiffness matrices of robotic hands and fingers. Int J Robot Res 19:835–847
13. Malvezzi M, Prattichizzo D (2013) Evaluation of grasp stiffness in underactuated compliant hands. In: Proceedings—IEEE international conference on robotics and automation
14. Cutkosky M, Kao I (1989) Computing and controlling the compliance of a robotic hand. IEEE Trans Robot Autom 5(2):151–165
15. Prattichizzo D, Malvezzi M, Gabiccini M, Bicchi A (2012) On the manipulability ellipsoids of underactuated robotic hands with compliance. Robot Auton Syst 60(3):337–346
16. Suárez R, Roa M, Cornellà J (2006) Grasp quality measures, Tech. Rep. IOC-DT-P 2006-10, Universitat Politècnica de Catalunya, Institut d'Organització i Control de Sistemes Industrials
17. Li Z, Sastry SS (1988) Task-oriented optimal grasping by multifingered robot hands. IEEE Trans Robot 4(1):32–44
18. Klein CA, Blaho BE (1987) Dexterity measures for the design and control of kinematically redundant manipulators. Int J Robot Res 6(2):72–83
19. Shimoga KB (1996) Robot grasp synthesis algorithms: a survey. Int J Robot Res 15(3):230–266
20. Kim BH, Oh S-R, Yi B-J, Suh IH (2001) Optimal grasping based on non-dimensionalized performance indices. In: Proceedings of the IEEE/RSJ international conference intelligent robots and systems, vol 2. IEEE, pp 949–956
21. Gioioso G, Salvietti G, Malvezzi M, Prattichizzo D (2013) Mapping synergies from human to robotic hands with dissimilar kinematics: an approach in the object domain. Robotics, IEEE Transactions on 29(4):825–837
22. Salvietti G, Malvezzi M, Gioioso G, Prattichizzo D (2014) On the use of homogeneous transformations to map human hand movements onto robotic hands. In: Proceedings of the IEEE international conference on robotics and automation, pp 5352–5357
23. Salvietti G, Gioioso G, Malvezzi, M, Prattichizzo D, Serio A, Farnioli E, Gabiccini M, Bicchi A, Sarakoglou I, Tsagarakis N, Caldwell D (2014) HANDS.DVI: a DeVice-independent programming and control framework for robotic HANDS. In: Gearing up and accelerating cross-fertilization between academic and industrial robotics research in Europe—technology transfer experiments from the ECHORD project. Springer tracts in advanced robotics. Springer, pp 197–215
24. Salvietti G, Meli L, Gioioso G, Malvezzi M, Prattichizzo D (2013) Object-based bilateral telemanipulation between dissimilar kinematic structures. Proceedings of the IEEE/RSJ international conference on intelligent robots and systems. Tokyo, Japan, pp 5451–5456
25. Salvietti G, Wimboeck T, Prattichizzo D (2013) A static intrinsically passive controller to enhance grasp stability of object-based mapping between human and robotic hands. Proceedings of the IEEE/RSJ international conference on intelligent robots and systems. Tokyo, Japan, pp 2460–2465

Chapter 13
Quasi-Static Analysis of Synergistically Underactuated Robotic Hands in Grasping and Manipulation Tasks

Edoardo Farnioli, Marco Gabiccini and Antonio Bicchi

Abstract As described in Chaps. 2–5, neuroscientific studies showed that the control of the human hand is mainly realized in a *synergistic* way. Recently, taking inspiration from this observation, with the aim of facing the complications consequent to the high number of degrees of freedom, similar approaches have been used for the control of robotic hands. As Chap. 12 describes SynGrasp, a useful technical tool for grasp analysis of synergy-inspired hands, in this chapter recently developed analysis tools for studying robotic hands equipped with *soft synergy* underactuation (see Chap. 8) are exhaustively described under a theoretical point of view. After a review of the quasi-static model of the system, the *Fundamental Grasp Matrix* (FGM) and its *canonical form* (cFGM) are presented, from which it is possible to extract relevant information as, for example, the subspaces of the *controllable internal forces*, of the *controllable object displacements* and the *grasp compliance*. The definitions of some relevant types of manipulation tasks (e.g. the *pure squeeze*, realized maintaining the object configuration fixed but changing contact forces, or the *kinematic grasp displacements*, in which the grasped object can be moved without modifying contact forces) are provided in terms of nullity or non-nullity of the variables describing the system. The feasibility of such predefined tasks can be verified thanks to a decomposition method, based on the search of the *row reduced echelon form* (RREF) of suitable portions of the solution space. Moreover, a geometric

E. Farnioli (✉) · M. Gabiccini · A. Bicchi
Research Center "E. Piaggio", Università di Pisa, Largo Lucio Lazzarino 1,
56122 Pisa, Italy
e-mail: edoardo.farnioli@iit.it

M. Gabiccini
e-mail: marco.gabiccini@iit.it

A. Bicchi
e-mail: antonio.bicchi@iit.it

E. Farnioli · M. Gabiccini · A. Bicchi
Department of Advanced Robotics, Istituto Italiano di Tecnologia,
Via Morego 30, 16163 Genoa, Italy

M. Gabiccini
Department of Civil and Industrial Engineering, Università di Pisa,
Largo Lucio Lazzarino 1, 56122 Pisa, Italy

M. Bianchi and A. Moscatelli (eds.), *Human and Robot Hands*,
Springer Series on Touch and Haptic Systems, DOI 10.1007/978-3-319-26706-7_13

interpretation of the FGM and the possibility to extend the above mentioned methods to the study of robotic hands with different types of underactuation are discussed. Finally, numerical results are presented for a power grasp example, the analysis of which is initially performed for the case of fully-actuated hand, and later verifying, after the introduction of a synergistic underactuation, which capacities of the system are lost, and which other are still present.

13.1 Introduction

The research in robotic hand design was directed for long time to increase the dexterity and the manipulation capabilities. To follow this line, the number of degrees of freedom, and, more in general, the complexity of the design are increased in the years. Remarkable examples of such hands are the UTAH/MIT hand [1], the Robonaut Hand [2], the Shadow hand [3] and the DLR hand arm system [4], just for citing a few of them, as discussed in Chap. 8.

However, a large number of degrees of freedom, often, bring to enlarge weights and costs of such prototypes. Moreover, the expected advantages in terms of manipulability are often difficult to exploit in a real scenario. Recently, in order to face the complexity of such systems, the human hand was considered as a source of inspiration (see Chaps. 2, 8 and 9) not just for the mechanical design, but also in order to simplify the control strategies.

In recent years, many neuroscientific studies such as, for example, the ones discussed in [5–11] (see also Chaps. 2–7), despite significant differences in the definitions and in the requirements of the investigated tasks, share a main observation: simultaneous motion of multiple digits, also called *synergies*, occurs in a consistent fashion, even when the task may require a fairly high degree of movement individuation, such as grasping a small object or typing.

As extensively discussed in the previous chapters, one of the main result is that a large variety of everyday human grasps is well described by just five synergies. Moreover, the first two human synergies can describe the 80 % of the variance in human grasp postures (see also Chap. 9). This suggested the idea to move the description base for grasping, from the joint space to the human-inspired postural synergy space, taking advantage from the *underactuation*. Between the first approach to this idea, we find [12, 13], that try to implement a synergistic control via software and via hardware, respectively. Despite each one is characterized by its own peculiarities, they share the common characteristics of rigidly controlling the joint movements, via the synergistic underactuation. As discussed in Chap. 8, in the *soft synergies* approach, proposed in [14], a virtual hand is introduced, attracting the real one via a generalized spring, allowing a certain adaptability of the hand during grasps and manipulations tasks. The influence of the synergistic underactuation, in terms of reducing the hand capabilities in object motion and contact force control, is investigated in [15]. Moreover, the contact force optimization problem was faced in [14], considering the limitations imposed by the underactuation.

The present chapter, mainly based on the results presented in [16–18], describes and studies the quasi-static model of a synergistic underactuated hand grasping an object. Considering the results of the above mentioned papers, despite the fact that the analysis is performed in the neighborhood of an equilibrium configuration, also some non local considerations can be done under a more general, nonlinear kineto-static interpretation. More in detail, in Sect. 13.2 the congruence and the equilibrium equations of the system are presented in quasi-static form. A compliant contact model is introduced between the hand and the object, in order to cope with the static indeterminacy of the contact force distribution problem. Finally, a quasi-static model for the soft synergy underactuation (discussed in Chap. 8) is provided. The treatise is general enough to consider the presence of hand/object contacts also in the internal limbs of the hand. Moreover, the derivative terms of the hand Jacobian and of the grasp matrix are considered, in order to properly take into account the effects of the contact force preload.

Both the presence of internal contacts and of underactuation can greatly affect the capabilities of the hand/object system, in terms of controllable system variations, e.g. limiting the controllability of the forces and/or the object displacements. This problem is faced in Sect. 13.3 where, after the *Fundamental Grasp Matrix* (FGM) has been defined, its *canonical form* (cFGM) is derived, from which relevant information on the system can be easily obtained, despite the difficulties introduced by the presence of the synergistic underactuation in the model. In fact, as we will discuss in Sect. 13.3.2, from the cFGM we can obtain information on the *controllable internal forces*, on the *controllable object displacements*, and on the *grasp compliance*, i.e. the compliance perceived at the object level. Moreover, from the cFGM, input-output relationships between the *independent variables* (i.e. the joint displacements and the external wrench variation) and the *dependent variables* of the system can be easily deduced.

In order to go beyond the information provided by the cFGM, a method to investigate the solution space of the system is presented in Sect. 13.4. Different types of system behaviors are defined in terms of nullity or non-nullity of some system variables, such as, for example, the *pure squeeze*, where the contact forces are modified without affecting the object configuration, or the *kinematic grasp displacement*, where, on the contrary, an object movement is allowed, without changing the contact forces. Finally, a decomposition method, based on the *row reduced echelon form* (RREF) is presented, in order to find out the feasibility of those predefined solutions.

In Sect. 13.5 a geometrical interpretation of the FGM is given. With a proper arrangement of the equations, the FGM takes the form of a first-order Taylor series approximation of the *equilibrium manifold* (EM) of the whole system, describing the kineto-static behavior both of the hand and of the object during their interaction. As explained in [18], some properties of the EM can be exploited, in order to steer the system, along a trajectory composed by a sequence of equilibrium configurations, toward a final one, characterized by the desired kineto-static properties.

Many of the observations and methods presented can be applied, with small modifications, also in case of different types of underactuation, and the Sect. 13.6 is dedicated to discuss this topic (see also Chap. 12).

To conclude, in Sect. 13.7, a numerical example is presented, for a power grasp case. The example was firstly studied as if the hand was completely actuated, discovering its manipulation capabilities. Then, a synergistic underactuation is introduced, and the methods presented in the chapter are used to verify which possibilities are lost and which others are still present.

13.2 System Modeling

In this section, we will present the equations describing the quasi-static behavior of the hand/object system, schematically represented in Fig. 13.1, and already introduced in Chaps. 8 and 12. For both the hand and the object, the quasi-static equilibrium equations will be considered, obtained as a first order approximation of the general, nonlinear, equilibrium equations. Moreover, in connection with the previous, by means of kineto-static duality considerations, the congruence equations will be introduced, describing the displacement of the contact points, corresponding to the hand/object displacements. A linear elastic model for the contact is also introduced, in order to properly describe how the contact forces change, during the execution of a manipulation tasks. Finally, the underactuation will be introduced in the system according to the soft synergy pattern.

For the sake of clarity, in the following we will briefly recall some of the notations already introduced in Chaps. 8 and 12, also summarized in Table 13.1.

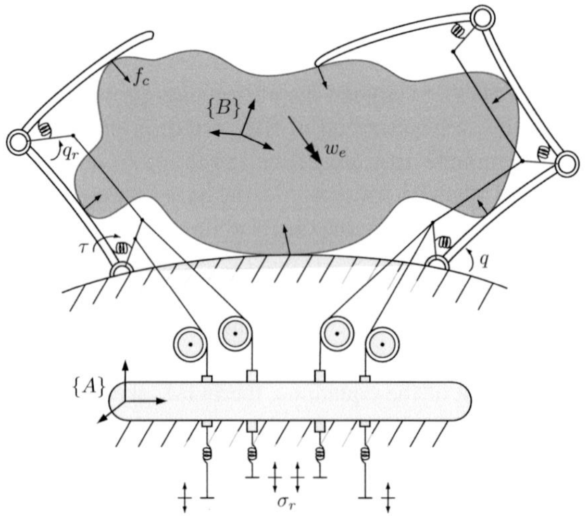

Fig. 13.1 Reference scheme for the analysis of compliant grasp by synergistically underactuated robotic hand

Table 13.1 Notation for grasp analysis

Notation	Definition
δx	Variation of the variable x
$\sharp x$	Dimensions of the vector x
$q \in \mathbb{R}^{\sharp q}$	Joint configuration
$q_r \in \mathbb{R}^{\sharp q}$	Joint reference
$\tau \in \mathbb{R}^{\sharp q}$	Joint torque
$\sigma \in \mathbb{R}^{\sharp \sigma}$	Synergy configuration
$\sigma_r \in \mathbb{R}^{\sharp \sigma}$	Synergy reference
$\eta \in \mathbb{R}^{\sharp \sigma}$	Synergy actuation (generalized) force
c	Number of hand/object contact constraints
$f_c \in \mathbb{R}^c$	Contact force/torque vector exerted by the hand on the object
$p_h^o \in \mathbb{R}^c$	Pose of the hand contact frame with respect to the object contact frame
$w \in \mathbb{R}^{\sharp w}$	(parametrized) External wrench acting on the object; $\sharp w = 6$ in 3D case, $\sharp w = 3$ in planar case
$u \in \mathbb{R}^{\sharp w}$	(parametrized) Object frame configuration
$J \in \mathbb{R}^{c \times \sharp q}$	Hand Jacobian matrix
$S \in \mathbb{R}^{\sharp q \times \sharp \sigma}$	Synergy matrix
$G \in \mathbb{R}^{\sharp w \times c}$	Grasp matrix
$\Phi^\star$	*Fundamental Grasp Matrix*, the coefficient matrix of the *Fundamental Grasp Equation* (13.14)
φ	*Augmented configuration*, vector collecting the kineto-static variables of the system

13.2.1 Object Equations

13.2.1.1 Equilibrium Equation of the Object

The grasped object is in equilibrium if the sum of all the contact forces/torques exerted by the hand, gathered in the contact force vector[1] $f_c \in \mathbb{R}^c$, and of a possible external wrench[2] $w \in \mathbb{R}^{\sharp w}$ is null, where the symbol $\sharp x$ indicates the

[1]The dimension of the contact force vector c is related to the number of contact points and to the local characteristics of the contacts. More details about this will be provided in Sect. 13.2.3.

[2]Strictly speaking, the vector $w \in \mathbb{R}^{\sharp w}$, in the present dissertation, represents a *parametrization of an external wrench*, abbreviated in the text simply as *external wrench*. Similarly, the object configuration is described by a parametrization vector $u \in \mathbb{R}^{\sharp w}$. As a consequence, the object velocity $\dot{u}$ in (13.3) is a parametrization of the object twist, and, for this reason, can be expressed as the time derivative of some physical variables. As an example, in a $3D$ case, a complete parametrization can be obtained considering a 6−DoF virtual kinematic chain describing the configuration of the object frame with respect to a fixed one. In this case, the vectors $\dot{u}$ and w will contain, respectively, the joint velocities and the joint torques of the virtual kinematic chain.

dimension of the vector x. In the present discussion, the contact forces are considered to be expressed in a local frame attached to the object. Before summing all the contributions, they have to be all expressed in a same reference frame, as for example the frame $\{B\}$ in Fig. 13.1, attached to the object. To this aim, it is usual in literature to introduce the *grasp matrix*, indicated as $G \in \mathbb{R}^{\sharp w \times c}$. Using the previous symbols, the equilibrium law for the object can be written as

$$w + Gf_c = 0. \tag{13.1}$$

It is worth observing that, despite the fact that the contact forces are described in a local frame attached to the object, the parametrization of the external wrench imposes that the grasp matrix becomes a function of the object configuration, as explained in [18]. In light of this, by means of a first-order Taylor series approximation, from (13.1) the quasi-static equilibrium equation for the object can be obtained in the form

$$\delta w + G\delta f_c + U_g \delta u = 0, \tag{13.2}$$

where the symbol δx expresses the variation of the variable x, the vector[2] $u \in \mathbb{R}^{\sharp w}$ describes a parametrization of the object configuration and $U_g := \frac{\partial G f_c}{\partial u}$.

13.2.1.2 Congruence Equation of the Object

From (13.1), by kineto-static duality considerations, it is possible to find that the transpose of the grasp matrix maps the object velocity[2], indicated as $\dot{u} \in \mathbb{R}^{\sharp w}$, into the velocities of the object contact frames, grouped into the vector $v_o \in \mathbb{R}^c$, as follows[3]

$$v_o = G^T \dot{u}. \tag{13.3}$$

The congruence equation, describing the displacements of the contact frames as a consequence of the object frame displacement, can be obtained from (13.3) by multiplying each member for an infinitesimal amount of time dt, obtaining

$$\delta C_o = G^T \delta u. \tag{13.4}$$

[3]More precisely, the vectors v_o and v_h contain the terms of the contact frame twists violating the (rigid) contact constraints between the hand and the object.

13.2.2 Hand Equations

13.2.2.1 Congruence Equation of the Hand

Let us define the *hand Jacobian matrix*, $J \in \mathbb{R}^{c \times \sharp q}$, as the map between the joint velocities, clustered in the vector $\dot{q} \in \mathbb{R}^{\sharp q}$, and the velocities of the hand contact frames[3] $v_h \in \mathbb{R}^c$, such that

$$v_h = J\dot{q}. \tag{13.5}$$

The displacement of the contact frames attached to the hand can be obtained by multiplying each member of (13.5) for an infinitesimal amount of time dt, obtaining

$$\delta C_h = J \delta q, \tag{13.6}$$

that describes the quasi-static form of the congruence equation of the hand.

13.2.2.2 Equilibrium Equation of the Hand

The equilibrium law for the hand comes from (13.5) by kineto-static duality considerations. As a result, indicating with the symbol $\tau \in \mathbb{R}^{\sharp q}$ the joint torque vector, the equilibrium law for the hand can be expressed as

$$\tau = J^T f_c. \tag{13.7}$$

The quasi-static equilibrium equation is obtained from (13.7), by means of a first order Taylor series expansion. To this aim, it is important to note that, since the fact that the contact forces are described in a local frame attached to the object, the Jacobian matrix, introduced in (13.5), is a function both of the joint parameters of the hand q, and of the object configuration parameters u, that is $J = J(q, u)$.

From these considerations, it follows that the quasi-static equilibrium of the hand can be expressed as

$$\delta\tau = Q_j \delta q + U_j \delta u + J^T \delta f_c, \tag{13.8}$$

where $Q_j := \frac{\partial J^T f_c}{\partial q}$ and $U_j := \frac{\partial J^T f_c}{\partial u}$.

13.2.3 Hand/Object Interaction Model

In the contact between the hand and the object, relative displacements of the contact frames are forbidden in some directions. In these directions, some reaction forces can arise. The dimension c_i of the ith reaction force vector depends by the nature of the materials involved. As an example, in the case of *contact point with friction*, or *hard finger* contact model, the force can be transmitted in any direction, but no moment is allowed, that is $c_i = 3$. Indeed, in the case of *soft-finger* contact type, also a moment about the normal to the contact can be transmitted, thus $c_i = 4$.

In most cases of interest, the total number of contact force elements $c = \sum_i c_i$ is greater than the number of the external wrench elements. For this reason, the problem of determining the contact force distribution is statically indeterminate.

This problem is generally faced in literature by *relaxing* the contact constraints. In other words, a relative displacement of the contact frames is allowed also in the directions nominally forbidden by the (rigid) contact constraint, and this is interpreted as the cause of the contact force variation. This behavior is modeled introducing a (virtual) linear spring between the two bodies in contact. Defining $K_c \in \mathbb{R}^{c \times c}$ as the contact stiffness matrix, i.e. a matrix collecting the stiffness values of all the contact springs, the constitutive equation of the contact can be, finally, expressed as

$$\delta f_c = K_c(\delta C_h - \delta C_o). \tag{13.9}$$

13.2.4 Soft Synergy Underactuation Model

As explained in Sect. 13.1, in this chapter we consider the problem of discovering the capabilities of *soft synergy* underactuated robotic hands in grasping, as already discussed in Chap. 8. Inspired by neuroscientific studies, the *soft synergy* underactuation model, can be seen as composed by two elements: (i) a virtual hand, which movement is governed by a synergistic correlation of the joints, and (ii) a set of virtual springs, connecting the virtual hand to the real one.

To mathematically describe this model, in each joint we introduce a compliant element by means of which the joint reference variables, collected in the vector $q_r \in \mathbb{R}^{\sharp q}$, transmit the motion to the real ones. Afterwards, the synergistic behavior of the hand is obtained imposing a correlation between the joint reference variables.

13.2.4.1 Elastic Joint Model

The equilibrium condition for the elastic joints requires that joint torques and the spring deflections, that is the mismatch between the reference joint variables and the real ones, are related by the joint stiffness. Considering this, by the introduction of the joint stiffness matrix $K_q \in \mathbb{R}^{\sharp q \times \sharp q}$, collecting all the joint stiffness values, it directly follows that the quasi-static equilibrium law for the elastic joints is described by the following

$$\delta \tau = K_q(\delta q_r - \delta q). \tag{13.10}$$

13.2.4.2 Introducing Synergies

The synergistic underactuation is imposed to the system by means of the *synergy matrix* $S \in \mathbb{R}^{\sharp q \times \sharp \sigma}$. In analogy to what seen in (13.6), the joint reference displacements can be expressed as

$$\delta q_r = S \delta \sigma, \tag{13.11}$$

where $\sigma \in \mathbb{R}^{\sharp \sigma}$ is the synergistic actuation vector.

Again, by virtue of the kineto-static duality, indicating with $\eta \in \mathbb{R}^{\sharp \sigma}$ the generalized actuation forces at the synergy level, with considerations similar to those that have led to (13.8), the quasi-static equilibrium for the synergistic underactuation level can be written as

$$\delta \eta = S^T \delta \tau + \Sigma \delta \sigma, \tag{13.12}$$

where $\Sigma := \frac{\partial S^T \tau}{\partial \sigma}$.

As already seen for the joints, an elastic model can also be introduced for the synergistic actuation by means of a synergy reference variable $\sigma_r \in \mathbb{R}^{\sharp \sigma}$, and the *synergy stiffness matrix* $K_\sigma \in \mathbb{R}^{\sharp \sigma \times \sharp \sigma}$. Thus, similarly to what seen in (13.10), the elastic actuation model for the synergy actuation can be described as

$$\delta \eta = K_\sigma (\delta \sigma_r - \delta \sigma). \tag{13.13}$$

13.2.5 The Fundamental Grasp Equation

Grouping together the equations for the object, the hand and the synergistic underactuation, that is considering the Eqs. (13.2), (13.4), (13.6) and (13.8)–(13.13), denoting with I an identity matrix of proper dimensions, we obtain the system

$$\begin{bmatrix} G & 0 & 0 & U_g & 0 & 0 & I & 0 \\ -J^T & I & 0 & -U_j & -Q_j & 0 & 0 & 0 \\ I & 0 & 0 & K_c G^T & -K_c J & 0 & 0 & 0 \\ 0 & I & 0 & 0 & K_q & -K_q S & 0 & 0 \\ 0 & -S^T & I & 0 & 0 & -\Sigma & 0 & 0 \\ 0 & 0 & I & 0 & 0 & K_\sigma & 0 & -K_\sigma \end{bmatrix} \begin{bmatrix} \delta f_c \\ \delta \tau \\ \delta \eta \\ \delta u \\ \delta q \\ \delta \sigma \\ \delta w \\ \delta \sigma_r \end{bmatrix} = 0, \tag{13.14}$$

where the contribution of (13.4) and (13.6) was considered in (13.9), as well as (13.11) was considered in (13.10).

Equation (13.14), also called *Fundamental Grasp Equation* (FGE), is a *linear* and *homogeneous* system, that can be written in compact form as $\Phi^\star \delta \varphi = 0$. The coefficient matrix of the system, $\Phi^\star \in \mathbb{R}^{r_\Phi \times c_\Phi}$ is the *Fundamental Grasp Matrix* (FGM), which matrix elements are evaluated in the reference equilibrium configuration of the system, and the variable vector $\delta\varphi \in \mathbb{R}^{c_\Phi}$ is the *augmented configuration*, collecting the variation of the system variables.

By direct inspection of (13.14), it is easy to verify that for the number of rows and columns of the FGM, that is for r_Φ and c_Φ respectively, it holds that

$$\begin{aligned} r_\Phi &= \sharp f_c + 2\sharp q + 2\sharp\sigma + \sharp w, \\ c_\Phi &= \sharp f_c + 2\sharp q + 3\sharp\sigma + 2\sharp w. \end{aligned} \tag{13.15}$$

In most cases of practical relevance the FGM is full row rank,[4] that is $\text{rank}(\Phi^\star) = r_\Phi$, and we will assume it in the rest of the dissertation. In these cases, Eq. (13.14) can be univocally solved when it is known a number of *independent variables*, or *inputs* for the system, equal to $c_\Phi - r_\Phi = \sharp w + \sharp\sigma$. In continuity with the grasp analysis literature, we consider to known, or to have a measure of, the external wrench variation δw. Moreover, the synergy references are supposed to be position-controlled, thus we consider to know[5] the variable $\delta\sigma_r$. The independent variables will be jointly indicates in next sections as $\delta\varphi_i \in \mathbb{R}^{c_\Phi - r_\Phi}$. We will refer to the set of all the other variables as the *dependent variables*, or *output* of the system, and they will be indicated as $\delta\varphi_d \in \mathbb{R}^{r_\Phi}$.

13.3 Controllable System Configuration Variations

13.3.1 The Canonical Form of the Fundamental Grasp Equation

Considering previous definitions, Eq. (13.14) can be also written as

$$\Phi^\star \delta\varphi = \begin{bmatrix} \Phi_d^\star & \Phi_i^\star \end{bmatrix} \begin{bmatrix} \delta\varphi_d \\ \delta\varphi_i \end{bmatrix} = 0. \tag{13.16}$$

Assuming the invertibility[4] of the matrix $\Phi_d^\star$, the so called *canonical form of the Fundamental Grasp Equation* (cFGE) can be obtained left-multiplying (13.16) for $\Phi_d^{\star^{-1}}$, thus obtaining

$$\begin{bmatrix} I & \Phi_i \end{bmatrix} \begin{bmatrix} \delta\varphi_d \\ \delta\varphi_i \end{bmatrix} = 0, \tag{13.17}$$

[4] Exceptions are analytically possible but they refer to pathological situations of poor practical interest.

[5] Other choices are possible, as for example considering to know the object displacement δu, instead of the external wrench δw, or the actuation force variation $\delta\eta$, instead of the synergistic displacement variable $\delta\sigma_r$. Many results of our analysis can be easily adapted to the above mentioned situations as well.

where $\Phi_i = {\Phi_d^\star}^{-1} \Phi_i^\star$. It is worth observing in passing that, since the matrix ${\Phi_d^\star}^{-1}$ is full rank, Eqs. (13.16) and (13.17) have the same solution space. In other words, all the vectors $\delta\varphi$ satisfying (13.16) are also a solution of (13.17).

The coefficient matrix of (13.17), characterized by the presence of an identity block corresponding to the dependent variables, is the *canonical form of the fundamental grasp matrix* (cFGM). From (13.17), it is easy to find that, once the variation of the independent variables is known, the value of the dependent variable variation can be directly computed as

$$\delta\varphi_d = -\Phi_i \delta\varphi_i, \tag{13.18}$$

which represents, in compact form, the relationship between the input and the output variables of the system.

13.3.2 Relevant Properties of the Canonical Form of the Fundamental Grasp Matrix

The cFGM can be further investigated, in order to find out some relevant information on the characteristics of the physical system. To this aim, let us consider again (13.17). More in detail, this can be written also as

$$\begin{bmatrix} I & 0 & 0 & 0 & 0 & 0 & W_f & R_f \\ 0 & I & 0 & 0 & 0 & 0 & W_\tau & R_\tau \\ 0 & 0 & I & 0 & 0 & 0 & W_\eta & R_\eta \\ 0 & 0 & 0 & I & 0 & 0 & W_u & R_u \\ 0 & 0 & 0 & 0 & I & 0 & W_q & R_q \\ 0 & 0 & 0 & 0 & 0 & I & W_\sigma & R_\sigma \end{bmatrix} \begin{bmatrix} \delta f_c \\ \delta\tau \\ \delta\eta \\ \delta u \\ \delta q \\ \delta\sigma \\ \delta w \\ \delta\sigma_r \end{bmatrix} = 0. \tag{13.19}$$

13.3.2.1 Controllable Internal Forces

From (13.19), we can extract the expression for the contact force variation, that is

$$\delta f_c + W_f \, \delta w + R_f \delta\sigma_r = 0. \tag{13.20}$$

In continuity with the literature, we define as *internal* the solutions of (13.19), or equivalently of (13.14), not involving the external wrench variation. From this definition, it immediately follows that the matrix R_f spans the subspace of the *controllable internal forces*, that is the subset of all the contact force variations that can be generated controlling the synergistic movement of the hand.

13.3.2.2 Contact Force Transmission Caused by an External Wrench

Again from (13.20), considering the hand actuation kept constant, the matrix W_f represents a map between the external wrench and the contact force variation.

In other words, $-W_f$ represents the *contact force transmission* caused by an external wrench variation.

Both controllable internal forces and contact force transmission have great relevance in some grasping problems, as e.g. in the force closure evaluation and in the contact force optimization problem.

13.3.2.3 Controllable Internal Object Displacements

In case of whole-hand grasp and/or of underactuated hands, it could be not easy to find out which motions can be imposed to the grasped object by the hand. The problem can be solved considering the fourth equation of (13.19), that provides a description of the object displacements as

$$\delta u + W_u \delta w + R_u \delta \sigma_r = 0. \tag{13.21}$$

Similarly to what discussed in Sect. 13.3.2.1, from (13.21) we can easily conclude that the matrix R_u spans the subspace of the *controllable internal object displacements*.

13.3.2.4 Grasp Compliance

Again from (13.21), we can find that the matrix $-W_u$ represents the *grasp compliance*. In other words, the matrix $C_g = -W_u$ is the compliance that a 6D spring should have in order to imitate the effects of the hand actuation on the object displacements, when an external wrench is applied.

13.3.3 GEROME-B: A Specialized Gauss Elimination Method for Block Partitioned Matrices

In Sect. 13.3.1, a numerical method to compute the cFGM was presented. Furthermore, the physical interpretation of some blocks composing the cFGM was discussed, providing relevant information on the hand/object system. However, since the relevance of these blocks, it may be helpful to have a symbolic form of the matrices W_j and R_j in (13.19), in order to better understand how some basic matrices of the system (such as the Jacobian matrix J, the grasp matrix G, the synergy matrix S, etc.) can affect the properties of the whole system (e.g. the controllable internal forces or the controllable displacement of the object). Moreover, the knowledge of such symbolic relationships can be profitably used e.g. in designing robotic hands or underactuation mechanism. An example can be found in Chap. 8, regarding the design of the underactuation of the Pisa/IIT SoftHand.

To achieve this goal, the typical *Gauss Elementary Row Operation Method* (GEROME) for linear and homogeneous systems was adapted to act on block partitioned matrices (GEROME-B), preserving the integrity of the initial blocks (see also Chap. 8).

The GEROME-B method can be applied by means of the following three elementary operations:

- exchanging the ith row-block with the jth row-block
- multiplying the ith row-block by a full-rank matrix Δ,
- adding the ith row-block with the jth row-block, possibly left-multiplied for a suitable matrix Λ to accord dimensions.

Each rule can be performed by left-multiplying the FGM for a suitable full-column rank matrix, thus without affecting the solution space of the initial system.

Let us consider a proper identity matrix I_p, initially partitioned such that the ith block on the main diagonal, indicated as I_{p_i}, has the same dimensions of the ith row-block of the FGM. From this, the three matrices, equivalent to the three elementary operations previously seen, can be written as

$$\begin{aligned}
M_{ij}^1 &= \mathrm{diag}(I_{p_1}, \ldots, I_{p_{i-1}}, I_{p_j}, I_{p_{i+1}}, \ldots, I_{p_{j-1}}, I_{p_i}, I_{p_{j+1}}, \ldots, I_{p_m}),\\
M_{ii}^2(\Delta) &= \mathrm{diag}(I_{p_1}, \ldots, I_{p_{i-1}}, \Delta, I_{p_{i+1}}, \ldots I_{p_m}),\\
M_{ij}^3(\Lambda) &= I_p \oplus \Lambda_{ij},
\end{aligned} \tag{13.22}$$

where the expression $I_p \oplus \Lambda_{ij}$ indicates the insertion of a suitable matrix Λ on the block on the ith row and jth column of the default partitioned identity matrix I_p, and where m is the number of row-blocks of the identity matrix I_p.

Moreover, similarly to the classical elimination method, to apply GEROME-B it is necessary to define and identify some *pivot* elements.

Definition 1 A block of the FGM can be a *pivot* if

- it is a full-rank square block,
- it is the only pivot in its row and column,
- it is not a coefficient of one of the input variables.

Without losing generality, describing the algorithm, we suppose to act on a matrix $\hat{\Phi}^\star$, such that all the pivots are on the main diagonal. The matrix $\hat{\Phi}^\star$ can be obtained from the initial $\Phi^\star$ by properly exchanging some rows and columns and/or using matrices of the type (13.22). Once the algorithm is completed, if desired, the permutation can be inverted, restoring the initial order. In our case, the desired new form of the FGM can be written as

$$\hat{\Phi}^{\star} = \begin{bmatrix} I & 0 & 0 & K_c G^T & -K_c J & 0 & 0 & 0 \\ -J^T & I & 0 & -U_j & -Q_j & 0 & 0 & 0 \\ 0 & -S^T & I & 0 & 0 & -\Sigma & 0 & 0 \\ 0 & 0 & 0 & U_g - G K_c G^T & G K_c J & 0 & I & 0 \\ 0 & I & 0 & 0 & K_q & -K_q S & 0 & 0 \\ 0 & 0 & I & 0 & 0 & K_\sigma & 0 & -K_\sigma \end{bmatrix}. \quad (13.23)$$

The three matrices seen in (13.22) can be used to describe the GEROME-B algorithm, able to bring to the cFGM acting on the new form of the coefficient matrix (13.23). The GEROME-B algorithm essentially operates through the following steps: (i) the ith block row is left-multiplied for the inverse of the ith pivot, thus the ith pivot becomes an identity matrix; (ii) the ith pivot is used to cancel out all the elements on its same column; (iii) the process is iterated for all the pivots. A formal description of these steps is presented in Algorithm 1.

Algorithm 1 GEROME-B

```
for h = 1 → m do
    Δ = Φ̂*_hh^{-1}
    Φ̂* = M²_hh(Δ)Φ̂*
  for k = 1 → m do
    if h ≠ k then
      Λ = −Φ̂*_kh
      Φ̂* = M³_kh(Λ)Φ̂*
    end if
  end for
end for
```

13.4 Solution Space Decomposition

Among all the possible solutions of the system, several are of greater practical interest. As a simple example, let us consider an object placement task. During the motion of the object, uncertainties of the model, as well as external disturbances, could bring one or more contacts close to the slipping condition. In order to increase the robustness of the grasp without affecting the performances of the positioning task, it is important to recognize the capability of the hand of redistributing internal forces, avoiding object movements. From this and other simple examples, it follows that some interesting behavior of the system can be described by defining proper (non-)nullity patterns of the system variables. In this way, in this section, some particular types of solutions will be defined, together with a method to discover their feasibility, by means of a numerical procedure acting on the solution space of the system, that is on the nullspace of the FGM.

13.4.1 Relevant Types of System Solutions

13.4.1.1 Internal System Perturbations

As discussed in Sect. 13.3.2.1, following the grasping literature, we will call *internal* the solutions in which an external wrench variation does not appear, that is in the cases in which $\delta w = 0$.

13.4.1.2 Pure Squeeze

We define the *pure squeeze* as the particular system behavior in which there is a contact force variation not caused by an external wrench, and do not involving any object displacements. In other words, a pure squeeze occurs if $\delta w = 0$, $\delta f_c \neq 0$ and $\delta u = 0$.

13.4.1.3 Spurious Squeeze

An internal contact force redistribution associated to a displacement of the object is defined as *spurious squeeze*. The definition correspond to a solution of the form $\delta w = 0$, $\delta f_c \neq 0$ and $\delta u \neq 0$.

13.4.1.4 Kinematic Grasp Displacement

The internal solutions in which the object is moved without changing the contact force distribution, that is do not violating the (rigid) kinematic contact constraints, are called *kinematic grasp displacement*. Such solutions have to verify the conditions $\delta w = 0$, $\delta f_c = 0$ and $\delta u \neq 0$.

It is worth observing that, considering the elastic model of the contact as descriptive of the deformations of the grasped object, requiring a null variation of contact forces implies a null variation of the object shape. In this interpretation the definition of *rigid object displacement* can be recovered.

13.4.1.5 External Structural Force

An *external* action causing a contact force variation without affecting the hand actuation level is defined as *external structural force*. If such kind of solution is possible, it is characterized by $\delta w \neq 0$, $\delta f_c \neq 0$ and $\delta \eta = 0$, $\delta \sigma_r = 0$. Considering Eq. (13.13), above conditions directly imply also that $\delta \sigma = 0$.

13.4.2 *Discovering (Non-)Nullity Patterns in the Solution Space*

In previous sections we showed how some relevant types of manipulation tasks can be defined in terms of nullity or non-nullity of some system variables. The feasibility of such solutions can be investigate by properly elaborating the solution space of the FGM. In this section, we briefly present a method to discover if the hand/object system is able to perform a task corresponding to a solution of (13.14), in the desired form. To this aim, we firstly recall some results from linear algebra, the details of which can be found in [19]. For the following discussion, it is useful to recall that from every matrix $C \in \mathbb{R}^{r_c \times c_c}$, with $\rho_c = \mathrm{rank}(C)$, its corresponding *reduced row echelon form* (RREF) can be obtained via a Gauss-Jordan elimination. The same result can be equivalently obtained by a suitable permutation matrix $\Pi \in \mathbb{R}^{r_c \times r_c}$, such that

$$\Pi C = \left[\frac{U}{0}\right], \tag{13.24}$$

where $U \in \mathbb{R}^{\rho_c \times c_c}$ is a staircase matrix, and the zero block has consequent dimensions. The RREF of a matrix, in (13.24), can be profitably used to discover the presence of desired (non-)nullity pattern in the nullspace base $\Gamma \in \mathbb{R}^{r_\gamma \times c_\gamma}$, that is in the solution space of (13.14). In later discussion, we will make the assumption to have access to a function $\mathtt{rref}(X)$ able to return the reduced row echelon form of its argument[6] X.

For the sake of simplicity, we consider the system variables divided in two groups, called $\delta\varphi_\alpha$ and $\delta\varphi_\beta$, and we will present the investigation method supposing that we are interested to find the solutions characterized by $\delta\varphi_\beta = 0$. In this case, all the solutions of the system can be written as

$$\delta\varphi = \begin{bmatrix} \delta\varphi_\alpha \\ \delta\varphi_\beta \end{bmatrix} = \begin{bmatrix} \Gamma_\alpha \\ \Gamma_\beta \end{bmatrix} x, \tag{13.25}$$

where $\Gamma_\alpha \in \mathbb{R}^{r_\alpha \times c_\gamma}$ and $\Gamma_\beta \in \mathbb{R}^{r_\beta \times c_\gamma}$, the portions of the nullspace relative to the variables just defined.

---

[6]This is a typical situation with the most popular computational platforms, e.g.: $\mathtt{rref}(X)$ in MATLAB and `RowReduce(X)` in Mathematica.

Considering (13.25), a suitable permutation matrix can be obtained running the function `rref(`$[\Gamma_\beta^T \mid I]$`)`, which result is a matrix in the form $\left[\begin{array}{c|c} \frac{U_\beta}{0} & \Pi_\beta \end{array}\right]$, where $U_\beta \in \mathbb{R}^{\rho_\beta \times r_\beta}$, and $\rho_\beta = \text{rank}(\Gamma_\beta)$. From the properties of the RREF, it is known that the block $\Pi_\beta \in \mathbb{R}^{c_\gamma \times c_\gamma}$ is the permutation matrix such that $\Pi_\beta \Gamma_\beta^T = U_\beta$. Using these results, it is possible to find a new form ${}^1\Gamma \in \mathbb{R}^{r_\gamma \times c_\gamma}$ for the solution space matrix such that

$$
{}^1\Gamma = \Gamma \Pi_\beta^T = \left[\begin{array}{c} {}^1\Gamma_\alpha \\ \hline U_\beta^T \,|\, 0 \end{array}\right], \tag{13.26}
$$

where ${}^1\Gamma_\alpha = \Gamma_\alpha \Pi_\beta^T$. From direct inspection of (13.26), it is evident that the last $c_\gamma - \rho_\beta$ columns of Γ_1 span all the solutions in which $\delta\varphi_\beta = 0$, while the first ρ_β columns of Γ_1 span all the solutions in which $\delta\varphi_\beta \neq 0$. The method explained can be easily extended, by a recursive application, to the case of searching (non-)nullity conditions for more than one variable. The reader can find more details about the above method in [17, 20].

13.5 Geometrical Interpretation of the Fundamental Grasp Equation

In Sect. 13.2, a model describing the local behavior of a grasp with a synergistic underactuated robotic hand was obtained, starting from both the differential kinematic and the equilibrium equations of the system. The quasi-static form of such equations was obtained considering the effects of the differential kinematic equations for an infinitesimal amount of time, and by means of a first-order Taylor series approximation of the equilibrium equations. Moreover, the constitutive equations of the contacts, as well as the compliance in the actuation (at different levels), were introduced via linear elastic models. All these equations were used to build the *Fundamental Grasp Equation*.

As we saw in (13.14), it is straightforward considering the contribution of the congruence equations into the other relationships. As a result, Eq. (13.14) can be seen as the first-order approximations of a suitable system of nonlinear equation. Without going into the details, we just mention that such system of equations, the Taylor series approximation of which correspond to Eq. (13.14), can be written as

$$
\begin{cases}
w + G(u) f_c & = 0 \\
\tau - J^T(q, u)\, f_c & = 0 \\
f_c - K_c p_h^o & = 0 \\
\tau - K_q(\psi(\sigma) - q) & = 0 \\
\eta - S^T(\sigma)\, \tau & = 0 \\
\eta - K_\sigma(\sigma_r - \sigma) & = 0,
\end{cases} \tag{13.27}
$$

where $p_h^o \in \mathbb{R}^c$ is a vector describing the configuration of the hand contact frames with respect to object ones, and where we introduce the function $\psi(\sigma) := q_r$, such that $\frac{\partial\psi(\sigma)}{\partial\sigma} = S(\sigma)$. We will refer to Eq. (13.27) as the *equilibrium manifold*[7] of the system. We note in passing that the FGE is the equation of the hyperplane tangent to the equilibrium manifold in a specific point, representing an equilibrium configuration of the system.

It is worth observing that, given the invertibility of the matrix $\Phi_d^\star$ in (13.16), the variables δq and δw can be considered a local parametrization of the equilibrium manifold in the neighborhood of a given equilibrium configuration of the system. As discussed more in detail in [18], this property can be exploited in order to steer the system toward a new equilibrium configuration characterized by different kineto-static properties, with respect to the initial one. Moreover, as explained in [21], the equilibrium manifold of the system can be used as the exploration space for planning algorithm for closed kinematic chains as e.g. in bimanual manipulation tasks, taking advantage of the compliance in the contacts for *relaxing* the geometric constraints imposed by the presence of the closed loop. In this case, the above discussed equilibrium manifold can be used for random sampling based technique in order to generate any-time paths for closed-loop robot manipulators.

13.6 Other Types of (Under-)Actuation

Despite the fact that the *soft synergy* (Chap. 8) is currently one of the most attractive and interesting underactuation approach, it is worth considering the possibility to apply the analytical tools presented in this chapter also in other cases. In literature, other underactuation approaches deserve attention, as e.g. the *eigengrasp*, presented in [12], the parallel structure based [22], or the recent *adaptive synergies* approach, described in [23] and in Chap. 8. Some parts of the previous discussions were strictly dedicated to the *soft synergy underactuation*, especially in Sect. 13.2. However, the methods presented in Sects. 13.3 and 13.4 can be easily recovered for other types of underactuation (as also discussed for the methods in Chap. 12). After the kinematic and static equations were obtained in quasi-static form for the particular underactuation mechanism in exam, the *Fundamental Grasp Matrix* directly follows. From this, a proper definition of the *dependent* and the *independent* variables bring to obtain the FGM in *canonical form*. Moreover, the GEROME-B algorithm can still be applied, obtaining the symbolic form of the block matrix composing the cFGM. These results can be used to study how the underactuation affects the main system characteristics. Many definitions of manipulation tasks by (non-)nullity patterns can be recovered, regardless of the particular type of underactuation. One remarkable exception is

[7]More precisely, the equations related to the elasticity do not describe an equilibrium law, and, for this reason, we should, more properly, talk about a *manifold describing the kineto-static behavior of the whole system*. For the sake of compactness, this definition will be left implicit in the rest of the discussion.

the subspace of the *external structural forces*. However, the definition provided in Sect. 13.4.1.5 can be generalized considering the conditions $\delta w \neq 0, \delta f_c \neq 0$, and $\delta\tau^\star = 0, \delta q^\star = 0$, where $\delta q^\star$ and $\delta\tau^\star$ are the generalized displacement and force variables at the underactuation level.

In Chap. 8, more space is dedicated to the application of some of the discussed methods to the case of the *adaptive synergies* undearctuation model.

13.7 Numerical Results

13.7.1 Power Grasp

As a test case, we consider a spider-like hand, composed by two fingers and 8 joints, grasping a square of side $2L$. Figure 13.2 shows the initial configuration of the system and the contact force preload. All the initial force components have unitary value along the directions depicted.

13.7.1.1 Perturbed Configuration for Fully Actuated Hand

The solution space of the system has dimension equal to $\sharp w + \sharp q = 11$. Elaborating the nullspace of the FGM, it is possible to find out that the *pure squeeze* subspace

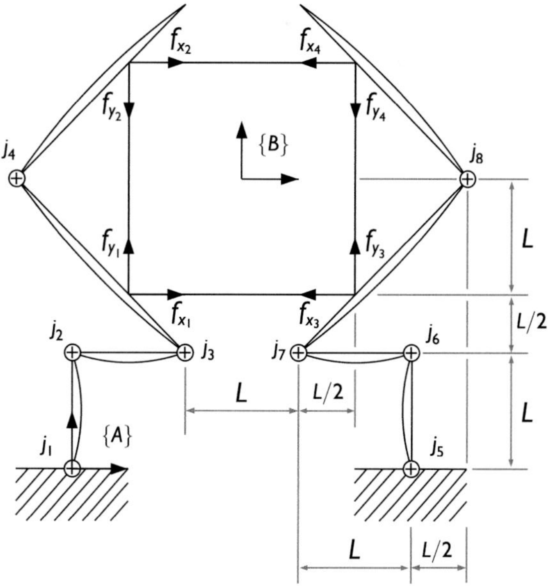

Fig. 13.2 Compliant grasp of a square object by a two fingered spider-like hand

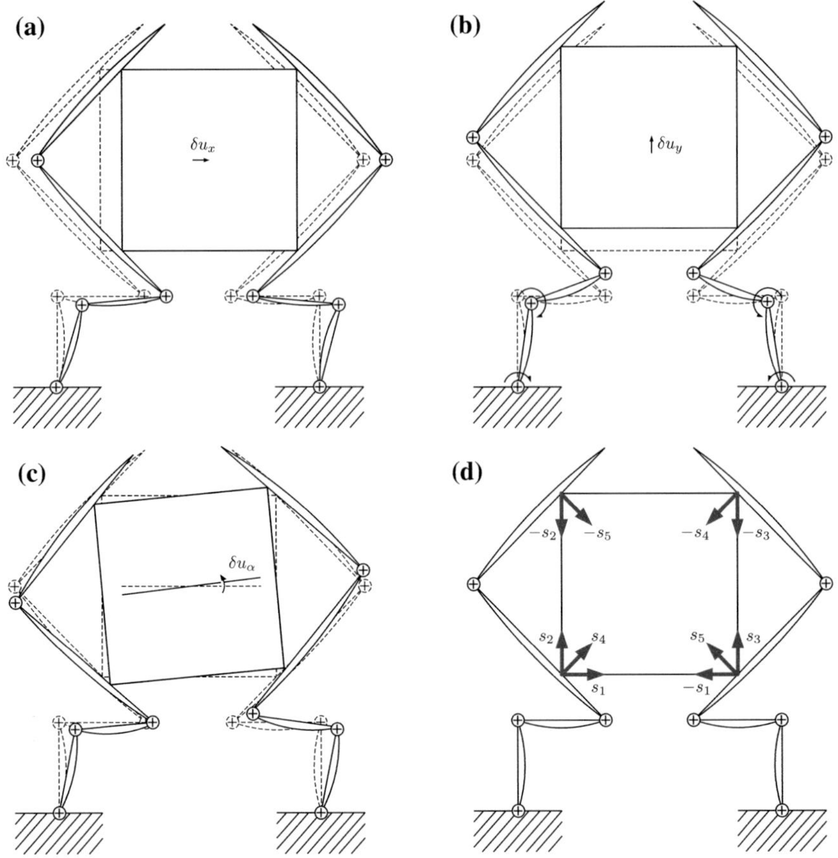

Fig. 13.3 Plates (**a**–**c**) represent the *kinematic displacements* of the grasped object, and plate (**d**) represents a basis for the *pure squeeze*

has dimension 5, the *kinematic grasp* subspace has dimension 3 and together they complete the internal solution subspace.

For the *kinematic grasp displacements*, simulation results show that it is possible to have a finite displacement of the object $\delta u_x = 0.001$, as in Fig. 13.3a, with no torque variations, but with the following joint angle displacements

$$\delta q = 10^{-3}\left[-1\ 1\ 0\ 0\ -1\ 1\ 0\ 0\right]^T. \tag{13.28}$$

For $\delta u_y = -0.001$, represented in Fig. 13.3b, the corresponding joint torques and joint angle variations are

$$\begin{aligned} \delta\tau &= 10^{-3}\left[-2\ -2\ 0\ 0\ 2\ 2\ 0\ 0\right]^T, \\ \delta q &= 10^{-3}\left[0\ 1\ -1\ 0\ 0\ -1\ 1\ 0\right]^T. \end{aligned} \tag{13.29}$$

To obtain an object rotation $\delta u_\alpha = 0.001$, without changing the contact forces, Fig. 13.3c, the necessary variations in the joint torques and joint angles are

$$\begin{aligned}\delta\tau &= 10^{-3}\left[3\;3\;0\;0\;3\;3\;0\;0\right]^T,\\ \delta q &= 10^{-3}\left[-1.5\;1\;1.5\;0\;-1.5\;1\;1.5\;0\right]^T.\end{aligned} \tag{13.30}$$

A basis for the pure squeeze is sketched in Fig. 13.3d, where the couple of forces s_i and $-s_i$ corresponds to the ith components of the basis. The numerical results for $\delta\tau$ and δq are omitted here for brevity.

13.7.1.2 A Synergy in the Power Grasp

Introducing in the system an underactuation characterized by a synergy matrix in the form

$$S = \begin{bmatrix} -0.6500 & 0 & -0.3200 & -0.4000 \\ 0.6500 & 0 & 0.3200 & 0.4000, \end{bmatrix}^T \tag{13.31}$$

in the solution space it remains a *pure squeeze* subspace of dimension 1.

In the absence of external disturbances, with an unitary synergistic actuation, $\delta\sigma_r = 1$, the contact forces and the object displacements become

$$\delta f_c = \begin{bmatrix} 0.5043 & 0.5043 & 0.5043 & -0.5043 \\ -0.5043 & 0.5043 & -0.5043 & -0.5043 \end{bmatrix}^T, \tag{13.32}$$

$$\delta u = \left[0\;0\;0\right]^T, \tag{13.33}$$

indicating that we are squeezing the object along both diagonals. It is worth noting that the above synergy was constructed by considering the contribution of two particular *pure squeeze* solutions, represented in Fig. 13.3d, for the fully-actuated system.

13.8 Conclusions

In this chapter, the basic concepts and methods for the quasi-static analysis of synergistically underactuated robotic hands were described. Moreover, compliance was integrated in the system at various levels, i.e. in the contacts between the hand and the object, and in the actuation mechanism, as discussed in Chap. 8. The derivative terms of the hand Jacobian and of the grasp matrix were also considered in the model, in order to properly take into account the effects of the contact force preload. Afterwards, the *Fundamental Grasp Matrix* (FGM) was defined, and a method for finding its *canonical form* (cFGM) was presented, both via a numerical and a symbolic approach. From the cFGM, relevant information on the system behavior can be easily extracted, as e.g. the *controllable internal forces*, the *controllable object displacements* and the *grasp compliance*.

Moreover, a method to investigate the solution space of the FGM was presented, able to point out the feasibility of relevant manipulation tasks, defined in terms of nullity or non-nullity of some system variables.

Despite the fact that the methods proposed provide information about local characteristics of the system around the initial equilibrium configuration, some results have also non-local relevance. In fact, it is possible to provide a geometrical interpretation of the FGE, for which this represents the tangential plane to the *equilibrium manifold* of the whole system. Exploiting the properties of the FGM, a local parametrization of the system can be found, which can be profitably used to steer the system over a continuum set of equilibrium configurations, until the desired kineto-static characteristics were fulfilled.

The generality of the proposed methods, as well as the technical tools described in Chap. 12, can be applied also in case of different types of underactuation, with small modifications.

Finally, in order to assess the validity of the proposed methods, an example of a power grasp has been presented showing the generality of the methods, capable of treating both the cases of fully actuated and synergistically controlled hands.

Acknowledgments This work was supported by the European Commission under the CP-IP grant no. 248587 "THE Hand Embodied", within the FP7-2007-2013 program "Cognitive Systems and Robotics", the ERC Advanced Grant no. 291166 "SoftHands: A Theory of Soft Synergies for a New Generation of Artificial Hands", and by the grant no. 600918 "PaCMan" - Probabilistic and Compositional Representations of Objects for Robotic Manipulation - within the FP7-ICT-2011-9 program "Cognitive Systems".

References

1. Jacobsen S, Wood J, Knutt D, Biggers K (1984) The Utah/MIT dextrous hand: work in progress. Int J Robot Res 3(4):21–50
2. Lovchik C, Diftler M (1999) The robonaut hand: a dexterous robot hand for space. In: Proceedings of 1999 IEEE International conference on robotics and automation, vol 2. pp 907–912
3. Shadow Robot Company Ltd. (2009) Shadow hand. http://shadowhand.com
4. Grebenstein M, Chalon M, Friedl W, Haddadin S, Wimbck T, Hirzinger G, Siegwart R (2012) The hand of the dlr hand ARM system: designed for interaction. Int J Robot Res 31(13):1531–1555
5. Fish J, Soechting JF (1992) Synergistic finger movements in a skilled motor task. Exp Brain Res 91(2):327–334
6. Angelaki DE, Soechting JF (1993) Non-uniform temporal scaling of hand and finger kinematics during typing. Exp Brain Res 92(2):319–329
7. Soechting JF, Flanders M (1997) Flexibility and repeatability of finger movements during typing: analysis of multiple degrees of freedoms. J Comput Neurosci 4(1):29–46
8. Santello M, Flanders M, Soechting J (1998) Postural hand synergies for tool use. J Neurosci 18:10105–10115
9. Latash, ML, Krishnamoorthy V, Scholz JP, Zatsiorsky VM (2005) Postural synergies and their development. Neural Plast 12:119–130, discussion 263–272

10. Thakur PH, Bastian AJ, Hsiao SS (2008) Multidigit movement synergies of the human hand in an unconstrained haptic exploration task. J Neurosci 28:1271–1281
11. Castellini C, van der Smagt P (2013) Evidence of muscle synergies during human grasping. Biol Cybern 107:233–245
12. Ciocarlie M, Goldfeder C, Allen P (2007) Dexterous grasping via eigengrasps: a low-dimensional approach to a high-complexity problem. In: Proceedings of the robotics: science and systems 2007 workshop-sensing and adapting to the real world. Electronically Published
13. Brown CY, Asada HH (2007) Inter-finger coordination and postural synergies in robot hands via mechanical implementation of principal components analysis. In: 2007 IEEE/RSJ international conference on intelligent robots and system, pp 2877–2882
14. Gabiccini M, Bicchi A, Prattichizzo D, Malvezzi M (2011) On the role of hand synergies in the optimal choice of grasping forces. Auton Robots [special issue on RSS2010] 31(2–3):235–252
15. Prattichizzo D, Malvezzi M, Bicchi A (2010) On motion and force controllability of grasping hands with postural synergies. Robotics: science and systems, vol VI. The MIT Press, Zaragoza, pp 49–56
16. Gabiccini M, Farnioli E, Bicchi A (2012) Grasp and manipulation analysis for synergistic underactuated hands under general loading conditions. In: International conference of robotics and automation—ICRA 2012, Saint Paul, MN, USA, pp 2836–2842, 14–18 May 2012
17. Gabiccini M, Farnioli E, Bicchi A (2013) Grasp analysis tools for synergistic underactuated robotic hands. Int J Robot Res 32:1553–1576
18. Farnioli E, Gabiccini M, Bonilla M, Bicchi A (2013) Grasp compliance regulation in synergistically controlled robotic hands with VSA. In: IEEE/RSJ international conference on intelligent robots and systems, IROS 2013, Tokyo, Japan, pp 3015–3022. 3–7 Nov 2013
19. Meyer CD (2000) Matrix analysis and applied linear algebra. Society for Industrial and Applied Mathematics, Philadelphia
20. Bicchi A, Melchiorri C, Balluchi D (1995) On the mobility and manipulability of general multiple limb robotic systems. IEEE Trans Robot Autom 11:215–228
21. Bonilla M, Farnioli E, Pallottino L, Bicchi A (2015) Sample-based motion planning for soft robot manipulators under task constraints. In: Accepted to international conference of robotics and automation—ICRA 2015
22. Birglen L, Laliberté T, Gosselin C (2008) Underactuated robotic hands. Springer tracts in advanced robotics, vol 40. Springer, Berlin
23. Catalano MG, Grioli G, Serio A, Farnioli E, Piazza C, Bicchi A (2012) Adaptive synergies for a humanoid robot hand. In: IEEE-RAS international conference on humanoid robots, Osaka, Japan

Chapter 14
A Simple Model of the Hand for the Analysis of Object Exploration

Vonne van Polanen, Wouter M. Bergmann Tiest and Astrid M. L. Kappers

Abstract When hand motions in haptic exploration are investigated, the measurement methods used might actually restrict the movements or the perception. The perception might be reduced because the skin is covered, e.g. with a data glove. Also, the range of possible motions might be limited, e.g. by wired sensors. Here, a model of the hand is proposed that is calculated from data obtained from a small number of sensors (6). The palmar side of the hand is not covered by sensors or tape, leaving the skin free for cutaneous perception. The hand is then modeled as 16 rigid 3D segments, with a hand palm and 5 individual fingers with 3 phalanges each. This model can be used for movement analysis in object exploration and contact point analysis. A validation experiment of an object manipulation task and a contact analysis showed good qualitative agreement of the model with the control measurements. The calculations, assumptions and limitations of the model are discussed in comparison with other methods.

14.1 Introduction

As discussed throughout the book, the investigation of human example can represent the winning approach to improve the design of artificial systems. The first step of all the neuroscientific studies on synergies and, consequently of their robotic applications, is the study of the human hand, and more specifically its kinematics. This motivates both the development of suitable kinematic models and sensing devices, like those described in Chap. 15, to investigate how human hands interact

V. van Polanen (✉)
Department of Kinesiology, KU Leuven, Movement Control and Neuroplasticity Research Group, Tervuursevest 101, 3001 Leuven, Belgium
e-mail: vonne.vanpolanen@kuleuven.be

W.M. Bergmann Tiest · A.M.L. Kappers
Department of Human Movement Sciences, MOVE Research Institute, Vrije Universiteit Amsterdam, Van der Boechorststraat 9, 1081 Amsterdam, BT, The Netherlands
e-mail: w.m.bergmanntiest@vu.nl

A.M.L. Kappers
e-mail: a.m.l.kappers@vu.nl

M. Bianchi and A. Moscatelli (eds.), *Human and Robot Hands*,
Springer Series on Touch and Haptic Systems, DOI 10.1007/978-3-319-26706-7_14

with or move within the external environment. The analysis of hand motions gives insights into how humans interact with objects (see e.g. Chaps. 2–8, 13 and 15). One of the purposes of hand motions might be to gain (haptic) perceptual information. Haptic perception is inherently an active process. While perception is important for performing an action, action also generates perceptual input. For the haptic modality, this consists of kinesthetic inputs from joint sensors, muscle spindles etc., but also cutaneous information benefits from the movement of an object against the skin. For instance, the object compresses the skin or rubs against it and this activates different sensors in the skin or in different ways compared to a static touch of the object. The hands are generally used to examine an object by touch. Therefore, it is of interest to see how the hand moves and how it manipulates the objects to perceive their properties.

Despite its importance for the haptic sense, the investigations of exploratory movements in haptic perception are scarce. Mainly, video analysis has been used to identify movement categories, as in the classic study of Lederman and Klatzky [15]. They defined exploratory procedures (EPs) that are optimal for the extraction of certain object properties. They define an EP as a "stereotyped movement pattern having certain characteristics that are invariant and others that are highly typical" [15, p. 344]. For instance, global shape can be perceived by the enclosure of an object and pressing an object is used for determining its hardness. These EPs were classified using video analysis where observers judge which EP is performed. This method has been used in the research of haptic exploration in healthy and blind individuals, as well as children [13, 14, 31].

Disadvantages of this procedure are that it is very time consuming and also subjective, because the judgement of the observer will depend on his or her interpretation of the exploratory procedure. In contrast, one study defined movement synergies from haptic exploratory movements, but did not distinguish between handling different object properties [26]. Another group of experiments were recently performed to quantify EPs by tracking the hand of the exploring participant [10, 11]. In those studies, only the index finger and the back of the hand were tracked. This might be enough for studying the exploration of large objects, where the hand is used as a whole. In the exploration of smaller objects, where the objects can be handled between several fingers or with multiple small objects in the hand, information of the movements of all fingers is important.

To measure the movements of all parts of the hand, the hand can be tracked by Data Glove-like systems (e.g. the CyberGlove and HumanGlove). In these systems, the observer wears a glove with sensors mounted inside. A review of these glove based systems can be found in [6] and a complete discussion of this topic can also be found in Chap. 15. The observer manipulates objects with the glove and because the sensors in the glove monitor the movements or positions of the hand, the exploration strategies can be studied. In perception research, besides the kind of movements that are made during exploration, it is also of interest which parts of the hand are used to extract the appropriate information. Some of these gloves measure the pressure an object exerts against the skin, so the contact points of the object with the hand can be determined. The combination of contact and position information would make a glove-based system ideal to investigate both exploratory procedures and hand contact information. However, an important limitation is that the glove reduces the cutaneous information, because it covers the skin. In addition, it might restrict the motions of participants.

An alternative to a glove would be to track the positions of the hand, where sensors are placed on the back of the hand. Since the palmar side of the hand is mostly used for extracting object properties, this leaves the skin free to explore. Such methods have been used in studies that analyzed grasping and finger movements (e.g. [3, 33]). They used reflective, passive markers that are captured by a camera. Using such a set of markers, a detailed model of the hand can be made [9, 24]. For instance, the model of Sancho-Bru et al. [24] calculates the contact forces and muscle forces that are used when holding an object. Their collision detection of the fingers on the object is based on a representation of the fingers as spherical polytopes. The spherical portrayal of fingers provide a more natural representation of the fingers than rectangular blocks. In addition, the algorithm assumes a soft contact model in the fingers that can deform on contact, although to determine contact with an object per se, the deformation of the fingers is not important.

However, there are also disadvantages in these approaches. In more complex exploration motions, the markers might suffer from occlusion. If a marker is not in view of the camera(s), it cannot be tracked. Some occluded markers might be reconstructed from others [5], but this might be more difficult with multiple occluded markers. Markers that do not suffer from occlusion, like magnetically instead of optically tracked sensors, might still limit hand motions due to the number of wires that are needed for these sensors. Moreover, the large number of markers might require a detailed calibration. In passive sensors, the sensors might also be confused with each other, especially if many are used and are placed close together. For these reasons, a smaller number of sensors might be desirable. Some approaches aim to ease the usage of all these systems by reducing the number of cumbersome sensors to be employed for the measurement process, leveraging the notion of synergistic reduction as discussed in Chap. 15. A balance must be sought between a small number of sensors and the amount of hand position information. With fewer sensors, the positions of the hand that are not measured, must be captured by a model (see Chap. 15 for a synergy-inspired reduction of the number of sensing elements and Chap. 7 for a more general discussion on sensory-motor synergies).

This chapter presents a way to model the hand in movement tracking analysis, where the position of all fingers and the hand palm is modeled from just 6 sensors with 6 degrees of freedom (DoFs). This small number of sensors poses few restrictions on the possible motions of the hand. In addition, the sensors are placed on the dorsal side of the hand, leaving the skin of the palmar side free to touch the object(s). In short, the hand model is calculated from positions of the fingertips and the hand palm. First, the joint positions are calculated for the fingers. Next, the phalanges are modeled as a series of connected rigid bodies. The hand palm is modeled as one rigid body. In this way, not only a kinematic chain of joints is made, but a 3D model of the hand is formed.

The model is not aimed to represent an actual hand, because a realistic hand is also very complex. The hand consists of 27 bones (including the wrist) and thus has many degrees of freedom. This is difficult to model with a few sensors and not necessary for most global movement descriptions. Therefore, the hand is simplified to fewer segments in the model. Furthermore, the model does not aim to serve as input for the control of robotic or artificial hands, although it might be used to extend the research needed to improve this challenging task (see also [17]). Instead, the model as presented in this chapter can be used as a tool to analyze the exploration and in-hand manipulation of objects. It is purposed to represent a hand that holds one or more objects and moves the fingers to explore and manipulate the object(s) in the hand. The model can be used to analyze contact points, but also the movements made by the individual hand parts. This chapter describes the calculation of the model and evaluates it for some possible applications.

14.2 Model of the Hand

The modeled hand consists of a series of rigid bodies: three for each finger, corresponding to the different phalanges and one for the hand palm (see Fig. 14.1 for an example). From fingertip to hand, the phalanges of the finger are called the distal, middle and proximal phalanx. The joints between the phalanges are called the distal interphalangeal joint, proximal interphalangeal joint and metacarpophalangeal joint (knuckle joint). The distal joint is the joint between the distal and middle phalanx, the proximal joint is the joint between the middle and proximal phalanx and the knuckle

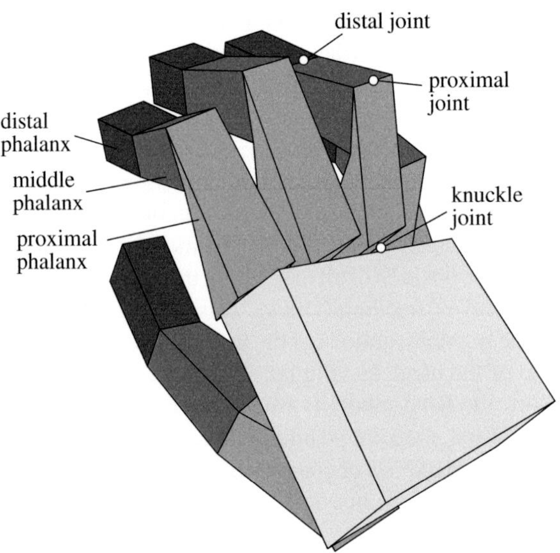

Fig. 14.1 An example of the hand model, the phalanges and joints are indicated. The joints were located in the fingers, but are drawn on top for illustration purposes

joint is the joint between the proximal phalanx and the hand palm. The thumb is modeled similar to the fingers, with 3 phalanges. The knuckle position of the thumb is then located more proximal in the hand, near the wrist.

In reality, the hand palm consists of a number of metacarpals. Here, the hand palm is modeled as a single rigid body. However, unless the hand is stretched out, the hand palm assumes a sort of bowl form where the knuckles lie on a slightly curved line. To be able to represent this curve, some rotations are applied to the hand palm and the fingers (see below). As a result, the proximal phalanges of the fingers form a slight bowl, where the proximal phalanx of the thumb lies on top of the hand palm.

To construct the model, 5 sensors are placed on the finger nails and one on the back of the hand. Next, the segments for the fingertips (distal phalanges) and the hand palm can be calculated from the sensor positions and orientations. For the other phalanges, first the joint positions need to be calculated. In the following sections, the data acquisition and calibrations that are necessary will be described. After that, the calculations of the joint positions and the construction of the segments will be explained step by step.

14.2.1 Sensors

The sensors that are used to measure the hand positions need to have 6 DoFs. This means that besides x, y, and z coordinates also the rotations azimuth (around z-axis), elevation (around y-axis) and roll (around x-axis) of the sensors are necessary. This is needed because not only the position of a sensor, but also its orientation in space must be known. Instead of 6 DoF sensors, also multiple (small) sensors in a grid might be used.

Sensors are placed on the nail of each finger, including the thumb. Another sensor is placed on the back of the hand, approximately in the middle in line with the knuckle of the middle finger (see Fig. 14.2). The centre of the finger sensors (corresponding to their measured position) is placed $\sim$5 mm from the fingertip. Sensors are attached with tape, but the palmar side of the hand is left free of tape, so the perceptual capabilities of the skin are not reduced. To further keep the sensors in place, tape is also placed over the wires on the second phalanx (not shown in Fig. 14.2).

14.2.2 Calibration

To be able to model the hand, some constant dimensions of the hand need to be determined. An overview can be found in Table 14.1. We propose an easy way to capture the dimensions of the finger by taking a picture of the hand that is placed on a grid, as shown in Fig. 14.2. The joint positions are marked on the observer's hand. The advantage is that the lengths of the fingers and joint positions can be calculated after the measurement, which reduces the time spent by the observer. The visible x

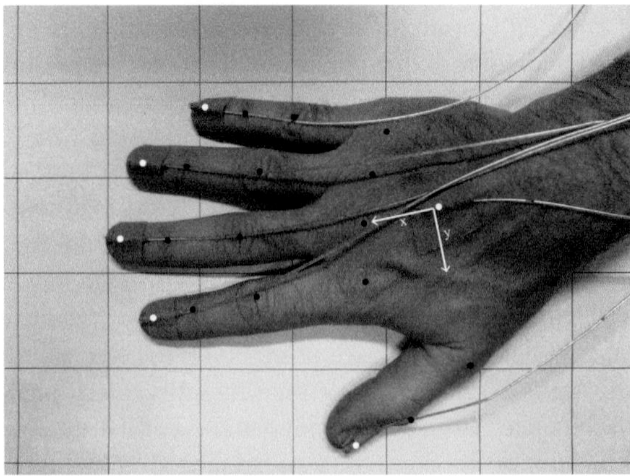

Fig. 14.2 Picture of the hand with the sensors (*white dots*) on the calibration grid. The joint positions are marked (*black dots*). The coordinate system drawn (*white*) is with respect to the hand, aligned with the hand sensor and the knuckle of the middle finger

Table 14.1 Overview of constant variables

Hand part	Variable	Meaning	Source
Finger phalanges	w	The width of the finger phalanx	Hand photo
	l	Length of the phalanx	Hand photo
	t_f [a]	Thickness of the finger	Calibration
	Roll	Initial roll of the distal phalanx only	Calibration
Finger knuckle	Δx	x-distance hand sensor to knuckle	Hand photo
	Δy	y-distance hand sensor to knuckle	Hand photo
Hand	t_h	Thickness of the hand	Calibration
	Roll	Initial roll of the hand	Calibration

[a]Proximal joint: 1.5 times as measured in calibration

and y grid coordinates of the photo are fitted to the known real-world coordinates of the grid with a 3rd order polynomial. In this way, the locations in the photo can be mapped to real-world coordinates. This mapping is then used to determine the coordinates of the marked joints in the photo. From these joint coordinates the lengths of the finger segments are calculated. The widths of the finger phalanges are also measured. For the thumb, the width of the proximal phalanx is chosen to be the same as the middle phalanx, as this cannot be measured in the photo. So, there are three lengths and widths for each finger. In addition, the positions of the knuckles with respect to the sensor on the back on the hand are measured as well. To do this, the knuckle of the middle finger and the hand sensor are aligned, so the y-distance of the hand sensor to the middle finger knuckle is 0.

Besides the finger lengths and widths, the thickness of the fingertips and of the hand need to be estimated. Therefore, a sensor is placed on a flat surface. The observer is asked to place each finger sequentially on the sensor, with the nail straight and horizontal. The z-distance between the sensor on the nail and the sensor on the surface is taken as the finger or hand thickness. The finger thickness is not assumed to be the same along the finger, but 1.5 times the measured thickness around the proximal joint and equal to the hand thickness around the knuckle joint.

In addition, in this calibration the initial roll rotation of each sensor is measured. To do so, the observer is asked to put his or her finger horizontally on the sensor, with the nail straight. This roll rotation is subtracted from the one measured in further analysis, to correct for the initial roll rotation of how the sensor was taped to the finger and hand. There is no need to correct for azimuth or elevation, because the sensor is placed in line with the finger on a flat surface (i.e. the nail or back of the hand).

14.2.3 Calculation of a Point from a Sensor

From the measurement of a certain sensor, not only the position, but also the orientation of the sensor is known. Hence, other points on the same rigid body can be easily calculated. The position of a point on the same segment as a sensor was calculated by transferring the relative position of the point with the rotations measured by the sensor. For example, to calculate the position of the distal joint from the sensor on the finger nail, the sensor coordinates need to be transferred a certain distance along the axis of the sensor. In Fig. 14.3, a distal finger phalanx is illustrated. Here, the coordinate system with respect to the sensor has its x-axis running along the sensor

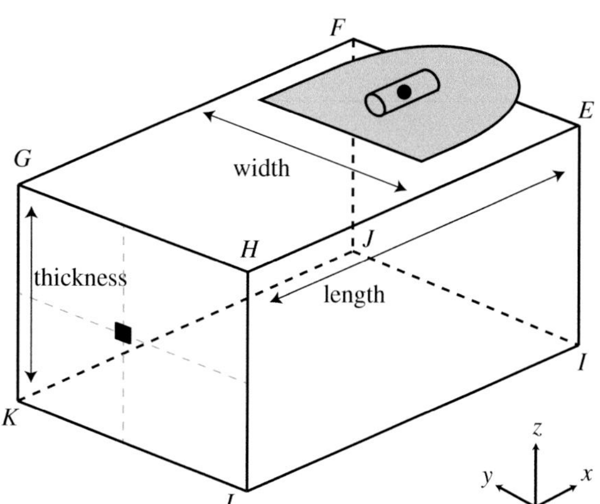

Fig. 14.3 Illustration of the phalanx of the finger. As an example the distal phalanx is pictured, with a *grey nail* to indicate the orientation. The sensor is placed on the nail, with the *black circle* to indicate the sensor position. The other phalanges connect to each other; so plane *GHKL* connects to plane *EFIJ* of the more proximal phalanx. The *black square* represents the distal joint position

from proximal to distal, the y-axis from right to left (ulnar to radial) and the z-axis from below to above the sensor (palmar to dorsal). The position of the distal joint is positioned in the middle of plane *GHKL* as a black square. This means that from the position of the sensor, the coordinates need to be transferred 'back' a distance equal to the length of the finger minus 5 mm (the sensor was placed ~5 mm from the fingertip) and 'down' half the thickness of the finger. So, with respect to the sensor, the distal joint is positioned at the coordinates $(-l + 5, 0, -\frac{1}{2}t_f)$. The length l and the thickness t_f of the finger are known from the calibrations. Next, a rotation transformation is applied to these coordinates, in which the measured azimuth, elevation and roll of the sensor are incorporated. Here, the initial calibrated roll rotation is subtracted from the measured roll. Lastly, the coordinates of the sensor are added to the result to obtain the coordinates of the distal joint in the original coordinate system. In short, the position of the distal joint is equal to:

$$\mathbf{R} * (-l + 5, 0, -\tfrac{1}{2}t_f) + (S_x, S_y, S_z) \tag{14.1}$$

where (S_x, S_y, S_z) are the coordinates of the sensor and $\mathbf{R}$ is a rotation matrix, calculated from the measured azimuth (around the z-axis), elevation (around the y-axis) and roll (around the x-axis, corrected for the initial roll) of the sensor. Other positions on the same rigid segment as the sensor can be calculated in a similar way.

As mentioned above, because the hand palm segment was modeled as a rigid block this does not the reflect the bowl form the hand makes when exploring objects in the hand. The assumed rigid form of the hand palm also affects the calculation of the positions of the knuckles, which are calculated from the hand sensor (see below), and thus the more proximal phalanges. To compensate for this, the hand sensor is extra roll rotated 40° (counterclockwise around the x-axis). The fingers knuckles are calculated from the hand sensor with an extra roll rotation in steps of 15°: 0, 15, 30, 45 and 60°, from thumb to little finger, respectively. This means that for every transformation of the hand sensor position, not only is there a correction for the initial roll rotation, but also the extra roll rotation is used. Pilot measurements verified that with these rotations, the calculated positions resembled the actual positions of the fingers and knuckles better than without these rotations.

14.2.4 *Joint Positions*

A simple illustration of a finger is shown in Fig. 14.4. To be able to construct the segments for each phalanx, first the joint positions of the fingers are calculated. Each finger (including the thumb) has three joints, the distal, the proximal and the knuckle joint. The position of the distal joint can be calculated from the position of the sensor on the nail with Eq. 14.1. The distal joint is positioned inside the finger, at half the thickness and width of the finger (point B in Fig. 14.4) and is calculated from the sensor on the finger, as described above. The knuckle position (A) can be determined at the relative position $(\Delta x, \Delta y, -\frac{1}{2}t_h)$ from the sensor on the back of

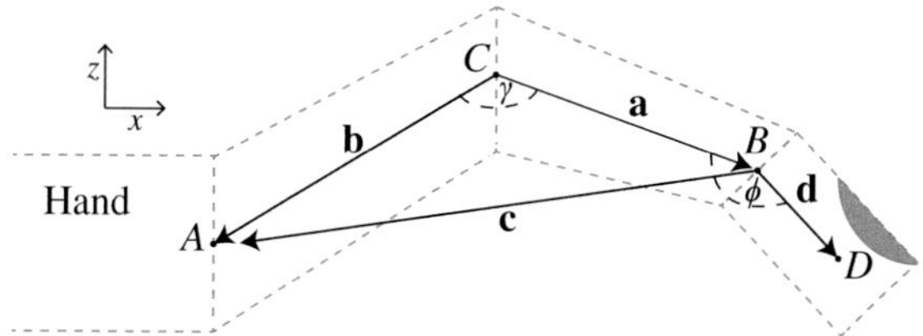

Fig. 14.4 Side view of the finger and the joints. A, C and B represent the knuckle, proximal and distal joints, respectively. D is the position in the finger underneath the nail sensor. **a**, **b**, **c** and **d** are the vectors between the joints and γ and ϕ the angles between the vectors that are used to calculate the proximal joint (C). The nail is colored *gray*

the hand. Δx and Δy are the distances to the knuckle as measured from the hand photo. The knuckle is placed at a depth half of the hand thickness t_h inside the hand. D is positioned at a relative position $(0, 0, -\frac{1}{2}t_f)$ from the nail sensor.

To calculate the position of C, the proximal joint, a triangle is assumed between the positions of the three finger joints, as illustrated in Fig. 14.4. The points A, B, C and D are assumed to lie in a single plane. This seems a valid assumption, since the distal and proximal joints have only one degree of freedom (flexion and extension). In Fig. 14.4, the positions of A, B and D are known. The lengths a and b of vectors **a** and **b** are known as the lengths of the finger phalanges that were measured in the hand photo. The vectors **c** and **d** can be calculated by subtracting B from A and B from D, respectively. γ can be determined by the cosine rule. Next, to calculate C, the following equalities must be satisfied:

$$\det([\mathbf{a}, \mathbf{b}, \mathbf{d}]) = 0 \tag{14.2a}$$

$$\mathbf{a} \cdot \mathbf{b} = ab\cos(\gamma) \tag{14.2b}$$

$$\sqrt{{a_x}^2 + {a_y}^2 + {a_z}^2} = a \tag{14.2c}$$

$$\sqrt{{b_x}^2 + {b_y}^2 + {b_z}^2} = b \tag{14.2d}$$

The first equality follows from the assumption that the vectors between the points A, B, C and D lie in a single plane. The other equalities are common vector rules. These equalities will leave two solutions, where C can lie 'above' (as shown in Fig. 14.4) or 'below' **c**. Of these two solutions, the one is chosen with the largest angle between the distal and middle joint, ϕ. This criterion is chosen because the distal finger phalanx cannot bend to very small angles. ϕ is defined from the following equality:

$$\mathbf{d} \cdot -\mathbf{a} = da \cos \phi \tag{14.3}$$

On some occasions, the triangle might not be closed. This is because with the bending and stretching of the fingers, the lengths between the joints do not remain constant in reality. In cases were no solution can be found because c is longer than $a + b$, the finger is assumed to be fully stretched. In these cases, C is calculated as the adjusted mean of A and B, positioning it on $\mathbf{c}$ but adjusted to the ratio of the lengths a and b:

$$C = B + \frac{a}{a+b}\mathbf{c} \tag{14.4}$$

Sometimes, the triangle cannot be closed, because c is smaller than $b - a$, which can happen with a fully bent finger. Here, C is positioned on $\mathbf{c}$, and extended with a:

$$C = B - \frac{a}{c}\mathbf{c} \tag{14.5}$$

Note that the direction of $\mathbf{c}$ is from B to A.

14.2.5 Distal (1st) phalanx

The first phalanx is calculated as a block around the coordinates of the sensor. Consider the phalanx model in Fig. 14.3. Here, the sensor position is illustrated as the black circle on the nail. Because the orientation of the sensor is known, the positions of the vertices of the phalanx block can be calculated. This happens in a similar way as in Eq. 14.1, as described earlier, using the length, width and thickness of the finger.

14.2.6 Middle (2nd) phalanx

The middle phalanx had no sensor on it, so the positions cannot be calculated with a rotation matrix, because the orientation of the phalanx was unknown. Because the phalanges are connected, the vertices E, F, I, and J of the middle phalanx are the same as the vertices H, G, L and K of the distal phalanx (see Fig. 14.3). The other vertices G, H, K and L of the proximal phalanx are calculated by summing vectors to the proximal joint as shown in Fig. 14.5. The points B_1 and B_2 can be calculated from the nail sensor in a similar way as the distal phalanx vertices. From the cross product of the vectors from the proximal joint to these points, a vector from the proximal joint to position M can be obtained ($\mathbf{m}$). Next, the cross product of $\mathbf{m}$ and

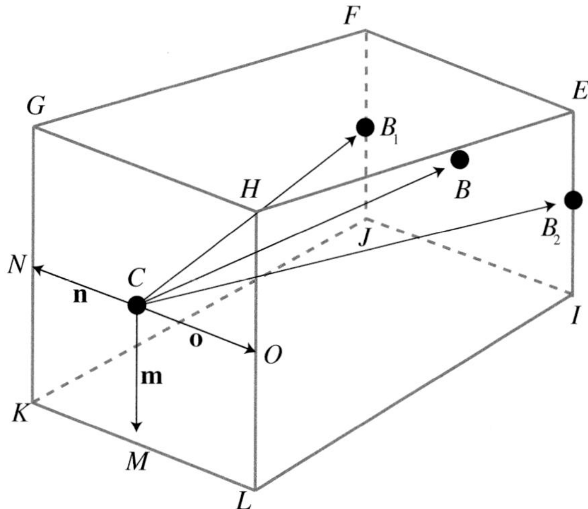

Fig. 14.5 Illustration of the calculation of vertices G, H, K and L around the proximal joint. In the figure, the proximal phalanx is shown with the proximal (C) and the distal (B) joints. See text for further explanation

the vector between the two joints gives the vectors perpendicular to **m**: **n** and **o**. The length of the vectors are adjusted to the thickness and width of the finger. By summation of these vectors to the proximal joint, vertices G, H, K, and L can be calculated:

$$\begin{bmatrix} G \\ H \\ K \\ L \end{bmatrix} = \begin{bmatrix} C + \mathbf{n} - \mathbf{m} \\ C + \mathbf{o} - \mathbf{m} \\ C + \mathbf{n} + \mathbf{m} \\ C + \mathbf{o} + \mathbf{m} \end{bmatrix} \quad (14.6)$$

where C are the coordinates of the proximal joint. Note that the thickness of the finger that is used in this plane is assumed to be 1.5 times the distal finger thickness. This means that the length of **m** is 0.75 times the calibrated finger thickness. The lengths of **n** and **o** are equal to half the width of the middle finger phalanx.

14.2.7 Proximal (3rd) phalanx

The vertices E, F, I and K of the proximal phalanx are the same as the vertices H, G, L, and K of the middle phalanx. The other vertices are calculated around the knuckle joint, using the sensor on the hand as reference point. The calculations are similar as those for the determination of the knuckle joint and use the proximal finger width and the hand thickness. The rotation matrix that is used includes the shifted roll rotations for each finger (0, 15, 30, 45 or 60°).

14.2.8 Hand

The hand part is a rigid body with eight vertices. The dorsal vertices are located at the dorsal positions above the little finger, index finger and thumb knuckle. The fourth dorsal vertex is the position on the ulnar side, at the same x-distance as the thumb knuckle and y-distance of the little finger. The palmar vertices were similar to the dorsal ones, but shifted in the z-direction by the thickness of the hand. The rotation matrix used here includes the extra roll rotation of $40°$. Therefore, the positions of the hand vertices will not be the same as the knuckle joint positions, since other roll rotations are used in those calculations.

14.3 Application Example: Contact Analysis

The model can be used for movement analysis. The sensors give, for example, the speed of the fingertips or their travelled distance. However, the model also offers extra parameters that can be considered. For instance, the angles of the various phalanges can be calculated from the joint positions. The angles reflect how much the finger is bent or stretched. This allows investigation of the closing and opening of the hand.

Another application of the model is an analysis of contact points with an object. In the following section, a mathematical procedure for determining which part of the hand comes in contact with a spherical object is explained in more detail.

The hand model consists of different hand parts. In this analysis we calculate which parts of the hand come into contact with ('touch') a specific object during exploration. Here, the position of the object must be tracked as well. For instance, a sensor might be placed inside or on the object. The model can then be used to determine the location of all parts of the hand with respect to the object. In this example, for each phalanx and for the hand palm we determine whether it contacts an object. The object is a sphere with a known radius and the position of the centre of the sphere is known.

To determine where the object touches the hand during the exploration, the distance from the object to the 6 planes of each segment (five distal phalanges, five middle phalanges, five proximal phalanges and the hand palm) is calculated. If one or more of the planes of a segment are touched, this is counted as a contact with this segment. An example situation of a plane and two spherical objects is shown in Fig. 14.6.

A plane of a segment can be represented with the equation $k_1x + k_2y + k_3z + k_4 = 0$. For coordinates (x, y, z) that lie in this plane, the equation holds. The constants k_1, k_2 and k_3 can be determined by calculating the cross product between two vectors that lie in the plane. For instance, $(V_1 - V_2) \times (V_4 - V_1)$ in Fig. 14.6 gives (k_1, k_2, k_3). Next, k_4 is calculated by solving the equation with a point that is known to be in the plane. For such a point, one of the vertices can be taken. The distance s from a point (x, y, z) to a plane can be found with the following equation:

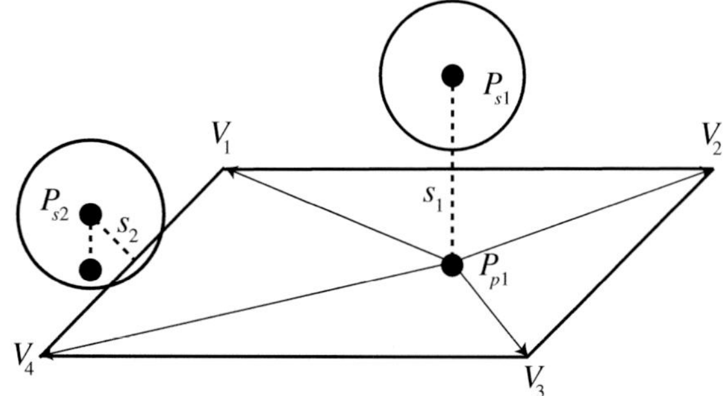

Fig. 14.6 Two examples of the calculation of the distance (s) from a sphere to a plane. The centre of a sphere is point P_s, which projects on the plane at point P_p. The vertices on the plane are labeled $V_{1...4}$

$$s = \frac{|k_1 x + k_2 y + k_3 z + k_4|}{\sqrt{k_1^2 + k_2^2 + k_3^2}} \tag{14.7}$$

If the point is the centre of the sphere (sphere sensor position, P_{s1}) and s_1 is smaller or equal to the radius of the sphere, the sphere touches the plane. However, the equation holds for an infinitely large plane. Thus, it must also be determined whether P_{s1} is close enough to the actual plane of the segment. s_1 is equal to the length of a vector that is perpendicular to the plane and runs from the plane to the point. This vector is a cross product of vectors in the plane and can be calculated from vectors on the edges of the segment plane (the same vector used to calculate k_1, k_2 and k_3 of the plane equation). Using this vector, the 'projected' point (P_{p1}) of P_{s1} on the plane can be calculated. If P_{p1} lies in the plane of the segment, the sum of the angles between the vectors running from P_{p1} to the vertices must be 360°. The angle of two vectors can be determined using Eq. (14.2b). So, if P_{p1} lies in the segment plane, the following equality must hold:

$$\begin{aligned} 360^\circ = \arccos\left(\frac{(V_1 - P_{p1}) \cdot (V_2 - P_{p1})}{|V_1 - P_{p1}||V_2 - P_{p1}|}\right) \\ + \arccos\left(\frac{(V_2 - P_{p1}) \cdot (V_3 - P_{p1})}{|V_2 - P_{p1}||V_3 - P_{p1}|}\right) \\ + \arccos\left(\frac{(V_3 - P_{p1}) \cdot (V_4 - P_{p1})}{|V_3 - P_{p1}||V_4 - P_{p1}|}\right) \\ + \arccos\left(\frac{(V_4 - P_{p1}) \cdot (V_1 - P_{p1})}{|V_4 - P_{p1}||V_1 - P_{p1}|}\right) \end{aligned} \tag{14.8}$$

where V_1, V_2, V_3 and V_4 are the coordinates of the vertices and P_{p1} that of the projected point.

A problem arrises at the border of a plane, where the projected point falls outside the plane, but the sphere contacts the side of the plane. This is illustrated in Fig. 14.6 by sphere P_{s2}. Therefore, the distance to the line segments on the border of the plane are calculated as well. For example, the distance s_2 of P_{s2} to the line segment $V_1 - V_4$ is:

$$s_2 = \frac{|(P_{s2} - V_1) \times (P_{s2} - V_4)|}{|V_4 - V_1|} \tag{14.9}$$

When P_{s2} is not positioned between V_1 and V_4, the distance is calculated to the endpoints (i.e., V_1 or V_4). If s_2 is smaller than the sphere radius, a contact point is also found. The distances to all four line segments are calculated. To summarize, the object only touches a part of the hand if (a) the distance s_1 is equal to or less than the sphere radius and Eq. (14.8) holds or, (b) s_2 is smaller or equal to the sphere radius.

14.4 Experimental Evaluation of the Model

To see how well the model could determine the finger positions, the model was evaluated in an object manipulation task. To do this, data from two control sensors were compared with the model calculations. Since there are two segments (middle and proximal phalanx) of each finger that are modeled and not tracked by a marker, two markers were placed on these segments to evaluate the models calculations. The position of the control sensors and the angle they made was used as a reference to which the model calculations were compared. Secondly, a contact analysis was performed on a task where a sphere was held between the thumb and another finger. This second task was chosen to evaluate the construction of the 3D phalanges. Since contact between the fingers and objects was known to occur, it should always be detected by the model.

14.4.1 Participants and Apparatus

Three right-handed participants took part in the experiment (2 females, 30 ± 4 years). Another participant was excluded due to technical errors. Participants provided written informed consent. A 3D Guidance TrakSTAR system (Ascension Technology Corporation) was used to measure the hand positions and orientations. Sensors were placed on the nails and hand as described above. The constant hand dimensions were measured as in the previously described calibration procedure. In addition, two extra sensors were placed on each finger sequentially. They were placed halfway between the joints on the middle and proximal phalanx. The objects used in the tasks were wooden beads, with a radius of 7.5 mm. They had a hole drilled in the center in which a sensor could be placed.

14.4.2 Task and Procedure

Two tasks were performed, an object manipulation task and a contact task. In the object manipulation task, five spheres were suspended from wires and were initially grasped. Then, the spheres needed to be dropped out of the hand one by one using the fingers. The two control sensors were placed on the thumb, index, middle, ring or little finger and participants performed three object manipulation trials. After that, the control sensors were placed on another finger and the process was repeated. The order of the control placement on the fingers was randomized.

The second task was a contact control task. Participants held a sphere, with a sensor placed inside, between the thumb and another finger. The experimenter indicated which finger was to be used. Three trials were performed for the four fingers, in a randomized order. While the participant held the sphere, the data was recorded for about 2 s. No control sensors were used in this task.

14.4.3 Analysis

For the object manipulation task, the positions and joint angles as determined by the model and the control sensors were compared. All parameters of the model as described above were computed. In addition, from these parameters, the positions halfway on the middle and proximal phalanx were calculated. These positions correspond to the locations of the control sensors and were directly compared with the measured positions from the control sensors. In addition, the angle between the control sensors was determined (around the proximal joint) and compared with the same angle as calculated from the model parameters. This angle was calculated from vectors running superficially on the dorsal side of the finger.

Comparisons were made by subtracting the positions or angles and average them across participants. In addition, the control and model values were correlated with each other. For the calculation of the correlation coefficients, one coefficient was determined for data of participants and trials grouped together. In the contact task, the contact analysis as described earlier was applied.

14.4.4 Results

The positions of the control sensors could be reproduced well by the model. An example of the x, y and z-positions of the control sensors and the calculated model positions is illustrated in Fig. 14.7. The qualitative representation was quite good, as correlations between the control and model positions were high (>0.95) for all fingers (Table 14.2). On average, the deviations were low, as shown in Table 14.3. The maximum absolute deviations could be high, though, up to a few cm.

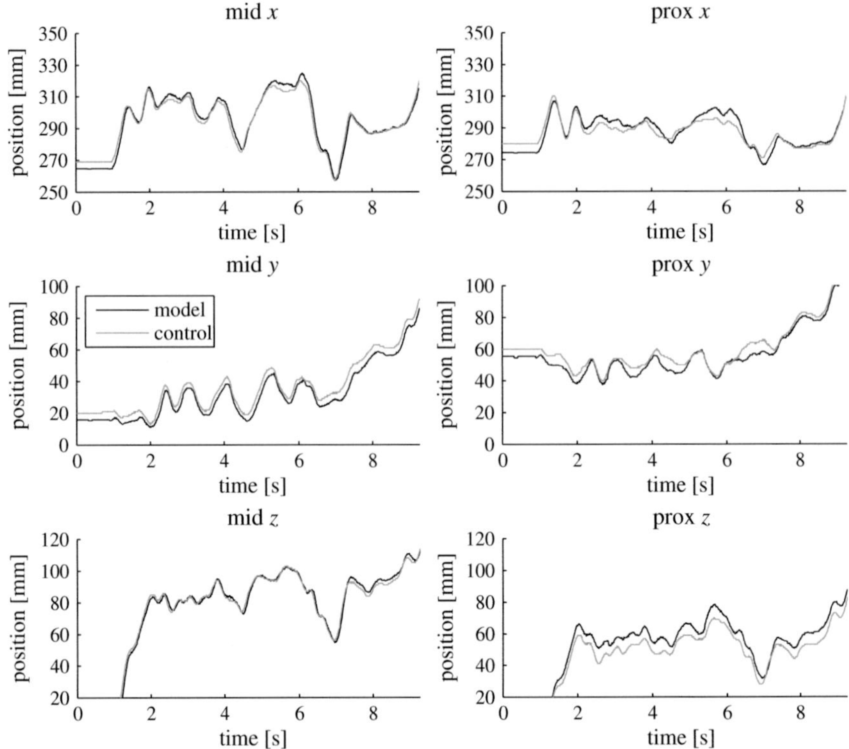

Fig. 14.7 An example of the comparison of the model and the control sensors for the *x*, *y* and *z*-positions. The positions of the middle (mid, *left panels*) and proximal (prox, *right panels*) phalanges of the middle finger are pictured. The *black solid line* represents the model, the *gray line* the control sensor

Table 14.2 Correlation coefficients between the model and control sensor positions for *x*, *y* and *z* directions and middle and proximal phalanges separately

Phalanx	Finger	*x*	*y*	*z*
Middle	Thumb	1.0	0.96	0.98
	Index	1.0	0.98	1.0
	Middle	0.99	0.99	1.0
	Ring	1.0	0.99	1.0
	Little	1.0	0.99	1.0
Proximal	Thumb	1.0	0.98	0.98
	Index	1.0	0.95	0.99
	Middle	0.99	0.98	0.99
	Ring	0.99	0.97	0.99
	Little	0.99	0.99	0.99

Table 14.3 Average deviations between the model and the control sensor positions for x, y and z directions and middle and proximal phalanges separately

Phalanx	Finger	x (mm)	y (mm)	z (mm)
Middle	Thumb	5.2 ± 0.7	-1.3 ± 1.8	-2.3 ± 1.6
	Index	−2.6±0.3	−4.4±1.3	−2.2±0.9
	Middle	−0.8±0.3	−4.5±0.2	1.3±1.3
	Ring	−0.1±0.2	−4.3±0.4	−0.2±0.6
	Little	1.1±0.1	−2.8±0.3	0.2±0.4
Proximal	Thumb	5.1±1.1	4.3±1.7	−6.8±1.5
	Index	−7.5±0.7	−10.5±2.5	1.4±1.5
	Middle	0.2±0.5	−5.8±0.9	5.9±1.8
	Ring	3.7±1.5	−2.1±0.9	4.3±1.4
	Little	4.8±2.4	1.8±0.7	6.3±0.8

Values are mean±standard error of the mean

The angles around the proximal joint were generally underestimated by the model compared to the angle as measured from the control sensors. Correlation coefficients were somewhat variable, but were still 0.85, 0.92, 0.91 and 0.89 for the index, middle, ring and little finger, respectively. For the thumb, the angles between the control and model calculations did not correlate (−0.06).

The alternative Eq. 14.4 had to be used in 34 % of the time samples for the thumb and in 0.4 % for the little finger in the object manipulation task. Equation 14.5 never had to be used in the present task.

The contact task indicated that contact of the thumb with the sphere was found on average on 99.6 ± 0.3 % (mean±sem) of the time. A contact was detected for $98.8 \pm 1.1, 79.1 \pm 19.1, 96.6 \pm 3.4$ and 100 ± 0 % of the time for the index, middle, ring and little finger respectively. Because with a light touch the calibrated thickness might be inaccurate, we also increased the tolerance by adding 1 or 2 mm to the finger thickness. Then, the percentages increased up to averages above 96 and 99 %, respectively.

In all trials but one, no other fingers contacted the sphere than the fingers used to hold the sphere. In one trial, a neighboring finger was also found to contact the sphere in 26 % of the trial time.

14.5 Discussion

A simple model of the hand was presented that constructs a hand with 16 segments and can be used as a tool to investigate the manipulation of objects by the hand. The measurement procedure that is presented requires only 6 sensors that are placed on the fingernails and the back of the hand. This procedure has two main advantages. First, the limited number of sensors poses few restrictions on the range of motion and allows the observer to move his or her hand in a natural way (see also Chap. 15 for a complete discussion of this topic). Secondly, because the sensors are placed on the dorsal side of the hand, the palmar side is left free to manipulate and explore the handheld object. In this way, the cutaneous input is not reduced in any way and finer object details can be perceived. Hence, the model is particularly suited for the investigation of the haptic perception and exploration of objects in psychophysical experiments.

The evaluation tasks indicated that the model represented the hand qualitatively well. In the contact task, where an object was held between thumb and finger, a contact with the object was usually well detected. Very light touches might not always be detected, because the finger thickness might be underestimated by the flat representation of the finger pad. Furthermore, the object manipulation task indicated that the model seems less suited for the investigation of angles, perhaps only in a qualitative way. In this task, the model represented the measured control positions qualitatively well. There were, however, also some deviations.

The differences between the model and the control measurement might arise from various sources, which can be roughly categorized into errors from calibration, measurement inaccuracies and model simplification. Calibration errors would include incorrect estimates of finger lengths and widths. Measurement inaccuracies consist of apparatus imprecisions and sensor displacements due to skin movements. Here, it must be noted that the control sensors are also subject to these inaccuracies and noise. Thus, deviations of the control measures and the model might also arise from these sources. This makes the validation experiment not a comparison of the model to correct hand positions, but to another measurement procedure. Nevertheless, this procedure provides a more direct measure of the validated positions because there are fewer calculations where errors can accumulate. Overall, the model leads to a good qualitative representation of the hand motions.

To represent the hand better, the hand model as presented is this chapter can be extended, but this will also make it more complex. In particular, the hand palm is now modeled as a single rigid body, whereas it consists of multiple parts in reality. Also the thumb segments are determined in a similar way as the other fingers, despite the fact that the thumb has more degrees of freedom than the other fingers and its ‘proximal phalanx’ is connected to the hand palm. The model could be further extended with more sensors on the back of the hand, to better measure the bowl form of the hand. The knuckle positions of the thumb and the index finger might for example be determined from an extra sensor close to the radial side of the hand, whereas the other knuckles are calculated from an extra sensor on the ulnar side. This requires that the hand palm segment consists of multiple parts, or a more complex form. However, considering the poor representation of the thumb joint angles and the difference in anatomy of the thumb compared to the other fingers (e.g. the degrees of freedom of the joints), an extra sensor to measure thumb kinematics might be a useful extension.

At this point, the model does not pose constraints on the the range of motion of the joints or in the degrees of freedom for the knuckle joint, as some other models do [1, 5, 16, 19, 24]. This makes the current model more flexible and it will find a solution more often, although reasonable constraints might provide more natural results. In the next part, the advantages and limitations of the model are discussed in comparison with the current state of the art of hand modeling.

14.5.1 Comparison with Other Models

One large advantage of the present model is that it requires a low number of markers. In contrast to gloves, this reduces restrictions on finger motions, makes the system easily adjustable to various hand sizes and leaves the skin uncovered to allow for cutaneous sensations. Many other models require the placement of many markers on the finger joints [5, 18, 33], but this also has limitations. For example, the wires of multiple markers might also restrict finger movements. In addition, there is the problem of occlusion with optically tracked markers. There are methods to reconstruct occluded markers [5] or ignore their input and rely on previous posture estimations (e.g. [8, 27]). However, with a lot of occluded markers or with occlusion for an extended time period, reconstruction might still prove difficult.

However, the disadvantage of few markers is the decreased accuracy, as multiple markers allow for a better estimation of the center of joint rotation. With just a few sensors, a small deviation (movement or displacement) of a sensor can have a large influence on the model calculations. The validation experiments showed a good qualitative representation of the hand, but the absolute deviations could be large. There are a few other methods that derive a hand model from a small marker set. Recently, a method for deriving index finger and thumb kinematics was presented by Nataraj and Li [19]. Their method for the calculations of the joint positions from two sensor sets is comparable to the one presented here, although different constraints are used. They showed that from a marker placed on the nail and on the back of the hand, similar results as with the measurements of more markers were obtained in a precision grasp task, which gave some validation for the use of fewer markers. Their model might be more accurate than ours, partly because they include a more extended calibration procedure (see below) and have a more realistic representation of thumb kinematics, especially for the metacarpophalangeal joint.

Most of the mentioned models are evaluated with simple grasping movements, usually power grasps. The evaluation of more complex movements is therefore still lacking. The current model is aimed at providing simple analysis of complex movements and information about object contact. It is therefore difficult to compare the current hand model to other biomechanical models aimed at grasping. Evidently, models that are based on more input data, such as more markers or calibration data, are more accurate.

When using few markers, some form of inverse kinematics is needed to calculate the joint angles of the finger phalanges. As described above, multiple techniques could be used to obtain these values. In methods that can be applied to a small marker set, iterative processes [2], geometrical models [19], optimization techniques [16] and filters [7, 27] are all procedures that can aid to find the unknown parameters. Our model uses only a geometrical model without optimal solution fitting. This makes the calculations simple and fast to perform, but it largely depends on the input values. As discussed earlier, small deviations will have a large influence on the outcomes. A problem with solution fitting that is absent in our model is that it can be trapped in local minima. However, in the present method sometimes a solution cannot be found, although this mainly occurs with fully extended finger and Eq. (14.4) can be applied.

To cope with outliers in the data, a model could be partly based on previous hand estimations. For instance, an extended Kalman filter could be used, where the current measure is the estimation and the previous posture is used as the prior [7, 8, 27]. The estimation and prior are then optimally implemented into the final model. Alternatively, a least median square algorithm could be used to determine the hand motions [32]. Because these methods to estimate the hand pose are partly based on the previous time frame, the influence of outliers is diminished.

A completely different procedure is applied by [1]. They use an iterative process (FABRIK, [2]) to estimate the joint position of a hand model from a minimal set of markers. Their hand model is constrained by the lengths between the joints and the possible rotation limits of each joint. Their methods are aimed at real-time hand tracking, and are mainly focused on limiting computation time. Unfortunately, they only provide a qualitative validation of their model, so it cannot be compared to ours.

Furthermore, besides occlusion and wearability, markers might not correspond to a stable skeletal position due to skin artifacts. The movement of the skin changes the position of the marker with respect to the joint dependent on the posture. This can be compensated by including skin deformation into a model [8] using measures from magnetic resonance imaging (MRI) [9]. In our presented procedure, the placement on the nail prevents sensor movement due to skin deformation. Therefore, inaccuracies due to skin deformation will be negligible.

Another aspect that will greatly influence the accuracy of the model are the calibrated measures. The present model does require some calibrations to be able to be fitted to the hand size of the observer. Calibration is often a time-consuming process, so we proposed to measure the lengths and widths of the fingers from a photograph. The various measurements can thus be calculated after the actual experiment, which reduces the time spent on calibration procedures. The calibration grid adjust distortions due to the perspective of the photo, but still the lengths depend on the accuracy of the marked positions of the joints. The hand is photographed in a stretched position, but with bending and stretching of the fingers, the distances between joints vary and this might cause problems with the model calculations. At this point, we think it is difficult to solve this problem without a more extensive calibration process and more measurements of joint positions during the exploration. Studies that compare their model to specific known postures in the calibration procedure require knowledge about all parameters in those postures, which might require extra measurements as well.

If a more accurate performance is required, there are many possible options to improve the calibrated measures. For instance, [19] partly resolved the problem of many markers by only including them in the calibration procedure and once the joint rotation centers are calculated, omitting them later in the experiment. Another way is to use a caliper to measure all finger lengths and thicknesses or even employ detailed MRI data.

The photo of the hand provides a simple way to scale the model to different hand sizes as individual measures can be easily acquired and put into the calculations. Although hand size scaling seems essential, not all models are equally flexible. For instance, a model might be partly fitted on experimental data. Lien and Huang [16] use inverse kinematics to calculate all possible solutions of the finger joints and the best solution is found through a regression search method. Constraints and relations between the joints are used in the calculations, where these are partly fitted onto experimental data. This makes this model difficult to apply to different hand sizes. Also Sancho-Bru et al. [24] base their model on experimental data, but scale it to the hand length and width of participants.

As hand models are only a representation of the actual hand, all models make assumptions. Even the more complex and detailed models (e.g. [24]) make assumptions that might differ from actual hand anatomy [4]. To gain insight into most hand movements, these constraints will often be adequate. In our model, the hand is also simplified, for instance, with a rigid palm and phalanges in the shape of blocks. In addition, in our model, the hand is assumed to hold one or more objects and thus to be shaped in a slight bowl form. In situations were the hand is mostly stretched, this might induce deviations. In such a task, the roll rotation corrections might be omitted. Currently, the roll rotations increase from the thumb to the little finger in steps of 15°. Further validation experiments could refine the best rotations for specific tasks.

Finally, another important feature of the current model is that it goes further than the above models by forming 3D phalanges in addition to the joint positions. Even if other minimal marker methods would be used to define joint positions, still the creation of a complete finger is very useful for psychophysical experiments. These 3D phalanges allow for a contact analysis, which is of high interest in haptic exploration research. A procedure for such a contact analysis was also described in this chapter. Whether a finger makes contact with an object determines if properties of that object could have been processed by that specific finger. In the last part of this chapter, the possible applications of the model are discussed.

14.5.2 Applications

The model as presented in this chapter is targeted at psychophysical studies where extensive technological devices, like many markers or data gloves, cannot be used as such systems would severely impair task performance. It is essential that cutaneous input is not hindered and movements are not restricted. Moreover, it is of relevance to have a method that is quick to execute, both in calibration and calculation as many psychophysical experiments have time restrictions or online updating is desired. In addition, the simple calculations make the model easy to use. This is a method that is specifically designed for psychophysical experiments, whereas most models are usually optimized to provide hand animations or investigate different grasping movements to objects, like power or pinch grips. So although our model cannot compete with other models in terms of accuracy, it will still be the method of choice in many psychophysical experiments. Of course, further development of the model and integration with optimal estimation and design techniques as described in Chap. 15 is needed and possible, but its present state provides a good start.

Our model is specifically aimed at the analysis of interaction with objects where haptic perception is critical and objects are held in the hand. Previous investigations of haptic perception have only analyzed a part of the exploring hand, and mostly objects were 2D [11, 20, 21, 25, 28]. However, this would not be enough to describe the movements where all the fingers are used to explore a 3D object, for instance, in the manipulation of small objects or multiple objects. Examples of such tasks are search tasks (e.g. [23, 30]), numerosity tasks (e.g. [22]) or the discrimination of small objects (e.g. [12]). Until now, the research into the exploration of 3D objects is rarely studied in a quantitative manner [10, 26]. The use of the model could facilitate this research. For instance, the object contact analysis is useful to investigate whether in specific explorations some parts of the hand are used more than others and at which moment in time.

A first example of the use of this model is the analysis of hand movements made in a haptic search task [29]. In this study, we varied object properties, like texture and shape, of the target object that had to be haptically searched for. Participants performed complex and detailed movements that would have been difficult to capture directly without restricting the finger motions. The model allowed us to investigate these unconstrained movements and we found differences in the use of the fingers and exploration strategies dependent on the specific target object property.

14.5.3 Conclusion

In conclusion, the current model provides a trade-off between the use of a minimal marker-set and simple calculations with accuracy. Improvements to the calibration procedure, marker-set and computations can make the hand model more realistic, but also increase the complexity of the calculations and measurements. For general purposes, the representation of the model in its current form will be enough. At the same time, integration with the synergy-inspired optimal estimation techniques and design of hand-pose reconstruction devices discussed in Chap. 15 can improve the experimental outcomes described in this chapter. Especially in experiments were the use of many sensors is undesirable, this model can be used to analyze whole hand motions even with a small number of sensors. For general movements, the model provides a basis to analyze these in an objective way. This is an advantage to procedures that use subjective measures, such as video analysis. The measurement with few sensors and simple calculations will then be a huge advantage.

Acknowledgments This work was supported by the European Commission with the Collaborative Project no. 248587, "THE Hand Embodied", within the FP7-ICT-2009-4-2-1 program "Cognitive Systems and Robotics".

References

1. Aristidou A, Lasenby J (2010) Motion capture with constrained inverse kinematics for real-time hand tracking. In: 4th International symposium on communications, control, and signal processing, ISCCSP 2010, IEEE, pp 1–5
2. Aristidou A, Lasenby J (2011) FABRIK: A fast, iterative solver for the Inverse Kinematics problem. Graphical Models 73(5):243–260
3. Braido P, Zhang X (2004) Quantitative analysis of finger motion coordination in hand manipulative and gestic acts. Hum Mov Sci 22(6):661–678
4. Bullock IM, Borras J, Dollar AM (2012) Assessing assumptions in kinematic hand models: A review. In: Proceedings of the IEEE RAS and EMBS international conference on biomedical robotics and biomechatronics, IEEE, pp 139–146
5. Cerveri P, De Momi E, Lopomo N, Baud-Bovy G, Barros RML, Ferrigno G (2007) Finger kinematic modeling and real-time hand motion estimation. Ann Biomed Eng 35(11):1989–2002
6. Dipietro L, Sabatini AM, Dario P (2008) A survey of glove-based systems and their applications. IEEE Trans Syst Man Cybern Part C: Appl Rev 38(4):461–482
7. Fu Q, Santello M (2010) Tracking whole hand kinematics using extended Kalman filter. In: 2010 Annual international conference of the IEEE engineering in medicine and biology society, EMBC'10, pp 4606–4609, doi:10.1109/IEMBS.2010.5626513
8. Gabiccini M, Stillfried G, Marino H, Bianchi M (2013) A data-driven kinematic model of the human hand with soft-tissue artifact compensation mechanism for grasp synergy analysis. In: IEEE International conference on intelligent robots and systems, pp 3738–3745, doi:10.1109/IROS.2013.6696890
9. Gustus A, Stillfried G, Visser J, Jörntell H, van der Smagt P (2012) Human hand modelling: kinematics, dynamics, applications. Biol Cybernet 106(11–12):741–755
10. Jansen SEM, Bergmann Tiest WM, Kappers AML (2015) Haptic exploratory behavior during object discrimination: A novel automatic annotation method. Plos One 10(2):e0117,017, doi:10.1371/journal.pone.0117017
11. Jansen SEM, Bergmann Tiest WM, Kappers AML (2013) Identifying haptic exploratory procedures by analyzing hand dynamics and contact force. IEEE Trans Haptics 6(4):464–472
12. Kahrimanovic M, Bergmann Tiest WM, Kappers AML (2011) Discrimination thresholds for haptic perception of volume, surface area, and weight. Atten Percept Psychophys 73(8):2649–2656
13. Kalagher H, Jones SS (2011) Young children's haptic exploratory procedures. J Exp Child Psychol 110(4):592–602
14. Klatzky RL, Lederman SJ, Reed C (1989) Haptic integration of object properties–texture, hardness, and planar contour. J Exp Psychol Hum Percept Perform 15(1):45–57
15. Lederman SJ, Klatzky RL (1987) Hand movements–a window into haptic object recognition. Cogn Psychol 19(3):342–368
16. Lien CC, Huang CL (1998) Model-based articulated hand motion tracking for gesture recognition. Image Vision Comput 16(2):121–134
17. Liu H (2011) Exploring human hand capabilities into embedded multifingered object manipulation. IEEE Trans Ind Inform 7(3):389–398
18. Miyata N, Kouchi M, Hurihara T, Mochimaru M (2004) Modeling of human hand link structure from optical motion capture data. Intell Robots Syst 3:2129–2135

19. Nataraj R, Li ZMM (2013) Robust identification of three-dimensional thumb and index finger kinematics with a minimal set of markers. J Biomech Eng 135(9):91002
20. Overvliet KE, Smeets JBJ, Brenner E (2007) Haptic search with finger movements: using more fingers does not necessarily reduce search times. Exp Brain Res 182(3):427–434
21. Plaisier MA, Bergmann Tiest WM, Kappers AML (2008) Haptic pop-out in a hand sweep. Acta Psychol 128(2):368–377
22. Plaisier MA, Bergmann Tiest WM, Kappers AML (2009a) One, two, three, many–Subitizing in active touch. Acta Psychol 131(2):163–170
23. Plaisier MA, Bergmann Tiest WM, Kappers AML (2009b) Salient features in 3-D haptic shape perception. Atten Percept Psychophys 71(2):421–430
24. Sancho-Bru JL, Mora MC, León BE, Pérez-González A, Iserte JL, Morales A (2014) Grasp modelling with a biomechanical model of the hand. Comput Methods Biomech Biomed Eng 17(4):297–310
25. Smith AM, Gosselin G, Houde B (2002) Deployment of fingertip forces in tactile exploration. Exp Brain Res 147(2):209–218
26. Thakur PH, Bastian AJ, Hsiao SS (2008) Multidigit movement synergies of the human hand in an unconstrained haptic exploration task. J Neurosci 28(6):1271–1281
27. Todorov E (2007) Probabilistic inference of multijoint movements, skeletal parameters and marker attachments from diverse motion capture data. IEEE Trans Biomed Eng 54(11):1927–1939. doi:10.1109/TBME.2007.903521
28. van Polanen V, Bergmann Tiest WM, Kappers AML (2012a) Haptic pop-out of movable stimuli. Atten Percept Psychophys 74(1):204–215
29. van Polanen V, Bergmann Tiest WM, Kappers AML (2014) Target contact and exploration strategies in haptic search. Sci Reports 4:6254. doi:10.1038/srep06254
30. van Polanen V, Bergmann Tiest WM, Kappers AML (2012b) Haptic search for hard and soft spheres. PLOS ONE 7(10):e45298
31. Withagen A, Kappers AML, Vervloed MPJ, Knoors H, Verhoeven L (2013) The use of exploratory procedures by blind and sighted adults and children. Atten Percept Psychophys 75(7):1451–1464
32. Wu YWY, Huang T (1999) Capturing articulated human hand motion: a divide-and-conquer approach. Proceedings of the seventh IEEE International conference on computer vision, IEEE 1:606–611
33. Zhang X, Lee SW, Braido P (2003) Determining finger segmental centers of rotation in flexion-extension based on surface marker measurement. J Biomech 36(8):1097–1102

Chapter 15
Synergy-Based Optimal Sensing Techniques for Hand Pose Reconstruction

Matteo Bianchi, Paolo Salaris and Antonio Bicchi

Abstract Most of the neuroscientific results on synergies and their technical implementations in robotic systems, which are widely discussed throughout this book (see e.g. Chaps. 2, 3, 4, 8, 10, 12 and 13), moved from the analysis of hand kinematics in free motion or during the interaction with the external environment. This observation motivates both the need for the development of suitable and manageable models for kinematic recordings, as described in Chap. 14, and the calling for accurate and economic systems or "gloves" able to provide reliable hand pose reconstructions. However, this latter aspect, which represents a challenging point also for many human-machine applications, is hardly achievable in economically and ergonomically viable sensing gloves, which are often imprecise and limited. To overcome these limitations, in this chapter we propose to exploit the bi-directional relationship between neuroscience and robotic/artificial systems, showing how the findings achieved in one field can inspire and be used to advance the state of art in the other one, and vice versa. More specifically, our leading approach is to use the concept of kinematic synergies to optimally estimate the posture of a human hand using non-ideal sensing gloves. Our strategy is to collect and organize synergistic information and to fuse it with insufficient and inaccurate glove measurements in a consistent manner and with no extra costs. Furthermore, we will push forward such an analysis to the dual problem of how to design pose sensing devices, i.e. how and where to place sensors on a glove, to get maximum information about the actual hand posture,

M. Bianchi (✉) · A. Bicchi
Department of Advanced Robotics (ADVR) – Istituto Italiano di Tecnologia, via Morego 30, 1613 Genova, Italy
e-mail: matteo.bianchi@iit.it

M. Bianchi · A. Bicchi
Research Center "E. Piaggio" – Università di Pisa, Largo Lucio Lazzarino 1, 56126 Pisa, Italy
e-mail: bicchi@centropiaggio.unipi.it

P. Salaris
Laboratoire d'Analyse et d'Architecture des Systèmes (LAAS) – CNRS, Avenue du Colonel Roche, 7, 54200 31031 Toulouse Cedex 4, France
e-mail: salarispaolo@gmail.com

M. Bianchi and A. Moscatelli (eds.), *Human and Robot Hands*,
Springer Series on Touch and Haptic Systems, DOI 10.1007/978-3-319-26706-7_15

especially with a limited number of sensors. We will study the optimal design of gloves of different nature. Conclusions that can be drawn take inspiration from and might inspire further investigations on the biology of human hand receptors. Experimental evaluations of these techniques are reported and discussed.

15.1 Introduction

The problem to achieve a correct and reliable hand pose estimation through Hand Pose Reconstruction (HPR) systems or "gloves" [10, 31] has gained an increasing importance for human-machine interactions in numerous applications such as robotics, rehabilitation, virtual reality and motion analysis. Furthermore, the study of human hand in psychophysical and neuroscientific studies requires accurate biomechanical and postural measurements together with refined kinematic models [15] to test and analyze theoretical motion control hypotheses [26], as it is widely discussed e.g. in Chaps. 2, 3, 4 and 14.

Unfortunately, all current HPR methods are limited due to non—idealities, such as an imperfectly known relationship between the measurements and the complexity of the mechanical Degrees of Freedom (DoFs) of the human hand as well as considerations that tend to discourage the usage of many sensors. Regarding this last point, economic motivations are crucial to determine the choice of both the technology solution in use and the number of sensing elements. Under this regard, a meaningful example is the CyberGlove (CyberGlove System LLC, San Jose, CA–USA), which is one of the most popular HPR glove-based systems: such a glove can come equipped with 18 or 22 piezoresistive sensors but its overall cost grows from 12,297 USD to 17,795 USD (2010 quotes). On the other side, the need of enabling mass diffusion has led to the development of more economic but inaccurate devices: e.g. Mattel's PowerGlove (Mattel Inc., El Segundo, CA–USA), which usually met with scarce acceptance due to their imprecision. Ideally, the goal is to have systems that are economic but effective.

This chapter, which is based on [1, 2], gives a global vision of the twofold problem of (i) optimally estimating human hand posture from partial and noisy HPR data—hence improving their accuracy at no extra costs—and (ii) how to optimally design pose sensing devices, i.e. how and where to place sensors on the human hand, to get maximum information about the actual hand pose despite limitations on their number and capabilities. This last point can be inspired from and offers interesting insights into biological investigations on human mechanoreceptors, as it will be discussed later in Sect. 15.2. Such a bi-directional relationship between natural and artificial side will be deeply analyzed in this chapter, representing the *leitmotif* of this work and all this book.

Indeed, the leading idea of our approach is the concept of "human hand synergies" [26–28] (see also Chaps. 2–6, 8, 10, 12 and 13): i.e. although very complex and possibly different in size and shape, human hands share many commonalities in how they are shaped and used in frequent everyday tasks. We will exploit such an information on the most frequent and probable hand postures to advance the state of art of hand sensing systems and robotics and, at the same time, to provide technical and theoretical tools to improve neuroscientific knowledge on human hand, in a mutual inspiration between biology and artificial sciences.

15.2 Biology and Artificial Systems: A Mutual Inspiration

As deeply discussed throughout this book, in recent years numerous studies have inquired in how the brain can organize the huge sensory—motor complexity of the human hand, with particular reference to grasping. One of the main findings is that there is a reduced number of coordination patterns, or *synergies*, related to both biomechanical [12] and neural factors [19], which correlate both joint motions and force exertions of multiple fingers [28] (see Chaps. 2–6, 8–13). Multivariate statistical methods over a grasping data set also revealed that a limited amount of so-called *principal components* (or *eigenpostures* [22][1]) can explain a great part of hand pose kinematic variability [26]. All these results suggest that it is possible to reduce the number of DoFs to be used according to a desired level of approximation.

Such an idea has been extensively used in robotics from a *controllability* point of view to define simplified strategies for the design and control of artificial hands [6, 7, 14] as it is discussed in Chap. 8. However, synergy concept can be also profitably exploited from the *observability* point of view, i.e. how to reduce the number of independent DoFs to be measured in order to obtain reliable hand pose estimations (cf. [23] for an application in hand avatar animation). Indeed, if the human hand moves according to patterns of most frequent use, it could be possible to exploit this information to improve hand pose reconstruction despite measurements, which are in general noisy and reduced in number. This observation suggests a strong relationship between sensory and motor side, which lays the foundations of the concept we defined as *sensory-motor synergies*, as discussed in Chap. 5 and in particular in Chap. 7.

In this chapter we will deal with such an observability problem. More specifically, in the first part we will provide Minimum Variance Estimation techniques to fuse synergistic kinematic information with partial and noisy glove measurements. In the second part, we will push forward such an analysis, wondering: “and if I were the designer, how could I choose and place the sensors on a glove to maximize hand postural information?”.

This last question is extremely important since it further reveals deep relationships between the artificial and natural side. Indeed, that the optimal distribution of sensitivity for HPR is not trivial is strongly suggested by the observation of the human example. Let us consider the role of cutaneous information and its relationship with proprioception and kinaesthesia of human hands and fingers, as it was investigated in [11] where the response to finger movements of cutaneous mechanoreceptors in the dorsal skin of human hand was studied. Two main classes of mechanoreceptors

[1]Here in after, in this chapter the terms *synergies* and *principal components* will be used as synonyms.

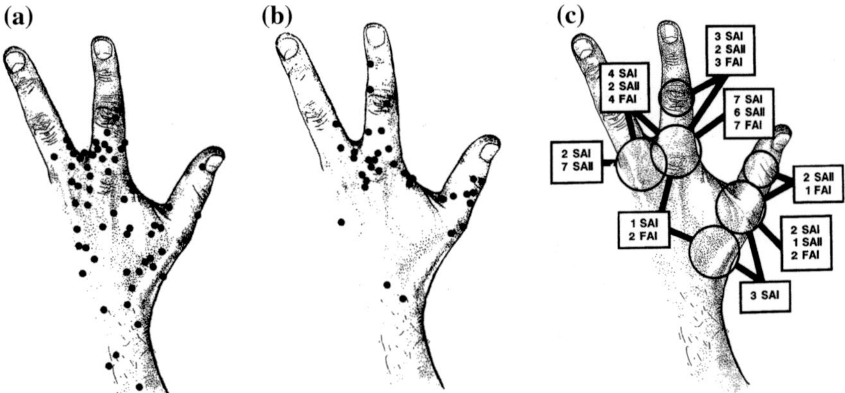

Fig. 15.1 Location of cutaneous mechanoreceptive units in the dorsal skin of the human hand. Adapted from [11], courtesy of the authors. **a** Slowly adapting (SA) units location. **b** Fast adapting (FA) units location. **c** Mechanoreceptor afferent units responding to $\geq$1 joint

involved in this response were roughly identified: Fast Adapting afferents of the first type (FAI), and Slow Adapting afferents, of both the first and second type (SAI and SAII, respectively). These two classes have non-uniform distributions as it is shown in Fig. 15.1a, b. Indeed, FA units, which have a more localized response to movements about one or, at most, two nearby joints, are primarily close to joints, while SA units, which respond to several joints at the same time, can be found more uniformly distributed (see Fig. 15.1c).

Conclusions that can be drawn suggest that in the human hand sensory system there are different typologies of proprioceptive sensors on the skin with different distributions and densities, thus producing a non-uniform map of sensitivities to joint angles. Nonetheless the functional motivations of these data is still unclear, a fascinating interpretation might be the different importance of different elementary percepts in building an overall representation of the hand pose. These biological results motivate our approach to deal with the problem of searching for a preferential distribution and density of different typologies of sensors, which optimize the accuracy of glove-based HPR systems, especially when restrictions on the production costs limit both the number and the quality of sensors. As kinematic synergies is the leading idea for our optimal estimation approach and, together with the observations on the biology of human mechanoreceptors, the motivation for the optimal design of hand pose sensing systems, results we have achieved on the artificial side might further inspire biological investigations, providing theoretical and technical tools to advance the study of human hand sensory—motor apparatus.

15.3 Performance Enhancement

The approach we propose to improve the reconstruction accuracy of existing sensing gloves can deal with noisy measured data and relies on classic *Minimum Variance Estimation* (MVE). To validate this technique, we used a set of grasp postures acquired with a low cost sensing glove, which provides few noisy measurements, and an optical tracking system, which represents the accurate ground—truth for pose reconstruction.

15.3.1 The Hand Posture Estimation Algorithm

Let us consider a set of measures $y \in \mathbb{R}^m$ given by a sensing glove. By using a n degree of freedom kinematic hand model, let us assume a linear relationship between joint variables $x \in \mathbb{R}^n$ and measurements y given by

$$y = Hx + \nu, \qquad (15.1)$$

where $H \in \mathbb{R}^{m\times n}$ $(m < n)$ is a full rank matrix, which represents the relation between measures and joint angles, and $\nu \in \mathbb{R}^m$ is a vector of measurement noise. The goal is to determine the hand posture, i.e. the joint angles x, by using a set of measures y whose number is lower than the number of DoFs describing the kinematic hand model in use. As a consequence, (15.1) represents a system where there are fewer equations than unknowns and hence is compatible with an infinite number of solutions, described e.g. as

$$x = H^{\dagger}y + N_h\xi, \qquad (15.2)$$

where $H^{\dagger}$ is the pseudo-inverse of matrix H, N_h is the null space basis of matrix H and $\xi \in \mathbb{R}^{(n-m)}$ is a free vector of parameters. Among these possible solutions, the least-squared solution resulting from the pseudo-inverse of matrix H for system (15.1) (hereianfter referred to as Pinv) is a vector of minimum Euclidean norm given by

$$\hat{x} = H^{\dagger}y. \qquad (15.3)$$

However, the hand pose reconstruction resulting from (15.3) can be very far from the real one. The goal is to improve the accuracy of the pose reconstruction, choosing, among the possible solutions to (15.2), the most likely hand pose, taking into account the fact that finger motions in grasping tasks are strongly correlated according to some coordination patterns, or synergies [26] (cf. Sect. 15.2 and Chaps. 2, 3 and 8).

To achieve this goal, we use as a priori information the synergistic information obtained by collecting a large number N of grasp postures x_i with n DoFs into a matrix $X \in \mathbb{R}^{n\times N}$. This information can be summarized by means of a covariance matrix $P_o \in \mathbb{R}^{n\times n}$, which is a symmetric matrix computed as $P_o = \frac{(X-\bar{x})(X-\bar{x})^T}{N-1}$, where

$\bar{x}$ is a matrix $n \times N$ whose columns contain the mean values for each joint angle arranged in vector $\mu_o \in \mathbb{R}^n$. We assume that the above described a priori information is multivariate normal distributed, and hence can be described by the covariance matrix P_o.

15.3.1.1 Minimum Variance Estimation

Minimum Variance Estimation (MVE) technique minimizes a cost functional that expresses the weighted Euclidean norm of deviations, i.e. cost functional $J = \int_X (\hat{x} - x)^T S(\hat{x} - x)dx$, where S is an arbitrary, semidefinite positive matrix. Under the hypothesis that ν has zero mean and Gaussian distribution with covariance matrix R, the solution for the minimization of J is achieved as $\hat{x} = E[x|y]$, where $E[x|y]$ represents the a posteriori probability density function (pdf) expectation value of the multivariate normal distribution. This function is expressed by [32] as

$$f(x) = \frac{1}{\sqrt{2\pi \|P_o\|}} \exp\left\{-\frac{1}{2}(x - \mu_o)^T P_o^{-1}(x - \mu_o)\right\}. \tag{15.4}$$

The estimation $\hat{x}$ can be obtained as in [16] by

$$\hat{x} = (P_o^{-1} + H^T R^{-1} H)^{-1}(H^T R^{-1} y + P_o^{-1}\mu_o), \tag{15.5}$$

where matrix $P_p = (P_o^{-1} + H^T R^{-1} H)^{-1}$ is the a posteriori covariance matrix, which has to be minimized to increase information about the system. This result represents a very common procedure in applied optimal estimation when there is redundant sensor information. In under-determined problems, it is only thanks to the a priori information, represented by P_o and μ_o, that Eq. (15.5) can be applied (indeed, $H^T R^{-1} H$ is not invertible).

When R tends to assume very small values, the solution described in Eq. (15.5) might encounter numerical problems. However, by using the Sherman-Morrison-Woodbury formulae,

$$(P_o^{-1} + H^T R^{-1} H)^{-1} = P_o - P_o H^T (H P_o H^T + R)^{-1} H P_o \tag{15.6}$$

$$(P_o^{-1} + H^T R^{-1} H)^{-1} H^T R^{-1} = P_o H^T (H P_o H^T + R)^{-1}, \tag{15.7}$$

Equation (15.5) can be rewritten as

$$\hat{x} = \mu_o - P_o H^T (H P_o H^T + R)^{-1}(H\mu_o - y), \tag{15.8}$$

and the a posteriori covariance matrix becomes $P_p = P_o - P_o H^T (H P_o H^T + R)^{-1} H P_o$. By placing $R = 0$ in (15.8), it is possible to obtain equation (15.7) and the a posteriori covariance matrix becomes

$$P_p = P_o - P_o H^T (H P_o H^T)^{-1} H P_o \tag{15.9}$$

Notice that, (15.8) with $R = 0$ can also be obtained by maximizing the pdf (15.4), that is equivalent to solving the following optimal control problem (see [2] for details):

$$\begin{cases} \hat{x} = \arg\min_{\hat{x}} \frac{1}{2}(x - \mu_o)^T P_o^{-1}(x - \mu_o) \\ \text{Subject to} \quad y = Hx. \end{cases} \tag{15.10}$$

It is interesting to give a geometrical interpretation of the cost function in (15.10), which expresses the square of the Mahalanobis distance [21]. The concept of Mahalanobis distance, which takes into account data covariance structure, is widely exploited in statistics, e.g. in Principal Components Analysis, mainly for outlier detection [18]. Accordingly, to assess if a test point belongs to a known data set, whose distribution defines an hyper-ellipsoid, its closeness to the centroid of data set is taken into account as well as the direction of the test point w.r.t. the centroid itself. In other words, the more samples are distributed along a given direction, the higher is the probability that the test point belongs to the data set even if it is further from the center.

15.3.2 Data Acquisition

To assess hand pose reconstruction effectiveness, without loss of generality, we used a 15 DoF model for the hand,[2] which was also considered in [14, 26] and reported in Fig. 15.2. We collected a large number of static grasp positions using 19 active markers and an optical motion capture system (Phase Space, San Leandro, CA, USA). More specifically, all the grasps of the 57 imagined objects described in [26] were performed twice by subject AT (M,26), in order to define a set of 114 a priori data. We characterized such an a priori information in terms of P_o and μ_o.

Moreover, 54 grasp poses of a wide range of different imagined objects were executed by subject LC (M,26).[3] The set of the latter poses will be referred hereinafter as *validation set*, since these poses can be assumed to represent accurate reference angular values for successive comparisons with the obtained hand pose reconstructions. For this reason, these data were recorded in parallel with the sensing glove, whose performance we wanted to optimize, as it will be described later in this Section, and the Phase Space system, in order to achieve also glove calibration. The processed hand poses acquired with Phase Space can be considered as reliable approximations of real hand positions, given the high accuracy provided by this optical system to detect markers (the amount of static marker jitter is inferior than 0.5 mm, usually

[2]The human hand, considering only fingers and metacarpal joints, has 23 DoFs [10]. Various models have been proposed in literature, which try to reproduce hand and wrist kinematics at different levels of approximation, e.g. [13, 15, 30].

[3]All these data and more information about hand pose acquisitions are available at http://handcorpus.org/.

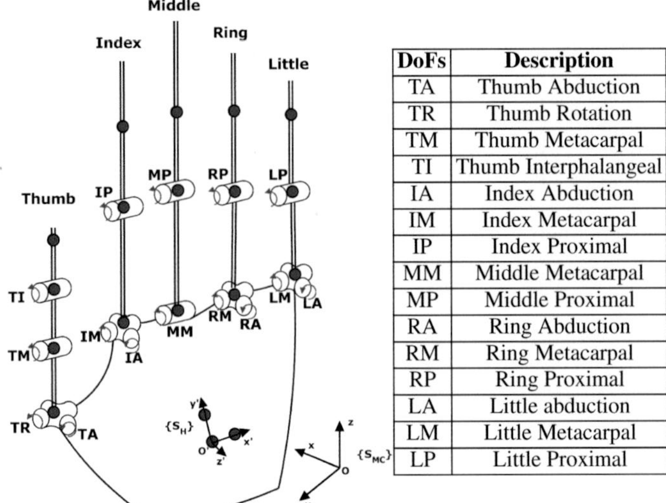

DoFs	Description
TA	Thumb Abduction
TR	Thumb Rotation
TM	Thumb Metacarpal
TI	Thumb Interphalangeal
IA	Index Abduction
IM	Index Metacarpal
IP	Index Proximal
MM	Middle Metacarpal
MP	Middle Proximal
RA	Ring Abduction
RM	Ring Metacarpal
RP	Ring Proximal
LA	Little abduction
LM	Little Metacarpal
LP	Little Proximal

Fig. 15.2 Kinematic model of the hand with 15 DoFs. Markers are reported as *red spheres*

0.1 mm) and assuming a linear correlation (due to skin stretch) between marker motion around the axes of rotation of the joint and the movement of the joint itself [37]. Since the sensing glove perfectly adapts to subject hand shape when it is worn, the latter assumption is still reasonable also in this case, even if departures from real reference configurations can happen. None of the subjects had physical limitations that would affect the experimental outcomes. Data collection from subjects in this study was approved by the University of Pisa Institutional Review Board. For the markerization protocol and additional details on the acquisition, the reader is invited to refer to [1].

15.3.3 Experimental Results

The reconstruction procedure was tested using a sensorized glove based on Conductive Elastomer (CEs) [33]. CE strips are printed on a Lycra®/cotton fabric in order to follow the contours of the hand, see Fig. 15.3.

Since CE materials present piezo-resistive characteristics, sensor elements corresponding to different segments of the contour of the hand length change as the hand moves. These movements cause variations in the electrical properties of the material, which can be revealed by reading the voltage drop across such segments.The sensors are connected in series thus forming a single sensor line while the connections intersect the sensor line in the appropriate points.

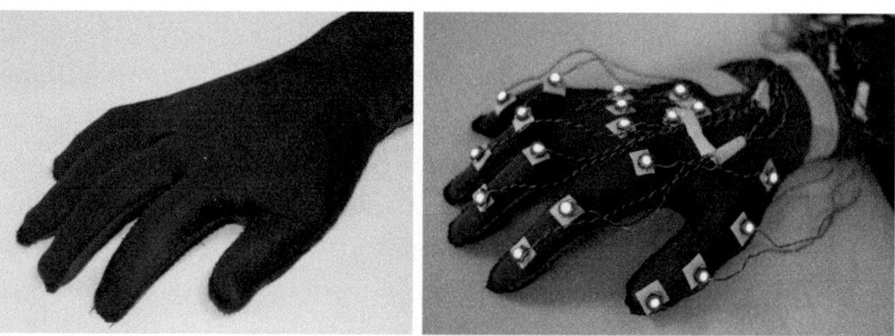

Fig. 15.3 The sensing glove used in our study (*on the left*) and the sensing glove with added markers (*on the right*)

In the present study, long finger flexion-extension recognition was obtained by means of an updated multi-regressive model having the metacarpophalangeal (MCP) flexion-extension angles of the five long fingers as dependent variables and the outputs of CE sensors covering MCP joints as independent ones. According to the hand kinematic model adopted in this work they are referred to as TM, IM, MM, RM, LM. The model parameters were identified by measuring the sensor status in two different position: (1) hand totally closed (90°), (2) hand totally opened (0°). For more information about the design and structure of the here described sensing glove and the signal processing system employed, the reader is invited to refer to [20, 33, 34].

Although this sensorized glove can be regarded as one of the most recent and inexpensive envisions in glove device literature, it is limited under several aspects that can reduce its performance, e.g. cloth support that affects measurement repeatability as well as hysteresis and non linearities due to piezo-resistive material properties. Indeed, although this kind of glove is suitable for general opening/closening hand movement measurement, it is not the best choice for sensing fine hand adjustments. Moreover, the assumptions done for data processing (the relationship between joint angles and sensors as well as the linearity between hand aperture and electrical property changes) and the calibration phase based only on two-points fitting can act like aditional potential sources of errors. To overcome this last point we performed a new calibration to estimate the measurement matrix.

15.3.3.1 Results and Discussions

First, we obtained an estimation of the glove measurement matrix H_g, i.e. $\hat{H}_g$. For this purpose, a calibration phase was performed by collecting a number of poses N in parallel with the glove and the optical tracking system. This number has to be larger or equal than the dimension of the state to estimate, i.e. $N \geq 15$. $X_c \in \mathbb{R}^{15\times 15}$ collects the reference poses, while matrix $Z_c \in \mathbb{R}^{5\times 15}$ organizes the measures from the glove. These measures represent the values of the signals referred to measured joints averaged over the last 50 acquired samples (@250 kS/s). Matrix $\hat{H}_g$ can be obtained by exploiting the relation $Z_c = \hat{H}_g X_c$ as $\hat{H}_g = Z_c((X_c^T)^\dagger)^T$. Measurement noise was characterized in terms of fluctuations w.r.t. the aforementioned average values of the measures, thus obtaining noise covariance matrix R. Noise level is less than 10 % measurement amplitude. However consistent errors in the measurement matrix estimation might be obtained due to intrinsic non-linearities and hysteresis of glove sensing elements. Once the measurement matrix $\hat{H}_g$ was obtained, we applied MVE estimation techniques to the measurements provided by the glove and compared the results with those achieved using simple pseudo-inversion (Pinv) (15.2).

Pose estimation errors (i.e. the mean of DoF absolute estimation errors computed for each pose $e_i = \frac{1}{n}\sum_{i=1}^{n} |x_i - \hat{x}_i|$), and DoF absolute estimation errors are considered and averaged over all the number of reconstructed poses.

Results clearly show that MVE outperforms Pinv in terms of estimation outcomes. Indeed, the average absolute pose estimation error with MVE is $10.94 \pm 4.24°$, while it is equal to $19.00 \pm 3.66°$ by using Pinv. Statistical difference was observed between the two techniques (p-value less than 10^{-4}). Notice that MVE exhibits best pose reconstruction performances also in terms of maximum errors (25.18° for MVE vs. 30.30° for Pinv). Absolute average reconstruction errors for each DoF are reported in Table 15.1. MVE produces the best results which are statistically different w.r.t. Pinv algorithm, see Table 15.1, except, respectively, for those DoFs which are directly measured (i.e. IM, RM and LM), for RA DoF, which exhibits a limited average estimation error ($\approx$6°), and finally TA. For TI the smallest average estimation is observed with Pinv; a possible explanation for this might be still related to the difficulties in kinematic modeling thumb phalanx kinematics. IA DoF presents the smallest absolute average estimation error with Pinv, although p-values from the comparisons between the two techniques for the estimation of this DoF are close to the significance threshold. Maximum DoF reconstruction errors for MVE are observed especially for those measured DoFs with potentially maximum variations in grasping tasks; this fact may be probably due to the non linearities in sensing glove elements leading to inaccurate estimation of H_g, and hence to inaccurate measures. Furthermore, MVE aims at minimizing the error statistics and guarantees that the mean squared norm of the joint error vector (i.e. the Mean Squared Error, $MSE = \frac{1}{N}\sum_{i=1}^{N} \|\hat{x} - x\|^2$, where N represents the number of predictions) is minimized, but not necessarily the value of each single component. For this reason, some worst-case sensing results can be found.

To conclude, except for some singular poses, the best estimation performance is provided by MVE for which a good robustness to errors in measurement process modeling is also observed. However, the latter errors are not taken numerically into account in these analyses. Moreover, as it can been seen in Fig. 15.4, reconstructed hand configurations obtained by MVE preserve likelihood with real poses, as opposed to pseudo-inverse based algorithm.

Table 15.1 Average estimation errors and standard deviations for each DoF [°], for the sensing glove acquisitions

DoF	Mean±Std		Max Error		p-values
	MVE	**Pinv**	**MVE**	**Pinv**	
TA	12.12±9.98	14.37±10.78	36.63	34.28	**0.28**
TR	9.20±7.13	26.46±10.49	26.34	46.43	0
TM∗	4.36±3.73	6.43±4.44	13.25	18.50	0.0093
TI	14.56±9.96	7.84±5.47	33.25	22.38	0.0008
IA	9.82±6.89	7.10±5.08	29.60	21.18	0.0381
IM∗	15.27±11.86	16.48±12.62	46.76	43.58	**0.58**
IP	9.60±7.65	31.47±14.70	27.40	61.11	0
MM∗	14.40±12.84	19.88±14.58	53.03	51.47	0.0232
MP	6.80±6.49	24.36±9.85	24.74	43.72	0
RA	6.20±4.31	5.69±4.72	15.72	20.90	**0.51**
RM∗	19.00±13.44	19.22±11.81	61.98	46.32	**0.67**
RP	8.98±8.91	31.51±13.98	32.24	60.62	0
LA	11.42±8.50	32.24±6.98	29.59	48.11	0
LM∗	17.37±12.51	17.98±11.81	58.40	45.05	**0.26**
LP	8.43±6.36	23.90±12.53	26.07	56.21	0

1 ← **p-values** → 0

∗ indicates a measured DoF.

MVE and Pinv methods are considered. Maximum errors are also reported as well as p-values from the evaluation of DoF estimation errors between MVE and Pinv. A color map describing p-values is also added to simplify result visualization. ◇ indicates that standard two-tailed t-test (T_{eq}) is exploited for the comparison. ‡ indicates a modified two-tailed T-test (Behrens-Fisher problem), T_{neq} test. When no symbol appears near the tabulated values, it means that Mann-Whitney U-test (U-test) is used. **Bold** value indicates no statistical difference between the two methods under analysis at 5 % significance level. When the difference is significative, values are reported with a 10^{-4} precision. p-values less than 10^{-4} are considered equal to zero

15.4 Optimal Design

In this part of the chapter, we extend the analysis to the optimal design of sensing gloves. The objective is to choose the optimal sensor distribution that maximizes the information on the actual posture. This information, used with the estimation method previously discussed, will lead to the minimization of the reconstruction error statistics.

As explained in Sect. 15.2, there is a strong biological evidence that the optimal distribution of sensitivity for a sensing glove should not be trivial. However, while most results in optimal experimental design [4, 8, 17] refer to the case where the number of measurements is redundant or at least equal to the number of variables

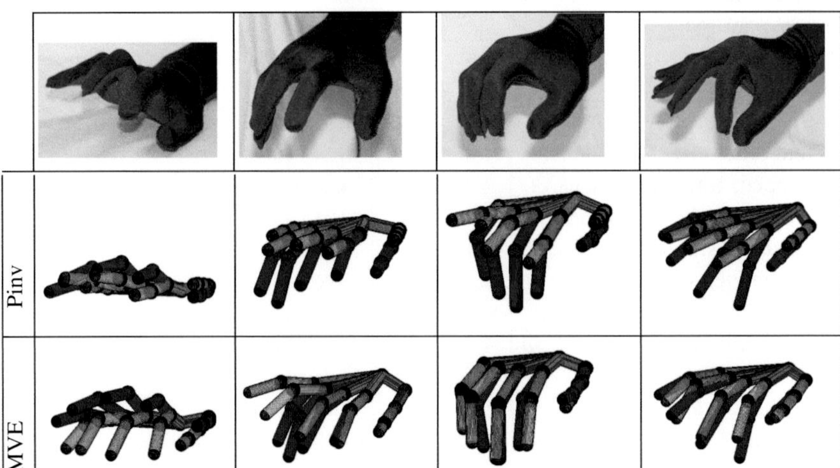

Fig. 15.4 Hand pose reconstructions with Pinv and MVE algorithms, with measures given by the sensing glove. In *blue* the "real" hand posture whereas in white the estimated one

to be estimated, the opposite case that fewer sensors are available than the hand variables is of main concern in our problem. To circumvent this limit, it is natural to think of exploiting synergistic a priori knowledge to disambiguate poses from scarce data.

In previous sections, synergistic prior knowledge on how humans most frequently use their hands is fused with partial and noisy data provided by any given glove device, to maximize reconstruction accuracy. Here, the goal is to characterize a design that enables for optimally exploiting—in a Bayesian sense—such an a priori information.

The optimization goals become particularly relevant when restrictions on the production costs limit both the number and the quality of sensors. In these cases, a careful design is instrumental to obtain good performance. Furthermore, different technologies and sensor distributions can be considered to realize the devices. At the physical level, sensors for gloves can be classified as either *lumped* (as e.g. a mechanical angular encoder about a joint or Hall-effect sensors, as in the Humanglove by Humanware s.r.l. (Pisa, Italy)) or *distributed* (e.g. a flexible optic fiber running along a finger from base to tip or conductive elastomeric strips as in the glove used in our experiments [33]). At the signal level, glove sensors can be *coupled* (if more than one hand joint angle influences the reading) or *uncoupled*. Of course, all distributed sensors are coupled, but also lumped sensors can exhibit cross-coupling.

Different sensor arrangements generate different measurement matrices H: the row corresponding to a lumped, uncoupled sensor has a non-zero element only in correspondence to the measured joint, hence (up to rescaling) it is a binary "selection" matrix. We will call such a matrix *discrete*, i.e. $H_{ij} \in \{0, 1\}$—a *discrete* set of values. Conversely, a coupled sensor with general weights, i.e. a distributed sensor or a lumped sensor with not negligible cross-coupling, produces a matrix whose row elements are real numbers, i.e., up to rescaling, $H_{ij} \in [-1,\ 1] \subset \mathbb{R}$—a *continuous* set of values. In the following, we will call such a matrix *continuous*. Finally, a glove employing both lumped (uncoupled and coupled) and distributed sensors will generate a *hybrid* measurement matrix, which consists of a continuous part and a discrete one.

Lumped, uncoupled sensing devices, which generate a discrete measurement matrix, are probably the easiest to be implemented, as they require to individually measure single joints according to the optimal measurement matrix. Common sensing strategies include Hall-effect or piezoresistive sensors (e.g. CyberGlove, by CyberGlove System LLC), directly placed on the joints to be measured, hence obtaining a lumped device. On the other hand, distributed sensors, which generate an optimal continuous matrix like the sensing glove used in the experiments described in Sect. 15.3.3, should provide measurements in terms of (optimally) weighted linear combinations of the contributions of different DoFs, e.g. using, among the different techniques, resistive ink printed on flexible plastic bends that follow the movement of hand joints (e.g. PowerGlove by Mattel Inc., El Segundo, CA–USA), or capacitive sensors (as e.g. in the Didjiglove by Dijiglove Pty. Ltd., Melbourne, AUS) [10]. Finally, the above discussed technologies (lumped, uncoupled and distributed) can be adopted and combined in an efficient manner to optimally realize devices that can be modeled by a hybrid measurement matrix. Notice that the human hand sensing distribution can be considered to belong to the latter glove class, as it will be discussed later in this chapter.

15.4.1 Problem Definition

In the ideal case of noiseless measures ($R = 0$), P_p becomes zero when H is a full rank n matrix, meaning that the available measures contain a complete information about the hand posture. In the real case of noisy measures and/or when the number of measurements m is less than the number of DoFs n, P_p cannot be zero. In these cases, the following problem becomes very interesting: find the optimal matrix H^* such that the hand posture information contained in a reduced number of measurements is maximized. Without loss of generality, let assume H to be full row rank and consider the following problem.

Problem 1 Let H be an $m \times n$ full row rank matrix with $m < n$ and $V_1(P_o, H, R)$: $\mathbb{R}^{m\times n} \to \mathbb{R}$ be defined as $V_1(P_o, H, R) = \|P_o - P_oH^T(HP_oH^T + R)^{-1}HP_o\|_F^2$, find

$$H^* = \arg\min_H \ V_1(P_o, H, R)$$

where $\|\cdot\|_F$ denotes the Frobenius norm defined as $\|A\|_F = \sqrt{\mathrm{tr}(AA^T)}$, for $A \in \mathbb{R}^{n\times n}$.

Frobenius norm has been already used in literature for optimization in measurement problem, e.g. [24]. Here the squared Frobenius norm is adopted to exploit its useful relation with matrix trace operator in order to simplify the derivation of the matrix gradient flow later defined. To solve Problem 1 means to minimize the entries of the a posteriori covariance matrix: the smaller the values of the elements in P_p, the greater is the predictive efficiency. Next sections will be dedicated to describe solutions to Problem 1 for different sensor distributions and hence measurement matrices, i.e. continuous, discrete and hybrid, the latter containing both lumped and distributed sensors.

Let us introduce some useful notations. If M is a symmetric matrix with dimension n, let its Singular Value Decomposition (SVD) be $M = U_M \Sigma_M U_M^T$, where Σ_M is the diagonal matrix containing the singular values $\sigma_1(M) \geq \sigma_2(M) \geq \cdots \geq \sigma_n(M)$ of M and U_M is an orthogonal matrix whose columns $u_i(M)$ are the eigenvectors of M, known as Principal Components (PCs) of M, associated with $\sigma_i(M)$. For example, the SVD of the a priori covariance matrix is $P_o = U_{P_o} \Sigma_{P_o} U_{P_o}^T$, with $\sigma_i(P_o)$ and $u_i(P_o)$, $i = 1, 2, \ldots, n$, the singular values and the principal components of matrix P_o, respectively.

15.4.2 Continuous Sensing Design

In this case, each row of matrix H is a vector in $\mathbb{R}^n$ and hence can be given as a linear combination of a $\mathbb{R}^n$ basis. Without loss of generality, we can use the principal components of matrix P_o, i.e. columns of previously defined matrix U_{P_o}, as a basis of $\mathbb{R}^n$. Consequently, naming H_c such a type of matrix related to a continuous sensing device, the measurement matrix can be written as $H_c = H_e U_{P_o}^T$, where $H_e \in \mathbb{R}^{m \times n}$ contains the coefficients of the linear combinations. Given that $P_o = U_{P_o} \Sigma_{P_o} U_{P_o}^T$, the a posteriori covariance matrix becomes

$$P_p = U \left[\Sigma_o - \Sigma_o H_e^T (H_e \Sigma_o H_e^T + R)^{-1} H_e \Sigma_o \right] U^T, \tag{15.11}$$

where, for simplicity of notation $\Sigma_o \equiv \Sigma_{P_o}$.

We will analyze the optimal continuous sensing design both under a numerical and analytical point of view. For this purpose, let us introduce the set of $m \times n$ (with $m < n$) matrices with orthogonal rows, i.e. satisfying the condition $HH^T = I_{m \times m}$, and let denote it as $\mathscr{O}_{m \times n}$.

15.4.2.1 Numerical Solution: Gradient Flows on $\mathscr{O}_{m \times n}$

A differential equation that solve Problem 1 is proposed. The following proposition describes an algorithm that minimizes the cost function $V_1(P_o, H, R)$, providing the gradient flow which can be used to improve the method of steepest descent.

Proposition 1 *The gradient flow for the function* $V_1(P_o, H, R)\colon \mathbb{R}^{m\times n} \to \mathbb{R}$ *is given by,*

$$\dot{H} = -\nabla \|P_p\|_F^2 = 4\left[P_p^2 P_o H^T \Sigma(H)\right]^T, \tag{15.12}$$

where $\Sigma(H) = (HP_oH^T + R)^{-1}$.

All the calculation to obtain the gradient can be found in the Appendix of [1].

Let us observe that rows of matrix H can be chosen, without loss of generality, such that $H_iP_oH_j^T = 0,\ i \neq j$ that implies that measures are uncorrelated, i.e. satisfying the condition $HH^T = I_m$. Of course, in case of noise-free sensors, this constraint is not strictly necessary. On the other hand, in case of noisy sensors, the minimum of $V_1(P_o, H, R)$ cannot be obtained since it represents a limit case that can be achieved when H becomes very large (i.e. an infimum) and hence increasing the signal-to-noise ratio in an artificial manner. Therefore, it is possible to use the constraint $HH^T = I_m$ to reduce the search space in order to find solutions.

To solve this constrained problem it is possible to use the Rosen's gradient projection method for linear constraints [25], which is based on projecting the search direction into the subspace tangent to the constraint itself.

Having the search direction for the constrained problem, the gradient flow is given by

$$\dot{H} = -4W\ \left[P_p^2 P_o H^T \Sigma(H)\right]^T \tag{15.13}$$

where $\Sigma(H) = (HP_oH^T + R)^{-1}$. The gradient flow (15.12) guarantees that the optimal solution H^* will satisfy $H^*(H^*)^T = I_m$, if $H(0)$ satisfies $H(0)H(0)^T = I_m$, i.e. $H \in \mathcal{O}_{m\times n}$.[4]

Notice that both $\mathcal{O}_{m\times n}$ and $V_1(P_o, H, R)$ are not convex, hence the problem could not have a unique minimum. To overcome this common problem in gradient methods, a multi-start search represents a classic procedure. The here described gradient-based technique can be useful to characterize optimal solutions also for discrete sensing design, in case of large dimension problem. Moreover, they can furnish interesting suggestions about a possible hybrid approach later discussed.

15.4.2.2 Analytical Solutions

Let us first consider the case of noiseless measures, i.e. $R = 0$. Let A be a non-negative matrix of order n. It is well known (see [24]) that, for any given matrix B of rank m with $m \leq n$,

$$\min_B \|A - B\|_F^2 = \alpha_{m+1}^2 + \cdots + \alpha_n^2, \tag{15.14}$$

[4] $H(0)$ indicates the starting point at t = 0 for the gradient flow.

where α_i are the eigenvalues of A, and the minimum is attained when

$$B = \alpha_1 w_1 w_1{}^T + \cdots + \alpha_m w_m w_m{}^T, \tag{15.15}$$

where w_i are the eigenvectors of A associated with α_i. In other words, the choice of B as in (15.15) is the best fitting matrix of given rank m for A. By using this result we can determine when the minimum of (15.11), and hence of

$$\| \Sigma_o - \Sigma_o H_e^T (H_e \Sigma_o H_e^T)^{-1} H_e \Sigma_o \|_F^2, \tag{15.16}$$

can be reached. Let us observe that the row vectors $(h_i)_e$ of H_e can be chosen, without loss of generality, to satisfy the condition $(h_i)_e \; \Sigma_o \; (h_j)_e = 0, \;\; i \neq j$, which implies that the measures are uncorrelated. As previously said, $\mathscr{O}_{m\times n}$ denotes the set of $m \times n$ matrices, with $m < n$, whose rows satisfy the aforementioned condition, i.e. the set of matrices with orthonormal rows ($H_e H_e^T = I$). By using (15.14), the minimum of (15.16) is obtained when (see [24])

$$\Sigma_o H_e^T (H_e \Sigma_o H_e^T)^{-1} H_e \Sigma_o = \sigma_1(\Sigma_o) u_1(\Sigma_o) u_1^T(\Sigma_o) + \cdots + \sigma_m(\Sigma_o) u_m(\Sigma_o) u_m^T(\Sigma_o). \tag{15.17}$$

Since Σ_o is a diagonal matrix, $u_i(\Sigma_o) \equiv e_i$, where e_i is the ith element of the canonical basis. Hence, it is easy to verify that (15.17) holds for $H_e = [I_m \,|\, 0_{m\times(n-m)}]$. As a consequence, row vectors $(h_i)_c$ of H_c are the first m principal components of P_o, i.e. $(h_i)_c = u_i(P_o)^T$, for $i = 1, \ldots, m$.

From these results, a principal component can be defined as a linear combination of optimally-weighted observed variables meaning that the corresponding measures can account for the maximal amount of variance in the data set. As reported in [24], every set of m optimal measures can be considered as a representation of points in the best fitting lower dimensional subspace. Thus the first measure gives the best one dimensional representation of data set, the first two measures give the best two dimensional representation, and so on.

In case of noisy measures, (15.15) cannot be verified since it represents a limit case that can be achieved when H becomes very large and hence increasing the signal-to-noise ratio. We hence describe an optimal solution for problem 1 in the set $\mathscr{A} = \{H \,:\, HH^T = I_m\}$. This problem was discussed and solved in [9], providing that, for arbitrary noise covariance matrix R,

$$\min_{H\in\mathscr{A}} V_1(H) = \sum_{i=1}^{m} \frac{\sigma_i(P_o)}{1 + \sigma_i(P_o)/\sigma_{m-i+1}(R)} + \sum_{i=m+1}^{n} \sigma_i(P_o), \tag{15.18}$$

and it is attained for $H = \sum_{i=1}^{m} u_{m-i+1}(R) u_i(P_o)$.

Hence, if $\mathscr{A}$ consists of all matrices with mutually perpendicular, unit length rows, the first m principal components of P_o are always the optimal choice for H

rows. As shown in [9] this situation changes under the Frobenius norm constraint, i.e. $\mathscr{A} = \{H : \|H\|_F \leq 1\}$ (see [9] for details).

Conclusions that can be drawn from this part is that in case of noise-free measures, the invariance of the cost function w.r.t. changes of basis, i.e. $V_1(P_o, H, 0) = V_1(P_o, MH, 0)$ with $M \in \mathbb{R}^m$ an invertible full rank matrix, suggests that there might exist a subspace in $\mathbb{R}^n$ where the optimum is achieved. Indeed, gradients become zero when rows of matrix H are any linear combination of a subset of m principal components of the a priori covariance matrix, or synergies [26]. Unfortunately, this does not happen in case of noisy measures and gradients become zero only for a particular matrix H which depends also on the principal components of the noise covariance matrix. In other terms, in case of continuous sensing gloves the sensing elements must be placed on the human hand in order to provide measurements that are related to the joints according to the first m PC weights.

15.4.3 Discrete Sensing Design

Let us consider now the case that each measure y_j, $j = 1, \ldots, m$ from the glove corresponds to a single joint angle x_i, $i = 1, \ldots, n$. The problem here is to find the optimal choice of m joints or DoFs to be measured.

Measurement matrix becomes in this case a full row rank matrix where each row is a vector of the canonical basis, i.e. matrices which have exactly one nonzero entry in each row: let H_d be such a type of matrix. The optimal choice H_d^* can be easily computed, by substituting all the possible sub-sets of m vectors of the canonical basis in the cost function $V_1(P_o, H, R)$. However, a more general approach to compute the optimal matrix is provided in order to obtain the solution also when a model with a large number of DoFs is considered, and eventually extended to all human body.

Let $\mathscr{N}_{m\times n}$ denote the set of $m \times n$ element-wise non-negative matrices, then $\mathscr{P}_{m\times n} = \mathscr{O}_{m\times n} \cap \mathscr{N}_{m\times n}$, where $\mathscr{P}_{m\times n}$ is the set of $m \times n$ permutation matrices (see lemma 2.5 in [36]). This result implies that if we restrict H to be orthonormal and element-wise non-negative, we get a permutation matrix. We extend this result in $\mathbb{R}^{m\times n}$, obtaining matrices which have exactly one nonzero entry in each row and the problem to solve becomes:

Problem 2 Let H be a $m \times n$ matrix with $m < n$, and $V_1(P_o, H, R) \colon \mathbb{R}^{m\times n} \to \mathbb{R}$ be defined as $V_1(P_o, H, R) = \|P_o - P_o H^T (H P_o H^T + R)^{-1} H P_o\|_F^2$, find the optimal measurement matrix

$$
\begin{aligned}
H^* &= \arg\min_H \; V_1(P_o, H, R) \\
s.t. \quad & H \in \mathscr{P}_{m\times n}.
\end{aligned}
$$

15.4.3.1 Numerical Solution: Gradient Flows on $\mathscr{P}_{m\times n}$

A solution for this problem can be obtained defining a cost function that penalizes negative entries of H. In [36] authors defined a function $V_2(P)$ with $P \in \mathbb{R}^{n\times n}$ that forces the entries of P to be as "positive" as possible. In this chapter, we extend this function to measurement matrices $H \in \mathbb{R}^{m\times n}$ with $m < n$ and hence, we consider a function $V_2 : \mathscr{O}_{m\times n} \to \mathbb{R}$ as

$$V_2(H) = \frac{2}{3}\mathrm{tr}\left[H^T(H - (H \circ H))\right], \tag{15.19}$$

where $A \circ B$ denotes the *Hadamard* or elementwise product of the matrices $A = (a_{ij})$ and $B = (b_{ij})$, i.e. $A \circ B = (a_{ij}b_{ij})$. The gradient flow of $V_2(H)$ is given by [36]

$$\dot{H} = -H\left[(H \circ H)^T H - H^T(H \circ H)\right], \tag{15.20}$$

which minimizes $V_2(H)$ converging to a permutation matrix if $H(0) \in \mathscr{O}_{m\times n}$.

Up to this point, we have introduced two gradient flows given by (15.13) and (15.20), both on the space of orthogonal matrices, that respectively minimize their cost function, while the second one also converges to a permutation matrix. By combining these two gradient flows a solution for Problem 2 can be achieved. Of course, we can combine the gradient flows in two different ways: by adding them in a convex combination or firstly ignoring the non-negativity requirement and switching to the permutation gradient flow when the objective function has been sufficiently minimized [36].

Theorem 1 *Let $H \in \mathbb{R}^{m\times n}$ with $m < n$ the measurement process matrix and let us assume that $H(0) \in \mathscr{O}_{m\times n}$. Moreover, we suppose that $H(t)$ satisfies the following matrix differential equation,*

$$\dot{H} = 4\,(1-k)W\,\left[P_p^2 P_o H^T \Sigma(H)\right]^T + k\,H\left[(H \circ H)^T H - H^T(H \circ H)\right], \tag{15.21}$$

where $k \in [0,\ 1]$ is a positive constant and $\Sigma(H) = (HP_oH^T + R)^{-1}$. For sufficiently large k (near one), $\lim_{t\to\infty} H(t) = H_\infty$ exists and approximates a permutation matrix that also minimizes the squared Frobenius norm of the a posteriori covariance matrix, $\|P_p\|_F^2$.

A proof for this theorem can be obtained directly by using results from [36] and further details can be found in [2].

15.4.4 Hybrid Sensing Design

In previous sections, optimal solutions for continuous and discrete sensing cases have been provided. However, in order to take advantage from both of them (the amount of information achievable vs low-cost implementation and feasibility) a hybrid sensing device which combines continuous and discrete sensors might represent a valid improvement, as it can be found also in biology. Indeed, human hand can be regarded—to some extent—as an example of hybrid sensory system. As previously discussed, among the cutaneous mechanoreceptors in the hand dorsal skin that were demonstrated to be involved in the responses to finger movements, and hence that possibly contribute to kinaesthesia, it is possible to find Fast Adapting (FA) type ones, which mainly respond to movements around one or at most two nearby joints and that can be regarded as “discrete” sensors, as well as the discharge rate of Slow Adapting (SA) afferents, which are influenced by several joints and can be regarded as “continuous” type sensors [11].

Up to re-arranging the sensor numbering, we can write a hybrid measurement matrix $H_{c,d} \in \mathbb{R}^{m \times n}$ as

$$H_{c,d} = \left[\frac{H_c}{H_d}\right],$$

where $H_c \in \mathbb{R}^{m_c \times n}$ defines the m_c rows of the continuous part, whereas $H_d \in \mathscr{P}^{m_d \times n}$ describes the m_d single-joint measurements of the discrete part, with $m_c + m_d = m$. Neither the closed-form solution valid for the continuous measurement matrix, nor the exhaustion method used for discrete measurements are applicable in the hybrid case. Therefore, to optimally determine the hybrid measurement matrix, we will recur to gradient-based iterative optimization algorithms.

By combining the continuous and discrete gradient flows, previously defined in (15.12) and (15.20), respectively, and constraining the solution in the sub-set $\mathscr{H}_{c,d} = \{H_{c,d} : H_{c,d} H_{c,d}^T = I_m\}$, we obtain

$$\dot{H}_{c,d} = 4\,(1-k)\,\left[P_p^2 P_o H_{c,d}^T \Sigma(H_{c,d})\right]^T\, W + k\,\bar{H}_d \left[(\bar{H}_d \circ \bar{H}_d)^T \bar{H}_d - \bar{H}_d^T (\bar{H}_d \circ \bar{H}_d)\right], \tag{15.22}$$

where $k \in [0,\ 1]$ is a positive constant, $P_p = P_o - P_o H_{c,d}^T (H_{c,d} P_o H_{c,d}^T + R)^{-1} H_{c,d} P_o$, $W = I_n - H_{c,d}^T (H_{c,d} H_{c,d}^T)^{-1} H_{c,d}$, $\Sigma(H_{c,d}) = (H_{c,d} P_o H_{c,d}^T + R)^{-1}$, and

$$\bar{H}_d = \left[\frac{0_{m_c \times n}}{H_d}\right].$$

Starting from any initial guess matrix $H_{c,d} \in \mathscr{H}_{c,d}$, the gradient flow defined in (15.22) remains in the sub-set $\mathscr{H}_{c,d}$ and, on the basis of Theorem 1, it converges toward a hybrid measurement matrix, (locally) minimizing the squared Frobenius norm of the a posteriori covariance matrix. Multi-start strategies have to be used to circumvent the problem of local minima.

When noise is not negligible, without constraining the solution in $\mathscr{H}_{c,d}$ by W, the gradient search method of (15.22) would tend to produce measurement matrices whose continuous parts, H_c, are very large in norm. This is an obvious consequence of the fact that, for a fixed noise covariance R, larger measurement matrices H would produce an apparently higher signal-to-noise ratio in (15.1).

15.4.5 Continuous and Discrete Sensing Optimal Distribution

Results we have described in the past Sections show that, in case of continuous sensing design, the optimal choice H_c^* of the measurement matrix $H \in \mathbb{R}^{m \times n}$ is given by the first m principal components of the a priori covariance matrix P_o. Figure 15.5 shows the hand sensor distribution for a number $m = 1, 2, 3$ of noise-free measures (for lack of space we have reported only the continuous case).

In case of discrete sensing, H_d^* does not have an incremental behaviour, especially in case of few measures. In other words, the set of DoFs which have to be chosen in case of m measures does not necessarily contain all the set of DoFs chosen for $m - 1$ measures (for further details the reader is invited to refer to [2]).

Figure 15.6 shows the values of the squared norm of the a posteriori covariance matrix for increasing number m of measures. In particular, in Fig. 15.6 values of V_1 for matrices H_c^* and H_d^* are reported, for noise-free measures.

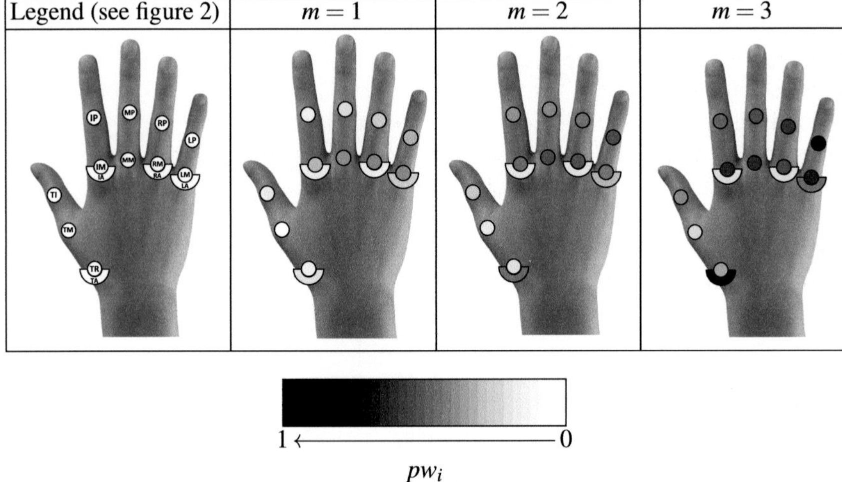

Fig. 15.5 Optimal continuous sensing distribution for $m = 1$, i.e. the first PC of P_o, for $m = 2$, i.e. the first two PCs of P_o and for $m = 3$, i.e. the first three PCs of P_o. The greater is the weight pw_i of the joint angle in the optimal measures, the *darker* is the *color* of that joint. We assume the weight of the ith joint in the optimal measures given as $pw_i = \sum_{k=1}^{m} |h_{k,i}|$, where $h_{k,i}$ is the (k, i)th entry of matrix H, normalized w.r.t. the maximum value of pw_i. For example, for $m = 3$, weight of *LA* joint is 0.53, whereas for *LM* joint is 0.74 and the maximum value is for *TA* joint

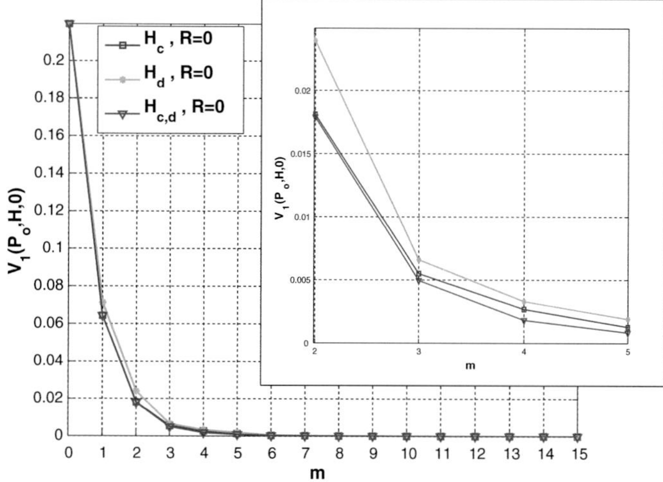

Fig. 15.6 Squared Frobenius norm of the a posteriori covariance matrix with noise-free measures in case of H_c^*, H_d^* and $H_{c,d}^*$ ($m_c = 1$) with an increasing number of noise-free measures. A zoomed detail of the graph is shown for $m = 2, 3, 4, 5$ measures

By analyzing how much V_1 reduces with the number of measurements w.r.t. the value it assumes for zero measures ($P_p \equiv P_0$), two types of observations can be done. First, the observed information quantified through V_1 (squared Frobenius norm of the a posteriori covariance matrix) is greatest for continuous case, while hybrid case provides better performance than the discrete one. Second, for the continuous case with noise-free measures, what is noticeable from the *observability* viewpoint is that a reduced number of measures coinciding with the first three principal components enable for $\simeq$97 % reduction of the squared Frobenius norm of the a posteriori covariance matrix. An analogous result can be found also under the *controllability* point of view. In [26] authors state that three postural synergies are crucial in grasp pre-shaping since they take into account for $\simeq$90 % of pose variability in grasping tasks.

15.4.6 Estimation Results with Optimal Discrete Sensing Devices

In this section, we compare the hand posture reconstruction obtained by H_s (which measured the joints *TM*, *IM*, *MM*, *RM* and *LM*) with the one obtained by using the optimal matrix H_d^* with the same number of measurements in case of noisy measures (*TA*, *MM*, *RP*, *LA* and *LM*), where an additional random noise was artificially added on each measure. A zero-mean, Gaussian noise with standard deviation 0.122 rad (7°)

Real Hand Postures

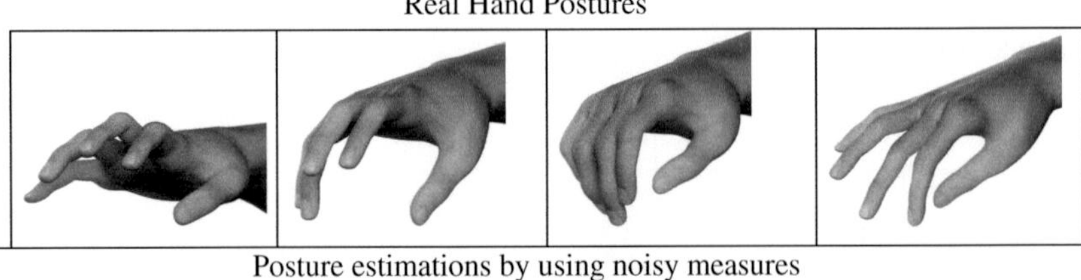

Posture estimations by using noisy measures

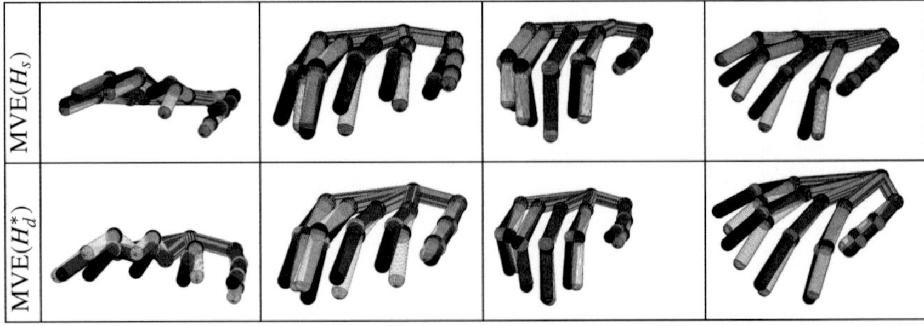

Fig. 15.7 Hand pose reconstructions MVE algorithm by using matrix H_s which allows to measure *TM*, *IM*, *MM*, *RM* and *LM* and matrix H_d^* which allows to measure *TA*, *MM*, *RP*, *LA* and *LM* (see Fig. 15.2). In color the real hand posture whereas in white the estimated one

was chosen based on data about common technologies and tools used to measure hand joint positions [29], thus obtaining a noise covariance matrix $R \approx \text{diag}(0.0149)$.

Measures were provided by grasp data from the *validation set*, where degrees of freedom to be measured were chosen on the basis of optimization procedure outcomes, while the entire pose was recorded to produce an accurate reference posture. In order to compare reconstruction performance achieved with H_s and H_d^* we used as evaluation indices the average pose estimation error and average estimation error for each estimated DoF. Maximum errors are also reported. In Fig. 15.7 a qualitative comparison of four hand pose reconstructions in case of noisy measurements is reported, with optimal and a non-optimal design.

In case of noise, performance in terms of average absolute estimation pose errors ([°]) obtained with H_d^* is better than the one exhibited by H_s (5.96 ± 1.42 vs. 8.18 ± 2.70). Moreover, maximum pose error with H_d^* is the smallest (9.30° vs. 15.35° observed with H_s). Statistical difference between results from H_s and H_d^* is found (p=0.001). In Table 15.2 average absolute estimation error with standard deviations are reported for each DoF. Also in this case, for the estimated DoFs, performance with H_d^* is always better or not statistically different from the one referred to H_s. Maximum estimation errors with H_d^* are usually inferior to the ones obtained with H_s.

Table 15.2 Average estimation errors and standard deviation for each DoF [°] for the simulated acquisition considering H_s and H_d^* both with five noisy measures

DoF	Mean Error [°]		H_s vs. H_d^*	Max Error [°]	
	H_s	H_d^*	**p-values**	H_s	H_d^*
TA⊗	6.7±5.62	4.87±3.57	**0.19**	23.35	15.93
TR	7.65±5.57	7.54±5.00	**0.91** ◇	27.46	22.73
TM○	2.81±1.75	2.63±1.90	**0.61** ◇	7.2	8.78
TI	6.08±4.63	5.42±4.74	**0.32**	19.6	19.10
IA	10.74±5.6	11.52±5.81	**0.32**	27.31	28.46
IM○	4.15±3.17	6.91±5.00	0.003	11.66	21.49
IP	14.61±7.93	6.61±6.01	0	31.85	38.07
MM○⊗	4.59±3.08	4.71±3.19	**0.77**	11.43	15.72
MP⊗	13.71±8.07	4.08±2.98	0 ‡	37.61	13.71
RA	3.12±2.37	3.28±2.45	**0.71**	9.18	9.37
RM○	4.03±3.07	6.30±4.72	0.01 ‡	12.94	12.91
RP	16.78±11.07	6.89±3.82	0 ‡	50.66	16.34
LA	8.97±5.11	9.86±5.45	**0.38** ◇	20.86	21.48
LM○⊗	3.82±3.05	4.82±4.30	**0.44**	11.33	14.26
LP⊗	14.64±9.68	3.94±2.95	0	48.61	11.03

1 ← ——— 0
p-values

○ indicates a DoF measured with H_s
⊗ indicates a DoF measured with H_d^*

Maximum errors are also reported as well as p-values from the evaluation of DoF estimation errors between H_s and H_d^*. ◇ indicates T_{eq} test. ‡ indicates T_{neq} test. When no symbol appears near the tabulated values, U test is used. **Bold** value indicates no statistical difference between the two methods under analysis at 5 % significance level. When the difference is significative, values are reported with a 10^{-4} precision. p-values less than 10^{-4} are considered equal to zero. Symbol "–" is used for those DoFs which are measured by both H_s and H_d^*. For further details on statistical tools, the reader is invited to refer to Table 15.1

15.5 Conclusions and Future Works

In this chapter we have dealt with the problem of achieving reliable hand pose reconstructions through sensing gloves. More specifically, we have exploited the synergistic information on how humans use most frequently their hands to optimize estimation performance and optimally design sensing systems, when constraints on the number and quality of the sensors can limit measurement outcomes. Results show that the exploitation of a priori information on kinematic synergies can be profitably used to advance the state of art of sensing devices, offering new insights to further investigate the biology of human hand, e.g. in conjunction with the techniques described in Chap. 14, in a bi-directional inspiration and relationship between neuroscience and robotic/artificial systems.

Future works will aim at physically realizing an optimal sensing glove. We are considering different technological solutions, e.g. knitted piezoresistive fabrics (KPF) textile goniometer technology that was developed by coupling two piezoresistive layers through an electrically insulating middle layer [35]. Such a technology was already used to develop an under-sensed glove whose measurements were completed through synergistic information for functional grasp recognition [3].

The driving idea will be the synergy-based strategy described in this chapter and protected by an Italian Patent [5], with the mid long-term of enabling mass production and commercialization for human-machine applications.

Acknowledgments This work was supported in part by the European Research Council under the Advanced Grant SoftHands "A Theory of Soft Synergies for a New Generation of Artificial Hands" (no. ERC-291166), and by the EU FP7/2007-2013 project (no. 248587) "The Hand Embodied (THE)".

References

1. Bianchi M, Salaris P, Bicchi A (2013) Synergy-based hand pose sensing: optimal glove design. Int J Robot Res 32(4):407–424
2. Bianchi M, Salaris P, Bicchi A (2013) Synergy-based hand pose sensing: reconstruction enhancement. Int J Robot Res 32(4):396–406
3. Bianchi M, Carbonaro N, Battaglia E, Lorussi F, Bicchi A, De Rossi D, Tognetti A (2014) Exploiting hand kinematic synergies and wearable under-sensing for hand functional grasp recognition. In: Wireless mobile communication and healthcare (Mobihealth), 2014 EAI 4th international conference on, IEEE, pp 168–171
4. Bicchi A, Canepa G (1994) Optimal design of multivariate sensors. Meas Sci Technol (Inst Phys J "E") 5:319–332
5. Bicchi A, Bianchi M, Salaris P (2014) A method for optimal hand pose reconstruction. Patent, 0001410855
6. Brown C, Asada H (2007) Inter-finger coordination and postural synergies in robot hands via mechanical implementation of principal component analysis. In: IEEE-RAS international conference on intelligent robots and systems, pp 2877–2882
7. Catalano MG, Grioli G, Farnioli E, Serio A, Piazza C, Bicchi A (2014) Adaptive synergies for the design and control of the pisa/iit softhand. Int J Robot Res 33:768–782. doi:10.1177/0278364913518998
8. Chaloner K, Verdinelli I (1995) Bayesian experimental design: a review. Stat Sci 10(3):273–304
9. Diamantaras K, Hornik K (1993) Noisy principal component analysis. Measurement'93, pp 25–33
10. Dipietro L, Sabatini AM, Dario P (2008) A survey of glove-based systems and their applications. Syst, Man, Cybern, Part C: Appl Rev, IEEE Trans 38(4):461–482
11. Edin BB, Abbs JH (1991) Finger movement responses of cutaneous mechanoreceptors in the dorsal skin of the human hand. J Neurophysiol 65(3):657–670
12. Fahrer M (1981) The hand. Saunders, chap Interdependent and independent actions of the fingers, Philadelphia, PA, pp 399–403
13. Fu Q, Santello M (2010) Tracking whole hand kinematics using extended kalman filter. In: Engineering in medicine and biology society (EMBC), 2010 annual international conference of theIEEE, pp 4606–4609

14. Gabiccini M, Bicchi A, Prattichizzo D, Malvezzi M (2011) On the role of hand synergies in the optimal choice of grasping forces. Auton Robots 31(2–3):235–252
15. Gabiccini M, Stillfried G, Marino H, Bianchi M (2013) A data-driven kinematic model of the human hand with soft-tissue artifact compensation mechanism for grasp synergy analysis. In: Intelligent robots and systems (IROS), 2013 IEEE/RSJ international conference on, pp 3738–3745. doi:10.1109/IROS.2013.6696890
16. Gelb A (1974) Applied optimal estimation. M.I.T Press, Cambridge, US-MA
17. Ghosh S, Rao CR (1996) Review of optimal bayes designs. In: Design and analysis of experiments, handbook of statistics, vol 13. Elsevier, pp 1099–1147
18. Hawkins DM (1980) Identification of outliers. Chapman and Hall, London
19. Kilbreath SL, Gandevia SC (2002) Limited independent flexion of the thumb and fingers in human subjects. J Physiol 543:289–296
20. Lorussi F, Rocchia W, Scilingo EP, Tognetti A, De Rossi DE (2004) Wearable, redundant fabric-based sensor arrays for reconstruction of body segment posture. Sens J, IEEE 4(6):807–818
21. Mahalanobis PC (1936) On the generalised distance in statistics. Proc Natl Inst Sci India 2(1):49–55
22. Mason CR, Gomez JE, Ebner TJ (2001) Hand synergies during reach-to-grasp. J Neurophysiol 86:2896–2910
23. Mulatto S, Formaglio A, Malvezzi M, Prattichizzo D (2010) Animating a synergy-based deformable hand avatar for haptic grasping. Int Conf EuroHaptics 2:203–210
24. Rao CR (1964) The use and interpretation of principal component analysis in applied research. Indian J Stat 26:329–358
25. Rosen JB (1960) The gradient projection method for nonlinear programming. part i. linear constraints. J Soc Ind Appl Math 8(1):181–217
26. Santello M, Flanders M, Soechting JF (1998) Postural hand synergies for tool use. J Neurosci 18(23):10,105–10,115
27. Santello M, Baud-Bovy G, Jörntell H (2013) Neural bases of hand synergies. Front Comput Neurosci 7
28. Schieber MH, Santello M (2004) Hand function: peripheral and central constraints on performance. J Appl Physiol 96(6):2293–2300
29. Simone LK, Sundarrajan N, Luo X, Jia Y, Kamper DG (2007) A low cost instrumented glove for extended monitoring and functional hand assessment. J Neurosci Methods 160(2):335–348
30. Stillfried G, van der Smagt P (2010) Movement model of a human hand based on magnetic resonance imaging (mri). In: Proc. ICABB
31. Sturman DJ, Zeltzer D (1994) A survey of glove-based input. Comput Graph Appl, IEEE 14(1):30–39
32. Tarantola A (2005) Inverse problem theory and model parameter estimation. SIAM
33. Tognetti A, Carbonaro N, Zupone G, De Rossi DE (2006) Characterization of a novel data glove based on textile integrated sensors. In: Annual international conference of the IEEE engineering in medicine and biology society, EMBC06, Proceedings., pp 2510–2513
34. Tognetti A, Carbonaro N, Dalle Mura G, Tesconi M, Zupone G, De Rossi DE (2008) Sensing garments for body posture and gesture classification. Tech Usage Text Mag 68:33–39
35. Tognetti A, Lorussi F, Mura G, Carbonaro N, Pacelli M, Paradiso R, Rossi D (2014) New generation of wearable goniometers for motion capture systems. J Neuroengineering Rehabil 11(1):56
36. Zavlanos MM, Pappas GJ (2006) A dynamical systems approach to weighted graph matching. In: Decision and control, 2006 45th IEEE conference on, pp 3492–3497
37. Zhang X, Lee S, Braido P (2003) Determining finger segmental centers of rotation in flexion-extension based on surface marker measurement. J Biomech 36:1097–1102

Printed in the United States
By Bookmasters